SSTWG Members

SETI INSTITUTE SCIENCE & TECHNOLOGY WORKING GROUP

MEMBERSHIP

1997-1999

BACKUS, Peter	SETI Institute
BILLINGHAM, John	SSTWG Executive Secretary, SETI Institute
BOCK, Douglas	University of California at Berkeley
BREGMAN, Jaap D.	NFRA, The Netherlands
BUTCHER, Harvey R.	NFRA, The Netherlands
CARLSTROM, John E.	University of Chicago
CHYBA, Christopher	SETI Institute
CULLERS, D. Kent	SSTWG Cochair, SETI Institute
CUTLER, Leonard S.	Hewlett-Packard Laboratories
DAVIS, Michael M.	Arecibo Observatory
DRAKE, Frank D.	SETI Institute
DREHER, John W.	SETI Institute
EKERS, Ronald D.	SSTWG Chair, Australia Telescope National Facility
FABER, Sandra	University of California at Santa Cruz
FISHER, J. Richard	NRAO, Green Bank, WV
HILLIS, W. Daniel	Walt Disney Imagineering
HOROWITZ, Paul	Harvard University
KELLERMANN, Kenneth I.	NRAO, Charlottesville, VA
KLERKX, Greg	SETI Institute

LAMPTON, Michael	University of California at Berkeley
LESYNA, Larry	Lockheed Martin Corporation
LIDDLE, David E.	Interval Research Corporation
MYHRVOLD, Nathan	Microsoft Corporation
PAPADOPOULOS, Greg	Sun Microsystems
PIERSON, Thomas	Chief Executive Officer, SETI Institute
POPOVIC, Zoya	University of Colorado
ROY, Alan	NRAO, Socorro, NM
SCHEFFER, Louis K.	Cadence Design Systems
SHOSTAK, Seth	SETI Institute
SMEGAL, Richard	SETI Institute
STARK, Anthony A.	Smithsonian Astrophysics Observatory
STAUDUHAR, Richard	SETI Institute
TARTER, Jill C.	SETI Institute
THORNTON, Douglas D.	University of California at Berkeley
TOWNES, Charles H.	University of California at Berkeley
van ARDENNE, Arnold	NFRA, The Netherlands
WEINREB, Sander	University of Massachusetts
WELCH, William J.	University of California at Berkeley
WERTHIMER, Daniel	University of California at Berkeley

Figure 1: The Hayes Convention Center in San Jose, California, was the SSTWG meeting venue.

Figure 2: Some of the SSTWG members gather for a group picture.

1. Richard Smegal	9. John W. Dreher	17. Sander Weinreb
2. Alan Roy	10. Daniel Wertheimer	18. Jaap D. Bregman
3. Frank D. Drake	11. Paul Horowitz	19. Harvey R. Butcher
4. Jill C. Tarter	12. Kenneth I. Kellermann	20. visitor
5. William J. Welch	13. Thomas Pierson	21. David E. Liddle
6. Ronald D. Ekers	14. Louis K. Scheffer	22. visitor
7. Vera Buescher	15. John Billingham	23. Anthony A. Stark
8. visitor	16. Greg Papadopoulos	24. Leonard Cutler

Contents

List of Tables

List of Figures

Prologue

Contented as I am–indeed, even awed–with the rich expertise of the fifty-plus contributors to *SETI 2020*, a prologue need not outline the book's diverse lore. The Preface and Summary provide a very good start-up. Perhaps an old-timer's personal remarks may strengthen an analogy or two, maybe only metaphors of use even in this high level of discourse.

I was one of very few five-year old radio listeners to the November 1920 election returns that reported the election of President Harding. That program came to me from a pioneer broadcast transmitter a mile or two away, using the one megahertz band (8XK, which became KDKA, East Pittsburgh, PA). My receiver was a birthday present from my father, who had days before bought me one of a couple of thousand Aeriola Jr. radios put on sale in the local department stores, each in its dark, polished hardwood box. With only the incoming RF for power, these crystal sets, with one tuned coil and a finicky galena detector, created a new headphone audience for music and voice, soon far outnumbering the code-using adepts of that day.

The symbol DX, in code or spoken as letters, was long the shorthand for distance. My childish interest was much less in the news or even the music than in the novel experience of hearing a new distant source, quite unlike the specific friend on the two-way wired telephone. As I recall it, within a matter of weeks I was tuning the single control to seek other novel signals from afar; my memory suggests Kansas City, one winter evening? DX held my affection for years before program content would pass mere path length as a goal.

The call CQ DX was a familiar part of the two-way standard in the high frequency amateur bands that had caught me by the fifth or sixth grade. It was clearly the announcement of a beacon, a general call to any distant station that heard the call. In fact, full antipodal distances were rare but memorable, once I switched on my own 25 W high frequency "beacon" at 14 MHz after grade school. The low transit time was the point we all took by assumption; nowhere would any answer once sent require a delay of even tenths of a second. There wasn't any longer DX to search out than that!

By graduate school days, quantum physics had drawn my attention away from the ham world. The opening of really grand distances came with Karl Jansky's radio astronomy. The wideband, noise-like, high frequency signals he found had traversed the Galactic medium for many thousands of years; astrophysical processes were the only plausible source. The key parameter is not simply the range, DX, but rather DX/c; it is time and not space that sets the firmest limits. Indeed, our Milky Way supports a hundred thousand light year range to its outskirts, but the round-trip transit time is longer than the full age of our articulate species. Historical time enters and rules. Human ability to signal across space beyond the hundred-mile horizon, open to the sunlit mirrors of mountaintop surveys or to ship's signal rockets by night, is no older than Marconi. Our SETI Milky Way surveys cannot rely on the guide of symmetry, for we are almost certain to be the juvenile in any dialogue.

It was the new Arecibo radio astronomy dish and its radar that supplied proof of concept to the first SETI microwave studies, independently for a theory paper and for Project Ozma, a real trial at 10 light years. Moreover, the inevitable transit delay implies a stretch of historical imagination we cannot be sure about. We mainly fall back upon the longest time limit we know, the geological time span that in fact saw radio and laser astronomers evolve from some microbial world, starting close to four gigayears ago. If that is the time span that underlies SETI everywhere, then our youthful asymmetry is not so serious; they will be well ahead of our signal capabilities anyway, and must have entered the hunt long ago. We optimists conclude that it is easier to receive than to send.

Receiving offers hope of saving half the round-trip time delay. More important perhaps, we save not only that delay, but can perhaps count on the far-off sender's lead by some far-extended Moore's Law of technology growth. But a corollary is the absence of much interest in extragalactic SETI today; a minimum round-trip of about five million years dampens all our societal extrapolations; how could anyone plan for an answer?

In fact the lifetime of an average species here on Earth is rather close to five million years. In the same way deep extragalactic space is splendid for radio astronomy, but not at all for SETI. Red shift ratios of 2 or 2.5 take us back too close to the time of the birth of galaxies, if four gigayears of evolution must be fitted in as well. Here there is only uncertainty. SETI 2020 still plans a megastar hunt, a million stars more or less Solar-type within one or two thousand light years. The modular phased array design reported in the text – all-the-sky, all-the-time – suggests we could see a bright enough source all the way out to the Galaxy edge, limited more by the time depth implied (L, in the Drake factors) and not by the physical attenuation of the signal by inverse-square or dust.

Here the Moore's Law exponential may be built in at both ends of the contact. That is a plausible gamble, if less compelling than recently in the time of 'dot com' exuberance. Will slowing-down first arise on the side of technological inventiveness, or in the overall human effort supported under some chillier investment climate a generation or two out?

For four decades Arecibo has been a wonderful 10 hectare tool. Proposed radio telescopes summing up to one hectare of surface, or even to a square kilometer in many-hectare bites, are fascinating, and current mass production both of TV dishes and of microwave modules lends them credibility. Infrared and optical laser pulse searching with large telescopes also seem fit for million-star surveys. Sending on our own remains a delayed option, perhaps to be considered at the close of a century of search, as even conventional habit might suggest.

The view of amateur radio as an utterly innocent harbinger to SETI, if by analogy alone, offers an encouraging sign that the cosmic context and insightful engineering of our SETI Working Group do tap hopes shared the world around ...and perhaps the Galaxy around as well.

One succinct lesson remains: search.

Philip Morrison

Professor of Physics, Emeritus
Massachusetts Institute of Technology
May 7, 2001

Figure 3: Dr. Philip Morrison is truly one of the founding fathers of SETI. His 1959 paper with Dr. Giuseppe Cocconi outlined a rationale and an approach to radio searches that is still germane today.

Preface

This book describes a plan for new searches for extraterrestrial intelligence to be carried out by the SETI Institute and its collaborators over the next twenty years.

These pages contain considerations synthesized from the work of over fifty scientists and engineers in a variety of disciplines. The shared conclusions rest upon common themes, namely, the vital importance of technology and its exponential growth. Though the wish to design a cost effective and growing search is common to all of the members of the SETI Science and Technology Working Group (SSTWG), viewpoints differ slightly.

The text of this document was reviewed by the group as a whole, but it would be overzealous to claim that every member reviewed and agreed to every word. Certain important contributions were individually authored, especially in the appendices.

Nonetheless, those named in the list of contributors aspire to a common purpose, to search the electromagnetic signatures of the stars for life. It is to this grand challenge, its design and implementation, that this document is dedicated.

Modern scientific investigations of intelligent life in the Universe had their genesis some 40 years ago. Then, after centuries of speculation about life on Earth's closest neighbors, researchers began to direct their attention to distant planetary systems. In 1959, two Cornell University physicists, Giuseppe Cocconi and Philip Morrison in a paper published in *Nature* [Coc59], concluded that radio waves would be a remarkably efficient means of communicating across the Galaxy. Independently, the following year, astronomer Frank Drake conducted the first radio SETI search. Drake used an 85 foot telescope at the National Radio Astronomy Observatory (NRAO) in Green Bank, West Virginia, to listen to the nearby solar-type stars Tau Ceti and Epsilon Eridani.

At about the same time a broader study of extraterrestrial life began. In 1959, the Nobel prize-winning biologist Dr. Joshua Lederburg, then at Stanford University, gave the name 'Exobiology' to this new discipline. In the same year, the first research organization dedicated to this subject was formally established at the National Aeronautics and Space Administration (NASA) Ames Research Center in California. It was called the Exobiology Division and was led by Dr. Harold P. Klein. A powerful stimulus to this new area of study was NASA's decision to send the Viking spacecraft to land on Mars and search for evidence of extant, extraterrestrial life.

The common threads in the search for primitive life and intelligent life were brought together in 1984 with the formation of the SETI Institute. It is a nonprofit research and educational organization dedicated to the Search for Extraterrestrial Intelligence and to the general study of "Life in the Universe". Located in Mountain View, California, the SETI Institute also conducts educational and public outreach programs in these two disciplines.

The story of SETI over the last 40 years, and the prominent role of the SETI Institute over the last 15 years, is recounted by Seth Shostak in Section 1.5 of this book, entitled *The Road from Ozma* (page 16). From its inception until October 1993, the SETI Institute participated with the NASA Ames Research Center and Jet Propulsion Laboratory in planning comprehensive searches for radio signals of extraterrestrial intelligent origin. Just as these projects were moving from planning to execution, Congress ordered the termination of government participation in SETI. Since that time, the mode of searching that focused on individual target stars has been continued and extended by the SETI Institute, as its Project Phoenix, with support provided by donors from the private sector.

In 1997, the SETI Institute decided to reexamine its SETI mission and objectives, in light of the many recent advances in science and technology. Experts from the relevant fields were invited to meet together as the SSTWG and to develop a plan for SETI for the first two decades of the new millennium. The SSTWG was chaired by Dr. Ronald D. Ekers, Director of the Australia Telescope National Facility (ATNF), and Dr. Kent Cullers, Director, SETI R&D, of the SETI Institute. The two-year study, which they most ably led, is summarized in Chapter 1, *Introduction and History*, in Section 1.2, *Goals and Objectives of the SSTWG* on page 7.

The Search for Extraterrestrial Intelligence is a combination of exploration, science and technology. That it is now largely accepted as a valid discipline among human intellectual pursuits is due in substantial part to the efforts of the SETI Institute's Project Director, Dr. Jill Tarter. In recognition of this and of her lifetime achievements, Dr. Tarter was recently appointed to the Bernard M. Oliver Chair for SETI, created in memory of Barney Oliver and endowed, as a result of his generosity.

In the university community, the first endowed Chair for SETI was established in 1999. It is the Watson and Marilyn Alberts Chair in the Search for Extraterrestrial Intelligence, established in the Department of Astronomy, University of California at Berkeley. It was created thanks to a generous gift by these two alumni who have had a long-standing interest in the field. The first holder of the chair is Professor William J. Welch, former director of the Radio Astronomy Laboratory, University of California at Berkeley, and the first Vice Chairman and current member of the SETI Institute's Board of Trustees. The rationale for, and wisdom of, having both Life in the Universe studies and SETI exploration housed within one institution are demonstrated by the Drake Equation.

The Drake Equation is usually written:

$$N = R_* \cdot f_p \cdot n_e \cdot f_l \cdot f_i \cdot f_c \cdot L$$

where:

N = The average number of civilizations in the Milky Way Galaxy whose radio emissions are detectable.

$R*$ = The rate of formation of stars in the Galaxy.

f_p = The fraction of those stars with planetary systems.

n_e = The number of planets, per extrasolar planetary system, with an environment suitable for life.

f_l = The fraction of suitable planets on which life actually appears.

f_i = The fraction of life bearing planets on which intelligent life emerges.

f_c = The fraction of civilizations that develop a technology that releases detectable signs of their existence into space.

L = The average length of time such civilizations release detectable signals into space.

Within the limits of our existing technology, any practical search for distant intelligent life must be a search for some manifestation of a distant technology.

Besides illuminating the factors involved in such a search, the Drake Equation is a simple, effective tool for stimulating intellectual curiosity about the Universe around us; for helping us to understand that life, as we know it, is the end product of a natural, cosmic evolution; and for making us realize how much we are a part of that Universe. A key goal of the SETI Institute is to further high quality research that will yield additional information related to any of the factors of this fascinating equation.

Today, the SETI Institute's "Life in the Universe" research program, dedicated to the discovery of more information about these Drake Equation factors, consists of more than 30 externally funded, peer-reviewed projects led by independent principal investigators.

Plans for the future of this program at the SETI Institute are now being formulated by the same type of team reviews that SETI has been subjected to in this

Figure 4: SETI Pioneer Dr. Frank D. Drake conducted Project Ozma, authored the Drake equation, and is now Chairman of the Board of the SETI Institute.

book. This process is being led by Dr. Christopher Chyba, the holder of the SETI Institute's new Carl Sagan Chair on the Study of Life in the Universe.

On behalf of the SETI Institute, we wish to acknowledge the many contributions and new initiatives, which have flowed from the members of the Working Group, and for their diligence and productivity over the course of the SETI 2020 study. They have provided the SETI Institute with a framework for a bold and exciting future.

Production of a report of this complexity requires a lot of hard work by many people. First, this report would not be possible without the outstanding leadership of the Working Group's Chair, Dr. Ron Ekers, who made numerous trips from Australia to supervise the entire working group process. In addition, we want to make special recognition of the work of the Cochair, D. Kent Cullers, in overseeing the final editing of *SETI 2020*, and transforming the draft report from the SSTWG into this final clear, cohesive document. Special thanks go to Dr. Lou Scheffer, a Cadence fellow, as well as Cadence itself, for Dr. Scheffer's time and significant editorial and technical contributions to this document, both during the meetings and afterwards, during his sabbatical with the SETI Institute. The quality of the end product is also due to the efforts of Ted Zajdel, the Technical Editor, who worked shoulder-to-shoulder with Dr. Cullers in the final production of this text. Finally, all SSTWG members agree that we would have been unable to carry out this enterprise without the administrative expertise and assistance of Vera Buescher and Chris Neller, who supported the effort from beginning to end.

John Billingham,
Executive Secretary,
SETI Science and Technology
Working Group

Thomas Pierson,
Chief Executive Officer,
SETI Institute

Figure 5: Dr. John Billingham, Executive Secretary of the Working Group, is clearly
pleased with the SSTWG's progress and offers a delighted smile.

Figure 6: The SETI Institute's Chief Executive Officer, Thomas Pierson, looks to the future.

Executive Summary

This book is the report of the Search for Extraterrestrial Intelligence (SETI) Science and Technology Working Group (SSTWG). The SSTWG met four times from August 1997 to May 1999, preparing a plan for new directions in SETI research and technological development, to be carried out over the years 2000 through 2020. The group was assembled and funded by the SETI Institute, a nonprofit research and educational organization dedicated to SETI and the general study of "Life in the Universe".

The SSTWG was made up of scientists and engineers from the SETI Institute and colleagues from academia and industry, working together to create the new SETI plan. Its meetings, utilizing the expertise of some fifty scientists and technologists from a wide variety of disciplines, were chaired by Ronald D. Ekers, Director of the Australia National Telescope Facility, and D. Kent Cullers, Director, SETI R&D, at the SETI Institute.

Since its modest beginning in 1960, the search for ETIs has grown by orders of magnitude. The SETI Institute's Project Phoenix is the most sensitive SETI research program ever developed. It uses the world's largest radio telescope to listen for microwave signals emanating from ETI civilizations. The signals are divided by spectrum analyzers into millions of frequency channels. The results are passed to high-speed computers, which examine cosmic noise for signals from alien technology. Searches of hundreds of nearby stars have been carried out in the Northern and Southern hemispheres. Project Phoenix will be completed in a few years, and the technology upon which its search systems are based is over a decade old. During this past decade, there have been new ideas for search strategies, improvements in telescope design, and a huge increase in the cost-effectiveness of computation. The SSTWG was charged to develop new approaches capitalizing on these changes and to develop a plan to implement the most promising among them over the next twenty years. Its report contains scientific recommendations to the whole SETI community, as well as strategic and political advice to the SETI Institute, which provides a large fraction of the funding and expertise for SETI.

The SSTWG report first establishes a set of strategic objectives. After addressing the historical background, the report re-examines the SETI philosophy and underlying logic in the section on Rationale (page 10). The whole of Chapter 5, *ETI Sources and Search Strategies* on page 111, is devoted to a review of possible search strategies, and options for new telescope and computer designs. The Drake Equation [Dra65] is re-examined in light of recent advances in astronomy and the planetary sciences, especially the discovery of extrasolar planets. The landmark 1970 Cyclops Design Study [Oli71] was used as a benchmark against which modern approaches to searches and systems were compared. It was used as a guide to set the standard for this analysis. By comparing its extrapolation of technology thirty years ago to technology today, it provided perspective when extrapolating current technology to thirty years hence.

The *Project Cyclops* report proposed an expandable phased array of 100 m telescopes listening for narrowband microwave signals from solar-type stars out to 1,000 light years (about a million stars). Some concepts in its proposed system, for example, optical Fourier Transforms, have been superseded by digital methods. However, many of the principles in the *Project Cyclops* report remain valid. Exotic searches for distant civilizations, such as massive particle or neutrino detection, and roundtrip interstellar travel, are beyond our technical capability. Intentional transmission from Earth of strong interstellar signals is considered a possible, long-term contact strategy. During the next twenty years, however, unintentional transmissions will still be very strong whatever SETI does. On the other hand, specialized listening techniques ready for development during the next two decades provide excellent opportunity for growth. The SSTWG Report concludes:

- All searches should be in the electromagnetic spectrum.
- Frequency ranges should be extended beyond the microwave to include the infrared/optical.
- Both targeted and sky survey systems should be considered.
- Multiple beams per telescope should be employed in microwave searches.
- Both continuous wave and pulsed signals should be sought.
- The focus should be on the detection of beacons without excluding the eavesdropping case.
- Multiple site detection will continue to greatly improve the efficiency and credibility of the search.

Included in the search strategy should be a new type of signal, of high power but with a low duty cycle, as might be produced by beacons illuminating many target civilizations sequentially. Overall, searches of *magic frequencies* are decreasing in relative importance because technology now permits very broadband, all-inclusive searches. Effective use of computing power is the key to geometric growth in searches for ETI technology. Dividing 1 MHz of spectrum into a million narrow channels in 1980 required special purpose computers and cost

$1 million. Today the same operation can be performed with general-purpose computers for about $1,000. Signals of interest are correspondingly inexpensive to detect.

Moore's Law (see Section 6.3, *Moore's Law* on page 152), which postulates that the power of computing systems doubles approximately every 18 months, is expected to remain valid during the next two decades. Designs in the SSTWG Report allow systematic upgrading of hardware during a project lifetime which may encompass decades. Signal detection systems, whose computing costs have typically been high, will henceforth be designed in a modular way to take advantage of Moore's Law through a gradual upgrading of capability over time. Since electronic beamforming is computationally intensive, it will benefit greatly as computer costs drop.

Our maturing microwave technology can now handle the broad band between 1 and 10 GHz. Over such bandwidths it is now possible, using phased arrays, to place individual beams on a large number of stars simultaneously. Since a receiving civilization may have telescopes distributed throughout its planetary system, it is reasonable to generate a patch of illumination some 10 Astronomical Units (AU) wide. Even relatively modest 1 MW transmitters give detectable fluxes over such areas.

To compare alternative search systems, the SSTWG calculated a figure-of-merit for each search and plotted it against the effective isotropic power of the beacon. They cautioned that the calculation neither incorporates interference rejection techniques, nor does it consider different signal types. Also, it is not very successful when comparing searches at very different frequency regimes.

Design options have been examined for a Square Kilometer Array (SKA), and they are now being considered by an international consortium. This proposed antenna has 30 times more collecting area than the largest existing radio telescope. It might consist of a few tens of antenna stations or clusters, each station having an effective diameter of the order of 300 meters. Also looked at were options for a smaller array, one hectare in collecting area (1hT), equivalent to a single dish 110 meters in diameter, as a proposed SETI telescope design. Individual antennas could be like those now widely available as satellite TV reception dishes – parabolic in shape, fully steerable, and inexpensive. A set of options for microwave sky survey instruments was considered:

- large single antennas;
- an array of identical 3 to 10 meter antennas comprising the one hectare array (1hT);
- an array of omnidirectional antennas as proposed in Dixon's Argus project [Dix95].

Initially, the array of omnidirectional antennas could be made of small elements with a total collecting area of only a few square meters. It would look at the whole sky at all times seeking signals of high power and low duty cycle. This concept was termed the Omnidirectional SETI System, and it was investigated in some depth. It was found that the costs were dominated by computation, especially for beamforming, and it was concluded that this approach should be explored further. If computing costs continue to be described by Moore's Law, which predicts a doubling of computer power every 18 months, then such an antenna, with a large collecting area, could be constructed in the not too distant future.

Because of cogent arguments which showed that direct photon detection eliminates quantum noise associated with narrowband, heterodyne techniques in the optical, it was decided to re-examine the possibility of infrared/optical SETI searches. The vast majority of the seventy-or-so searches carried out to date have been in the microwave region of the spectrum. Notable exceptions were: the infrared search carried out by Betz [Bet86] and the optical search currently being done by Kingsley [Kin96].

Townes has long pointed out that the rise of quantum noise with frequency, which bedevils heterodyne devices, disappears if they are eliminated from the system and replaced by direct photon detection [Tow83]. Such detectors are now available and improving every year. It has been demonstrated that lasers of enormous power can be constructed. A 1 MJ/ns laser pulse with 10 meter telescopes at the sending and receiving ends can be easily detected at three thousand light years. Particular attention was therefore given to searches for continuous wave signals in the infrared/optical regions with high resolution spectrometers, and for nanosecond pulses of intrinsically wide bandwidth.

Recommendations

On the basis of these considerations, the Working Group was divided into three teams to develop preliminary designs for the 1hT, the OSS, and for optical SETI. A fourth team was formed to examine the best strategy for implementing these designs. Together, the findings of the four teams represent the recommendations of the Working Group.

One Hectare Telescope

The Working Group recommends that the SETI Institute undertake the development and construction of a One Hectare Radio Telescope (1hT) to carry out targeted searches of candidate stars and some sky surveys in the galactic plane over the next twenty years in the 1 to 10 GHz region of the spectrum.

The telescope will be a phased array of identical 3 to 10 meter, fully steerable, commercially available parabolas. The antennas will be equipped with broad-band feeds and cryogenically cooled, low-noise amplifiers. The array will be capable of simultaneously forming multiple steerable beams. Processing of the data from each beam will allow forming up to one billion 1 Hz channels, detection of drifting continuous wave and pulse signals, radio frequency interference rejection, and data archiving. The array will be controllable from a central station.

Feasibility of the concept should be examined on a small-scale version of the array, composed of 10 antennas. Hardware and software should be modular and upgradable to allow for advances in computer capability. Maintenance costs must be low. Development, construction, test, and operation should be performed according to a detailed project plan. The plan should incorporate parametric cost estimates for expanding the collecting area of the telescope well beyond one hectare during the twenty year period. It should also include multiple arrays, located around the equator, to give continuous signal coverage.

Collaborations with other organizations should be explored, especially with the international consortium on the Square Kilometer Array (SKA). Mutually beneficial arrangements to use the telescope for radio astronomy should be made with the scientific community. An examination should be made of the possible value of the telescope as a test-bed for future Deep Space Network antennas communicating with interplanetary spacecraft. It is expected that the 1hT can be built in five years for a cost of $25 million. The collecting area of the telescope and its computational power should then both be expanded at the same annual cost rate until the year 2020.

The 1hT was a consensus favorite of the Working Group because of increased performance over existing facilities in many areas. The antenna cost will be reduced by a factor of 10 per unit area; its bandwidth will cover three octaves; it can beam perhaps 10 stars simultaneously; and it can perform SETI all the time. It may be the prototype for the SKA.

Omnidirectional SETI System

The Working Group recommends that the SETI Institute undertake extensive feasibility studies and test bed demonstrations of an omnidirectional sky survey telescope, designated the Omnidirectional SETI System (OSS), designed to search for strong, low-duty cycle microwave signals in the one to three gigahertz region of the spectrum.

If these tests are successful, the first OSS should be constructed over a period of fifteen years, providing a gradual increase in capability for reduced costs according to Moore's Law. The telescope would have 1,024 elements and a collecting area of a few square meters. It would cover one third of the sky at once and would be sensitive enough to detect intermittent ETI beacons out to 1,000 ly. It is expected that the total cost, dominated by signal processing, would be approximately $30 million over the period 2000 through 2015.

It should then be possible to further expand the OSS to 4,096 elements by the year 2020, with a proportional increase in sensitivity.

The main reason for selecting the OSS telescope was the ability to explore new dimensions of search space for transient microwave signals of high power, and for the first time to achieve a time-continuous, all-sky coverage for the low end of the microwave window.

Infrared/Optical Experiments

The Working Group recommends that the SETI Institute set aside funds for small-scale experiments to detect infrared/optical signals of ETI origin using existing telescopes and direct photon detection techniques.

A diverse set of inexpensive experiments should be carried out. Several good detectors exist which can be used with existing telescopes to search for beacon pulses at optical wavelengths. Nanosecond pulses, containing several photons, have very low probability as natural events, and thus the entire visible spectrum can be searched without requiring sophisticated spectrometers. Searches for continuous wave signals can use existing, high resolution spectrometers to look for sharp emission lines. In some bands, even existing data could be mined.

While infrared and visible band optical systems can be used for pulsed or continuous wave signals, infrared is preferred for sources at distances greater than 300 ly because of interstellar extinction. Thus, it is recommended that the SETI Institute send out requests for proposals (RFPs) to the astronomical community to fund a number of $10,000 to $50,000 optical SETI projects, up to a total of $1.3 million over the years 2000 through 2005. New or upgraded observing facilities will be constructed, if justified by these initial projects. At a later date, the SETI Institute should assess the potential for building a large, low cost collecting area for improving infrared and optical search sensitivity.

The main reason for selecting infrared/optical experiments is the ability to open up, at a very reasonable cost, frequency dimensions of search space that have previously been little explored.

Strategic Implementation

The Working Group recommends that the SETI Institute develop a strategic plan to address all questions pertinent to the implementation of the science and technology plan described in this book.

In addition to science and technology, the plan should incorporate strategies for continuing close interactions with the scientific and engineering communities, for organization and management, development, education, public programs, and studies of interactions between SETI and society. (See Chapter 9, *Epilogue*, on page 277). The strategic plan should include studies of how to proceed with an expanded SETI enterprise after the first signal is detected, and what to do if, after many decades, no signal has been detected. It is also recommended that the SETI Institute extend the life of the SSTWG, or at least part of it, to continue to provide assistance to the SETI Institute as the new programs get underway.

Of major importance is the acquisition of funds. The annual cost of the programs described above is $8 million, for a total of $169 million over the years 2000 through 2020. The Development Office of the SETI Institute should strive to raise these funds from private sources, and there are clear opportunities for capital campaigns for the large telescopes.

The SETI Institute should engage in discussions with the National Aeronautics and Space Administration (NASA) and the National Science Foundation (NSF) to encourage them to join in a public/private partnership for direct participation in these projects, or for technology development, or for other joint ventures such as using the 1hT for traditional radio astronomy research.

There is an exciting new NASA program called Astrobiology, and it seems widely agreed in the scientific community that a modestly funded SETI effort should be a component of this program. Since SETI is inherently an international endeavor, collaborations should also be sought with interested organizations in other nations.

A concerted attempt should be made to encourage the US Congress to support the visionary nature of the SETI endeavor, which is of such profound importance to science and indeed to people from all walks of life everywhere on Earth. It is worth remembering that Congress did appropriate $78 million for SETI over the years 1975 to 1992. A few million dollars a year is a small price to pay for the chance to find another civilization among the stars.

Figure 7: Dr. Ronald D. Ekers, from the Australia Telescope National Facility, was the inspiring Chair of the SSTWG.

Figure 8: SSTWG Cochair, Dr. D. Kent Cullers of the SETI Institute, the master editor of this book, offers insightful analysis and a smile. (Photo courtesy of Lisa Powers)

Figure 9: The headquarters of the SETI Institute, Landings Drive, Mountain View, California.

Chapter 1

Introduction and History

This chapter describes the historical background of the SETI Institute and its activities. It introduces and outlines the detailed discussions described in succeeding chapters. It emphasizes that all the projects proposed in this book are based on experience and extend current activities based upon recent and anticipated technological developments, especially in computing. Discussions of the ways to search for concrete evidence of the existence of extraterrestrial intelligence (ETI) are summarized. Possible and actual choices are outlined, and the requirements for each choice are stated.

The goals and objectives of the SETI Science and Technology Working Group (SSTWG) were to plan for the period 2000 through 2020, building on experience with Project Phoenix (which is near the end of its life), and taking advantage of developing technology and increasing knowledge. The value of the search and instruments it uses extends into other areas of research, so they will benefit other astronomical disciplines. This is illustrated by a list of various instruments that made discoveries outside their original purposes.

The history of the Search for Extraterrestrial Intelligence (SETI) is outlined starting with the 1959 Cocconi and Morrison paper, Drake's Project Ozma and the Drake Equation. It continues with the Cyclops Design Study and the role of the National Aeronautics and Space Administration (NASA). It concludes with discussions of the subsequent development of smaller and less expensive antennas, and with current searches and surveys. Historical and current target strategies are outlined and justified.

1.1   Overview

Are we the only creatures to view the Universe with understanding? In a cosmos filled with billions of galaxies containing trillions of stars, is it possible that

Earth, a world of inconsequential size and ordinary position, is alone in housing life that can discern the natural order? It is deeply incongruous to suppose that our enormous Universe is only sparsely occupied. To do so requires the belief that humans are exceedingly special. In view of astronomy's history, such a view is clearly suspect.

There is no lack of speculation about the existence of alien intelligence. Much of it has been optimistic, positing that the general processes that led to sentient life on this planet were repeated elsewhere. While this optimism has been based on the simple fact of the great extent of the Universe, new discoveries bolster confidence in its likely truth. We now know that planets, the most probable progenitors of life, are common around mid-sized stars. There is also suggestive evidence that biology itself may be a universal phenomenon, in much the same way that physics and chemistry are universal. Many researchers dare to envision a widely inhabited cosmos. Consequently, it hardly seems radical to hypothesize that, in at least a few places, life has evolved to intelligence. This book describes technology that could prove that hypothesis.

SETI is not a new idea. In its modern incarnation, the search for narrowband radio signals, SETI research has been pursued for nearly four decades. However, schemes to detect sophisticated beings *in situ* by eavesdropping on their signals are not the only methods with which we might search. We could, for example, expend effort in looking for alien artifacts, interstellar spacecraft, or large-scale, ETI engineering projects. Instead, SETI has sought signals, not because these other approaches are without merit, but simply because in electromagnetic signaling the speed is very high and the cost is very low. We expect communications, inadvertent or deliberate, to be commonplace in an inhabited Universe.

The title of this book hints at its mission – an analysis to identify the most fruitful SETI projects to undertake during the next two decades. These mostly focus on new instrumentation, although there is some consideration of strategies as well. In order to forestall undue provincialism, the SSTWG responsible for this book is comprised of experts from a wide range of disciplines including people outside of SETI. Participants in existing SETI programs are joined by colleagues from radio astronomy and the computer industry. Nonetheless, the projects described here inevitably owe much to SETI history. They extend current programs in the directions encouraged by foreseeable improvements in technology, and begin new ones (such as optical SETI) at a clearly practical level. They are also commensurate with attainable levels of funding. The approaches discussed in this book are exciting. They are also realistic.

Much of the impetus for SETI at the start of the 21st century derives from the galloping speed of computer technology. Moore's Law states that a doubling of computer chip densities occurs every 18 months. Rather than simply noting this exponential growth in computational power, this book incorporates it into its projects. The consequence is that by 2020, SETI instrumentation could be ten thousand times faster than that used in the best current experiments.

This is an exciting prospect, for it offers the hope that our grasp may soon equal our reach. Scrutiny of hundreds of thousands, or even millions of nearby stars will be possible.

At many wavelengths, the vast tracts of the Universe are quiet and dark and it is against this subdued backdrop that faint signals could be found. A success would resolve our enduring speculation about the uniqueness of humans and it would break the bubble of isolation that has surrounded Earth since its birth.

SETI attempts to answer the age-old question "Are we alone?" by searching for concrete evidence of other intelligent life in the Universe, and particularly, in our Galaxy. As is often the case for exploration, SETI is based on fragmentary knowledge. It is fraught with uncertainties and assumptions. For what evidence are we searching? When and where should we look? How should we conduct the search? The answers to these questions come from a wide range of disciplines. Some are based on science – What kinds of signals travel well across the Galaxy? Some are based on technology – What is the relative effort of sending spacecraft as compared to signals? Some are based on the supposed behavior of the postulated alien civilization – Should we look for something they have deliberately sent us, or for unintentional by-products of their technology? All these questions require at least provisional answers before we can search.

The Decision Tree in Figure 1.1 on page 4 summarizes the parameters of SETI, most of which are considered in further detail in this book. Each path through the tree, from the top to the bottom, corresponds to one set of choices for a possible SETI program.

Starting at the top of the Decision Tree, we need to decide if we will even try to answer the question: "Are we alone?" This is discussed in the section *Rationale*, and this book is evidence of the assumption that we wish to try.

Once we elect to answer the question, we need to decide on how one galactic civilization could learn of another. The main types of evidence appear to be physical artifacts, electromagnetic waves, and particles. We conclude that while scientists should keep their eyes open for particles (e.g., propulsion residues) and artifacts, the best conveyor of evidence still seems to be electromagnetic radiation. This is covered in detail in Chapter 5, *ETI Sources and Search Strategies* on page 111.

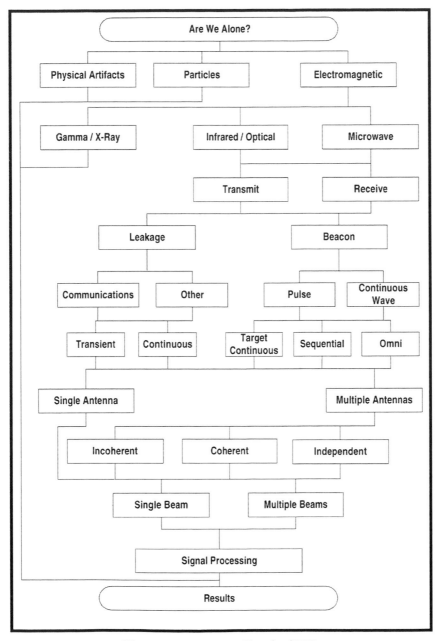

Figure 1.1: Decision Tree for SETI.

Electromagnetic radiation spans an enormous range of wavelengths, from the kilometers of very low frequency radio waves to the nanometers or less of gamma radiation. What frequencies should we use? The main criteria for choosing wavebands are:

- They should propagate well between the stars.
- They should suffer little interference from natural astrophysical sources.
- They must be easy to generate and detect.

The propagation and backgrounds are features of our Galaxy and are discussed in Chapter 2, *The Science of SETI* on page 37. The main conclusion is that the region from the low end of the microwave band (wavelength 30 cm) through the ultraviolet (UV) region of the spectrum (around 200 nm) seems plausible.

This is a very wide band. In Chapter 3, *Interstellar Communication Engineering* on page 67, we consider which wavelengths are easy to generate and detect. We conclude that although techniques in the optical and the radio differ considerably, both have advantages and disadvantages, and both should be pursued. Those who have followed the SETI enterprise will note that this is a considerable change from the conventional wisdom that the microwave region is highly favored for communication between the stars. This change is driven by improved optical technology, new understanding of interstellar propagation, and consideration of the detectability of short pulses in the optical compared with narrowband signals in the microwave. Since the optical techniques are new to SETI, whereas the microwave techniques are well understood, we look at optical technologies in considerable detail in Chapter 6, *Technology for SETI* on page 147.

Having chosen a band or bands for communication, we need to decide whether to receive, transmit, or both. This very interesting question involves philosophy, economics, ethics, game theory, and assumptions about the technology and motivation of other civilizations in our galaxy. We approach this question from several points of view in Chapter 5, *ETI Sources and Search Strategies* on page 111. Our basic conclusion is that deliberate transmission does not yet make sense – The Earth is already quite bright with radio leakage. Transmitting enough to improve this significantly is expensive, and any transmitting strategy cannot pay back for many years. However, as our technology improves, the balance tilts in favor of transmitting for two reasons. First, the Earth will produce less leakage as our own technology improves – high definition television, for example, is much weaker and harder to detect than our older analog television technology. Second, transmitting becomes easier and less expensive as time goes on. Our conclusion, therefore, is that we should continue just to listen for now, but plan to re-examine this question as the benefit becomes greater and the cost less.

As this discussion implies, there are two kinds of signals for which we can search:

- Beacons are signals deliberately sent to us.

- Leakage consists of signals generated by another civilization for their own purposes, but which we might detect.

If they exist, beacons should be much easier to detect, since they are designed for that purpose. Beacon signals could be pulses or continuous wave (CW). Both are technically possible for sending or receiving in either the microwave or the optical bands. In terms of overcoming the background noise and reducing the parameter space that needs to be searched by the receiver, the best choice seems to be CW in the microwave and pulses in the optical. This conclusion is far from certain, and we should continue to look for both signal types in both bands.

Leakage seems like an unavoidable byproduct of a technical civilization, but will be weak since it is intended for nearby receivers and not aimed at us. Worse yet, if we on Earth are a good example, leakage from radio communication becomes both weaker and harder to detect as technology improves. It is getting weaker because newer systems use directed transmissions where possible, and can assume low-noise receivers. It is getting harder to detect because simple signals are a waste of the valuable electromagnetic spectrum, and are being replaced by complex signals that are more efficient but depend on the transmitter and receiver agreeing on the exact signal format. To any other receiver, these signals look like noise. Consequently, as our technology becomes more sophisticated, and as new systems replace the old, detectable leakage from our communications decreases. This development seems likely for other civilizations as well. On the other hand, microwaves can be used for many other purposes, such as radar, power transmission, and navigation, and these have different constraints. So the amount of leakage we might expect is very uncertain.

The possible noise-like nature of signals concerned the SSTWG as a whole. Generally, for Gaussian noise, the bit rate is proportional to $log_2(1+SNR)(bandwidth)$, where SNR is the signal-to-noise ratio. Since the power is also proportional to the bandwidth (BW), it is generally useful to increase bandwidth rather than to increase signal-to-noise ratio. This is true, of course, if the SNR ratio is high. For weak signals, it does not hold.

This implies that both frequency and time will be fully occupied. However, it does not mean that the signal will be Gaussian. One can occupy all bandwidth with many signals that are highly coherent or with one signal that is noise-like. If spectrum use is not perfectly coordinated, if coherence leads to low cost reception, if low signal levels are expected, then coherence is an advantage.

Beacons may likely be coherent. However, most leakage power is not likely to be coherent. Even if a few percent is, however, it may be the most detectable signal of a civilization. We therefore advise that we look for complex signals as technological opportunity permits but expend energy in traditional coherent signals as well.

The exact nature of leakage is very uncertain. It might consist of sources that are usually on, such as our television broadcasts, or much more powerful but occasional signals, such as when a big radar happens to be aimed in our direction. The signals might be CW, as our carriers, or pulses, as our radars.

Since we choose to listen, the next question is to listen for what? According to our best understanding of science and engineering, and assuming that ETIs, like ourselves, are limited in their resources, we might expect deliberate signals to be weak but continuous (either aimed at a subset of all stars or sent omnidirectionally), or very strong signals that come in intermittently as they are beamed at star after star. Leakage, as discussed earlier, may well have the same characteristics.

Our existing searches look for relatively weak but continuously present signals. For this we need more sky coverage, better sensitivity, lower cost, and wider bandwidths. We consider these systems in Section 6.7 *Next Generation Sky Survey* on page 170, and Section 6.8 *Next Generation Targeted Search* on page 180. Powerful but occasional pulses in the radio band would be missed by all existing radio telescopes. An instrument to detect such pulses would open new fields of science as well as serving SETI. Such an instrument, considered in Section 6.6 *Omnidirectional SETI System* on page 162, is very computationally intensive, but could be built in the next two decades as computer technology continues to improve.

Once we have decided what to look for, we need to make the search as sensitive as possible. The key to good sensitivity is a large collecting area, which can be built as one large antenna or many smaller antennas. In the case of many antennas, we can combine them coherently, incoherently, or use the antennas independently. We can form one or many simultaneous windows on the sky. The choices made among these options will determine the speed, cost, and thoroughness of any search, and so deserve careful consideration.

Finally, having formed our window or windows, we need to examine the output for the signals we are looking for. We need to reject interference of all types, and detect the weakest possible signals without being overcome by false alarms. These topics are treated in Chapter 3, *Interstellar Communication Engineering* on page 67.

1.2   Goals and Objectives of the SSTWG

Early in 1997, the SETI Institute decided to establish a Science and Technology Working Group (SSTWG) to examine future opportunities for SETI and make recommendations for specific directions in which the science and technology of the discipline should proceed. The title of "SSTWG" was chosen because of the desire to make the entire process a joint enterprise between scientists and engineers from both inside and outside the SETI Institute, and particularly between SETI and the astronomy community.

The Chair was Ronald D. Ekers, Director of the Australia Telescope, and the Cochair was D. Kent Cullers, Director, SETI R&D, at the SETI Institute. The SSTWG drew its ideas from a broad cross-section of the scientific community. It was composed of astronomers, physicists, SETI experts and innovators in Silicon Valley. The excitement and freshness experienced by the group as a whole is drawn in large measure from interaction of these diverse groups. At the same time, some of the tendency of the report to contain many views is likely due to this. Members of the SSTWG are listed inside the front cover. The SSTWG held four three-day meetings during 1997, 1998 and 1999.

The major goal of the SSTWG was the development of a SETI science and technology plan for the years 2000 to 2020. The plan is embodied in this book, which is the final report of the SSTWG. The rationale for developing the plan was based on the many different needs identified and discussed by the SSTWG. Foremost among these was the realization that the existing Project Phoenix of the SETI Institute is based on science and technology of a decade ago (see Section 1.5, *The Road from Ozma* on page 16), and that there has since been substantial new thinking about SETI searches, combined with major developments in the relevant engineering and computation fields.

Project Phoenix will actually wind down over the next few years, so it needs to be replaced by next-generation searches and systems. It is important that the Science and Technology Plan be widely available, so it can answer questions about the new SETI enterprise from many different quarters. It must be available to the scientific and engineering communities, to the SETI community itself, to the "Astronomy Survey Committee for the 2000s", and as a background document for future international cooperation and agreements. It will be invaluable as a progress report for past donors to the SETI Institute. It will be indispensable as a cornerstone for the private fund-raising necessary to implement the plans, and for any discussions that may take place on funding by government agencies. Accordingly, this book, published by the SETI Press, is available to anyone.

1.2.1   Objectives

The following list contains the major objectives of the SSTWG.

1. Re-examine in the broadest way the fundamental philosophy and rationale of SETI.

2. Lay out the science requirements for the design of searches and systems for the years 2000 through 2020.

3. Explore computing and signal processing options for systems which would meet the science requirements.

4. Examine microwave search and technology options for systems which would meet the science requirements.

5. Investigate infrared (IR) and visible band search strategies for systems which would meet the science requirements.

6. Describe design drivers and logic rules for selecting from all of the options.

7. Select the preferred technical approaches from all of the options.

8. Develop specific conceptual designs for systems to implement these approaches during the years 2000 through 2020.

9. Prepare approximate costs and schedules for the selected projects.

10. Analyze major questions concerning strategies, advocacy, resources, and implementation for the Science and Technology Plan.

11. Achieve the widest dissemination of the report by publication in the open literature at minimal cost.

These objectives were followed approximately in sequence in the deliberations of the SSTWG. They are also expanded in the same sequence in the body of this book.

While the recommendations in this book were developed primarily for the SETI Institute, they could represent a broad prescription for the next phase of the evolution of SETI as a whole, and they may be of interest to practitioners of this young discipline around the world. It will surely lead to collaborative ventures among them. Since scientists and engineers from the SETI Institute were intimately involved in the preparation of the Plan, it is probable that the SETI Institute's projects will follow the recommendations closely, always depending on the resources that become available.

SETI has important implications for the future of society on Earth. The actual discovery of a signal will lead to sociological, intellectual and practical changes in our civilization. At least there would be immediate questions to answer, for example: "Should we respond to the signal?" "What type of follow-up SETI projects should be implemented?" Although this book primarily deals with the science and technology of SETI over the next twenty years, the SSTWG felt that these questions are so important that they should be mentioned in this book. It also should reference the limited amount of work currently addressing these questions, and include recommendations for further studies. These matters are discussed further in Chapter 9, the *Epilogue* on page 277.

1.3   Rationale

In a letter to Sky and Telescope in March 1990, Barney Oliver wrote:

> "To me SETI is a search for proof that natural selection and evolution are ubiquitous and that they frequently lead to beings as complicated as humans. We SETI buffs are enthralled by the knowledge that on this little planet the wonderful laws of physics have, in a few billion years converted the ravening chaos of the Big Bang into the most delicate and complex of structures; into spider webs and apple blossoms and leaping trout; and above all into brains capable of modeling the exterior world and puzzling out its origin. We want to know if this astonishing transformation is a local freak event or an inherent property of the Universe. We very much suspect the latter." [Oli90].

"We want to know..." Those four words sum up the rationale for an observational SETI program. "Where did we come from?" and "Are we alone?" are two of the oldest unanswered questions of the human race.

We live at a time when new instruments deployed on the ground, in space, under the oceans, and within the polar ice caps are helping us to piece together the story of our origins. We cannot yet make life in a test tube, but we can make artificial life in a computer and we can tailor individual molecules to perform tasks necessary for life as we know it and for alternative life as we do not yet know it. We do not yet fully understand how galaxies condensed so rapidly in the early Universe, but we can begin to model the process in our computers. The unified physics needed to compute the outcome of the first instants of the birth of our Universe still eludes us. Our instruments and computational models, however, have shown us that there was a beginning and have allowed us to formulate strategies for teasing out the details.

Finally, we live at a time when it is possible to contemplate an experimental approach to answering the second unanswered question "Are we alone?"; an approach that does not rely on appeal to a higher authority or to a prevailing philosophy. Our computational resources are beginning to be sufficient to allow us to look for cosmic neighbors in a systematic way. The search can be made in many ways. It is impossible to say whether any of these searches will be adequate, or when any particular search strategy might yield an answer. At the inception of SETI as a new scientific discipline, Cocconi and Morrison, the authors of the first refereed paper [Coc59] stated: "The probability of success is difficult to estimate; but if we never search, the chance of success is zero." That situation remains unchanged.

Figure 1.2: Dr. Bernard (Barney) M. Oliver in front of an artist's conception of the proposed Cyclops antenna array. (Photo of Dr. Oliver courtesy of Hewlett-Packard Corp.)

Because we *want* to know, the SETI Institute is attempting to structure an organizational entity whose longevity and fiscal soundness will be adequate to the task of searching, perhaps for many generations. As a species, we have a poor track record for completing multigenerational projects, particularly ones that are driven by curiosity. In the past, we have taken generations to build monuments such as the great cathedrals of Europe, but these tasks have been dictated from a strong central authority. How long will we persevere out of curiosity? When, if ever, should we give up? How should we proceed?

We have wondered about our place in the cosmos for so long, across so many cultures, and through so many cultural evolutionary changes, that it seems safe to assume that this question will remain of importance for the indefinite future. Thus, if there is a way to continue the search, there will likely be a will to do so. Whether there is a way will depend on how many resources are expended on the search.

Searching the Universe, or even just the Milky Way Galaxy, for signs of other technological species is an enormous task, but not one on which enormous resources should be concentrated. With no end in sight, and no knowledge of what is demanded for success, a vast expenditure of resources would soon be stopped by public consensus. This would occur if other, more immediate, problems went without solution because of a diversion of those resources. Even when the search was conducted as a government project under NASA, the resources devoted to it were less than 0.1% of NASA's budget. However, from the viewpoint of an individual researcher in another discipline, whose grant was denied for lack of funds, this represented a large expenditure.

When evaluated as an exploratory scientific project, dealing with a topic of enduring interest, the allocation of resources to SETI was appropriate. Why then did it not continue? The annual appropriations of funds in a political arena are, in hindsight, a poor venue for support for an open-ended quest. We need to fall back on an historical precedent, and primarily rely on philanthropy to fund the search. Thus, we conclude that the magnitude of the search should be scaled so as to be commensurate with the philanthropic capabilities of the world's visionary individuals of great wealth.

As was much of astronomy and science in the past, so too should SETI be conducted through patronage. Where possible, public/private partnerships should be developed to allow SETI to implement and share observational facilities with the astronomical community. This approach probably will not permit the largest scale searches that could be conceived during any technological epoch. Until some technology or strategy can promise a definitive answer to the question, this is the right way to enable SETI.

Given our state of ignorance, how should we proceed? As technology changes, it is necessary to re-examine and perhaps alter search strategies of the past. Yet sound strategies should not be abandoned simply because they are old, and have not yet borne fruit. This book is an attempt to chart the course of this multigenerational endeavor for at least the next generation. We shall feel gratified if this book correctly identifies the principles that will guide the strategies and decisions to be made by the SETI Institute for the next two decades.

1.4   Value

It has taken four billion years of tortuous evolution to evolve sentient beings on Earth, and centuries of hard-won scientific knowledge to set the stage for an endeavor whose positive outcome would surely be the most dramatic and profound moment in human history. It is only at the turn of the millennium that the SSTWG finally is undertaking the *coordinated* effort needed to search for purposeful signals from distant life. Yet, that search is likely to occupy decades. Years of effort may go unrewarded, and experiments will fail. Or will they?

Consider one of the recommendations of this book, the 'One Hectare Telescope'. It is an inexpensive radio telescope, radical in design, able to collect wide-band data from multiple celestial sources simultaneously at any band from 1 to 10 GHz, with extraordinary rejection of interference. Bold and elegant in design, this project is exploring new methods of signal processing, using a modular architecture of mass-produced dishes and electronics that invites future growth in size and performance. Its observational capabilities will rival, and indeed exceed, many of the world's best observatories. Its data, much of which will be at previously unexplored wavelengths, will be accessible to all. It explores new space, and is a project that generates tremendous excitement.

What is the lesson here? The 1hT, designed particularly for the specialized enterprise of SETI, is at the same time important to the broader scientific community. Both the innovative design of the instrument and the data collected will benefit the worldwide scientific enterprise. This is deliberate. For example, if an additional investment of 10% could double the productivity of traditional (non-SETI) radio astronomy, such an enhancement would be eagerly embraced by the SETI-based instrument designers. Such ideas are actively solicited.

We conclude with the observation that, historically, the greatest discovery of nearly every astronomical instrument has been different from the goal for which it was built. Table 1.1 on page 14 contains a modest list of examples.

Although we cannot predict what will come from the SETI science instruments and observations, it is certain that the technological spin-off will be of great value to science and technology. Every quality research endeavor is ultimately valuable, both to those who run it and to the larger community, regardless of whether it delivers the preconceived goal.

1.4.1   How the SSTWG reached their Conclusions

As the SSTWG participants began their deliberations, they had the benefit of the exploration of SETI possibilities, which was laid out decades ago in the

Instrument	Discovery	Original Purpose
Herschel 20 foot	Extragalactic Nebulae	New Galactic Catalog (NGC), to find comets
Mt. Wilson, 100 inch	Cepheids in Galaxies	Stellar Spectroscopy
Jansky Antenna	Radio Astronomy	Noise in Radiotelephony
Jodrell Bank Dish	Gravity-Lensing, Quasars	Meteor Trails
Palomar Mountain, 200 inch	Quasar Redshifts	Extragalactic Astronomy
20 foot Holmdel Horn	Primordial 3 K Radiation	Satellite Ground Station
Vela Satellite Array	Gamma Bursters	Detect Nuclear Tests
Cambridge Array	Pulsars	Interplanetary Scintillation
Arecibo	Gravitational Radiation	Ionospheric Backscatter
36 foot NRAO Dish	CO, etc., in Interstellar Medium	Far-Infrared Continuum
WSRT	Flat Rotation Curves	Source Counts
VLA	AGN Jets	Imaging

Table 1.1: Some Instruments, their Discoveries and Original Purposes.

Project Cyclops report[1]. For the reader unfamiliar with the *Project Cyclops* report, Appendix A, *Cyclops Revisited* on page 283, recapitulates its premises and conclusions. This section attempts to succinctly summarize how our knowledge base and technology have changed since then. It discusses the reasons why this book draws a somewhat different set of conclusions about how to proceed for the next two decades.

The initial intention was to be as inclusive as possible, considering a wide innovative range of technologies. At first, all possible technologies and search strategies would be considered. Additional studies would be conducted as needed, finally becoming selective only after comparative assessment. In reality that did not happen. Many ideas were discussed in the initial meeting of the SSTWG. By its close, the direction for the rest of the endeavor had been set, thanks to the synergy and expertise of the participants. During the meeting, it became clear

---

[1]In this book, we are using the following terms related to "Cyclops": 1. Cyclops Design Study – The "1971 Summer Stanford University / NASA Ames Faculty Fellowship Program in Engineering Systems Design" that resulted in the publication of the *Project Cyclops* report; 2. *Project Cyclops* report – the 1972/1973 published report resulting from the Cyclops Design Study; 3.Cyclops system – the radio telescope array and supporting equipment for SETI research, proposed by the Cyclops Design Study.

that there were new technologies which could provide immediate opportunities. Further, there was sufficient expertise among the participants to make the top-level design decisions. Microwave searches over the past 40 years had hardly begun to make a dent in the search, primarily because time was far too scarce on telescopes belonging to others. Dedicated facilities were needed for both targeted searches and sky surveys.

Since 1997, there has been ongoing work at the SETI Institute in collaboration with the radio astronomy community. This work indicates that a phased array of parabolic dishes, of modest size (about 30 m) is the desired configuration for a dedicated telescope for a targeted search. It would also enable some sky survey work. The key concept – using consumer commodity components, such as small satellite television (TV) dishes rather than 30 m antennas, to make the array affordable – surfaced during the first meeting of the SSTWG. Subsequent meetings refined the initial concept for the 1hT, and gave great consideration as to how it might leverage the international Square Kilometer Array (SKA) a decade hence. The SSTWG found themselves emulating the Project Cyclops strategy of building a large array in stages. Then, when detection occurs, it will be achieved with the minimum investment in search facilities.

At the first meeting, the review of the Project Cyclops premises and conclusions quickly pointed out that the rationale for ignoring the infrared/optical regime had largely been overtaken by events and technologies. Before the second meeting, several participants had conducted simple experiments using photon-counting detectors in anti-coincidence mode. The first infrared/optical search programs, in someone's backyard, were well under way before the final meeting. Deliberations at the later meetings concerned how such searches could be replicated, moved onto larger collecting areas, and turned into sky surveys.

The idea of a radio telescope that could look at much of the sky simultaneously to seek out strong transient signals, has been discussed for the past few years. The technologists who attended the meetings seized upon this idea as the ideal way to benefit from the continuing exponential growth anticipated for computational capacity. The discussions illuminated how to build a system in a way that should take maximal advantage of this growth.

In their task of creating a roadmap to guide SETI into the future, the SSTWG participants took a much more pragmatic and less academic approach than was envisioned by the meeting organizers. Does that imply that this book is likely to be too shortsighted, and to serve the community less well than the *Project Cyclops* report? On the contrary – because it forecasts technology trends, and considers costs and the ability to leverage other efforts, it may in fact be more effective for the community than was the *Project Cyclops* report.

The *Project Cyclops* report was inspirational, and many of the scientists and engineers working in SETI can trace their association back to it. However, by focusing on an ultimate Cyclops system that was so unaffordable, the intermediate, achievable stages were overlooked by most readers. During the subsequent three decades, no dedicated SETI observing facilities were built, and the signal processing advances we made are the result of technology not envisioned by the Cyclops Design Study. The recommendation of this book is to support new and innovative searches, thus helping to ensure that new technologies are embraced and utilized where appropriate.

1.4.2   Figure-of-Merit

It is always difficult to compare the merits of proposed research projects, since by definition they do not allow the comparison of results. The SSTWG participants tried to devise a quantitative figure-of-merit to compare all the suggested search strategies. While a specific figure-of-merit could be found to compare searches within a given wavelength regime, the SSTWG eventually concluded that there are no appropriate figures-of-merit for comparing searches using different parts of the spectrum (see Appendix F, *SETI Figure-of-Merit* on page 339).

The committee as a whole felt there were two main avenues to be pursued, since there is merit in diversity. First, the existing searches, optimized for continuous microwave sources, were still a sound idea. Modern technology could greatly improve these searches, making them less expensive, faster, with wider bandwidth and more sensitivity. Since all these goals are shared with the radio astronomy community, this seems like the right time for a combined effort in this area. The technology is ready, the SKA needs development effort, and the SETI community needs dedicated collecting area. This convergence of ends and means made the 1hT project a consensus favorite of the committee.

The second avenue of research is to pursue, or initiate, other approaches to interstellar communication. In particular, pulsed microwave beacons and pulsed optical beacons (see *Pulsed Beacons* on page 89), look at least comparable to continuous wave microwave beacons, and might well be chosen by ETs wishing to communicate. Furthermore, these represent unexplored corners of parameter space, and might reveal new natural phenomena. Therefore, an effort in these areas was warranted.

1.5   The Road from Ozma

It is nearly four decades since the remarkable scientific ideas that inspired this book burst upon the scene like a hot summer dawn. The year was 1959, and radio astronomers were grappling with the possibilities afforded by a new generation of low-noise amplifiers. Physicists Giuseppe Cocconi with Philip Morrison at Cornell, and astronomer Frank Drake at the National Radio Astronomy Observatory (NRAO) in Green Bank, independently realized that these improved devices were capable of catching artificially produced radio signals from other star systems. Mounted on large radio telescopes, these amplifiers could detect signals transmitted by equipment no more sophisticated than that which humans could both conceive and fashion.

This realization, and particularly the first implementation of these ideas in Drake's Project Ozma, created the possibility of finding intelligent beings far beyond the limited compass of our solar system. For a dozen years thereafter, a series of studies fleshed out the conceptual and practical aspects of the newborn discipline, soon christened SETI. These first efforts led to what might be called the conventional wisdom of SETI.

A list of important events in this early period would surely include the following:

- In 1959, Cocconi and Morrison [Coc59], published an article in *Nature*, which quantitatively established the feasibility of interstellar signaling by radio, and suggested that 1,420 MHz, the spectral location of the famous hyperfine line of neutral hydrogen, was the natural hailing frequency.
- Drake's two-month experiment in 1960, known as Project Ozma, [Dra61a], [Dra61b] in which Drake scrutinized a pair of local stars, as described in Section 1.5.1 starting on page 18.
- The Drake Equation was introduced in 1961 and first published in 1965 [Dra65]. This set the agenda for the first Green Bank Conference on SETI and for most SETI thinking since.
- Two highly influential and accessible books on the subject of SETI were published in the early 1960s – A. G. W. Cameron's collection of articles on *Interstellar Communication* [Cam63], and Iosef Shklovskii and Carl Sagan's *Intelligent Life in the Universe* [Shk66].
- The 1971 *Project Cyclops* report [Oli71], by Bernard Oliver and John Billingham, is a *tour de force* of engineering studies.
- In 1977, NASA published [Mrp77] a report of a two-year series of Workshops chaired by Philip Morrison, bearing the straightforward title, "The Search for Extraterrestrial Intelligence – SETI ."
- The *SETI Science Working Group Report* [Dra83], edited by Frank Drake, John Wolfe, and Charles Seeger, was published in 1983. This small volume was the blueprint for the NASA SETI program, later canceled by a Senate amendment in October, 1993.

The documents itemized above show a steady consolidation of ideas about how we might most effectively search for evidence of cosmic company. Now, with the perspective of two score years, the SSTWG has been convened to reconsider the conventional wisdom of SETI, to decide what strategies are worth continued effort, and which should be replaced. In our attempts to plot a course that will lead to future discovery, it pays to examine where we have been, and why we went there.

The remainder of this chapter describes the road from Project Ozma to the present. The description is certainly incomplete, inevitably biased, and of necessity omits many things that others will believe important. The authors have simply tried to note a few milestones that they hope, will delineate the broad contours of the topography thus far traveled.

Figure 1.3: Dr. Seth Shostak, of the SETI Institute, writes his own caption.

1.5.1 Project Ozma

While pioneering experiments are frequently canonized by science, it is remarkable when this occurs for an experiment that failed. Project Ozma did not detect transmissions from either of its target stars Tau Ceti (11.9 light years) or Epsilon Eridani (10.7 light years). It did, however, open a new field of inquiry, and bounded the parameter space of many subsequent observations. In 1985, Frank Drake gave a talk that eloquently described his experiment. The following paraphrase of those remarks emphasizes the aspects that are most relevant to the present work.

In 1959, the fledgling facility of the National Radio Astronomy Observatory (NRAO) in Green Bank, West Virginia, was looking for projects. In a move designed to get up and running quickly, the observatory decided to buy a commercially available 85 foot telescope from the Blaw-Knox Corporation. This would serve as an interim instrument until the completed construction of the far more ambitious 140 foot antenna. Drake, who already had an interest in the possibility of ETI life, calculated the range at which the 85 foot telescope, outfitted with a new, low-noise amplifier, could hear the strongest terrestrial transmitters. The distance turned out to be a few tens of light years. This surprising result encouraged Drake to propose a search for ETI broadcasts, and his idea was favorably reviewed by an observing committee of one – the NRAO director, Otto Struve.

Although his project was based on solid calculations, Drake recognized that the subject matter was sensational, and possibly controversial. Consequently, he kept his idea quiet, and did the experiment 'on the cheap'. The choice of frequencies, for instance, was made on the basis of the receiver's ultimate usefulness for studying the 21 cm neutral hydrogen Zeeman effect. Most of the $2,000 out-of-pocket cost of Project Ozma was used to purchase the narrowband filters used in the back end.

About six months into the project, the paper by Cocconi and Morrison [Coc59] appeared in *Nature*. Struve worried that the two Cornell professors would get all the credit for the SETI idea, and he broke the story of Project Ozma. The equipment was hurriedly completed, and observations began on 11 April, 1960. As an historical perspective, note that this was less than eight decades since the discovery of radio waves by Heinrich Hertz.

An immediate benefit of Struve's publicity was an offer by Microwave Associates, a commercial firm, to loan the project a parametric amplifier. The noise temperature of the parametric amplifier was a respectable 350 K. This was only achieved after a laborious tuning process, which Drake remembers vividly because he was obliged to undertake it at the focal point in the numbing chill of the early morning. The instantaneous bandwidth, set by the narrowband filters, was 100 Hz, and this signal band was differenced against a 1,000 Hz comparison band.

Project Ozma's single channel was slowly scanned up and down the dial by a motor attachment to a commercial receiver used as the back end. The scan rate was 100 Hz every 100 seconds, and the total bandwidth covered was about 360 kHz, corresponding to 76 km/s at a wavelength of 21 cm. This amounts to a serial examination of 3,600 channels, centered at 1,420 MHz. The total observing time expended on Project Ozma was about 150 hours, and the sensitivity achieved was approximately 10^{-21} W/m^2. All receiver output was sent to a chart recorder.

The only positive result of this effort was a pulsing noise source detected while the 85 foot was pointed at Epsilon Eridani. Later observations with a small antenna showed that this source appeared in other directions, thus confirming its terrestrial origins.

While Project Ozma failed to detect ETIs, it stimulated interest for new searches and for further refinement of SETI strategies. Within a year, a conference was held at Green Bank, and the Drake Equation was first offered to the research community.

Despite the considerable progress since Project Ozma, it is noteworthy how much experimental territory was reconnoitered by Drake. Project Ozma was a targeted search of nearby, solar-type stars, a strategy being pursued today as Project Phoenix. His integration time of 100 seconds is remarkably close to that used by contemporary targeted searches. The observed band, near the 21 cm line of hydrogen, continues to be the spectral region of choice for SETI. Even Project Ozma's use of a second antenna to sort out terrestrial interference has a modern analog. Was Drake prescient, or did subsequent experiments simply follow established tracks? That, of course, is one of the underlying questions being considered by the current SSTWG.

1.5.2   Cyclops Design Study

In 1971, the NASA Ames Research Center funded a summer faculty fellowship program whose objective was "to assess what would be required...to mount a realistic effort using then current, or near-term future, state-of-the-art techniques, aimed at detecting the existence of extraterrestrial intelligent life." The result was the *Project Cyclops* report [Oli71], a comprehensive, detailed, and remarkably durable description of the SETI enterprise.

In thinking of Project Cyclops today, the image that all-too-often comes to mind is a daunting array of 100 meter antennas splayed across a tract of real estate 16 km in diameter. However, this was the ultimate system, estimated to cost tens of billions of (1971) dollars. What is sometimes overlooked is that the intention was to start with a small group of antennas, increasing the size of the array only as needed. With this strategy, if an ETI were detected, it would be achieved with the minimum instrument capable of success. It is important to note that the economics of antennas at the time of this study favored individual elements that were as large as possible, and antenna cost dominated the expense of the system.

Figure 1.4: Dr. Frank D. Drake in front of the now-defunct 300 foot radio telescope in Green Bank, West Virginia. The 85 foot telescope used for Project Ozma (1960) is located approximately one mile away, in the direction of Drake's gaze. (Photo courtesy of Green Bank Observatory, NRAO.)

The Cyclops antenna array was designed to focus its antennas on nearby solar-type stars – a targeted search that was an obvious extension of Project Ozma. A simple model fitted to astronomical data allowed a statistical description of the data. The stellar density of F0 through K7 stars within 500 ly of the Sun can be approximated as:

$$n = \rho \frac{4}{3} \pi R^3 \qquad (1.1)$$

where R is in light years and the local density $\rho = 5.4 \times 10^{-4} ly^{-3}$.

This allows a straightforward estimate of the maximum distance of targets. If the probability of detection per stellar target is 10^{-3}, then stars out to about 100 light years (ly) must be scrutinized. If the probability is only 10^{-6}, then the hunt must extend to 1,000 ly. This assumes, of course, a thorough search.

It is noteworthy that if the latter distance limit proved inadequate, the authors of the report felt less hopeful of success: "Beyond 1,000 ly the situation becomes rather bleak. Not only does the cube law [describing the density of stars] fail, but also the radiative epoch becomes shorter than the round-trip delay making two-way exchanges unlikely." The first bit of pessimism is only geometry, and can be remedied by searches that concentrate on the Galactic plane. The latter is founded on the dystopian opinion that technological civilizations are rather short-lived, or at least that specific transmitting projects are.

Note that the arguments summarized here have an important corollary. If the likely SETI targets are stars within 1,000 ly, and a microwave search is undertaken, then at 3 GHz even a small, 10 meter dish will only have about six targets within its field of view. To scan many stars simultaneously is difficult if the instrument is constructed of highly directional elements, whether it is a single instrument or an array of instruments.

The Cyclops system was intended to cover frequencies from 0.5 to 3 GHz, with an instantaneous bandwidth of 100 MHz and spectral resolution of 1 Hz. The data processing envisioned first subdividing the signal using filters into bands 1 to 10 MHz wide. The time signal of each subband would be "DC-shifted" (to avoid negative values) and used to modulate a light beam recorded onto film as a two-dimensional raster. Illuminating the film with coherent light and passing it through a lens would accomplish the Fourier Transform, and the latter was to be read out by a high-resolution TV vidicon and written to disk. Within a half-decade, the appeal of this analog approach was trumped by the possibility of digital computation.

Because it was a report of a design study, much of the *Project Cyclops* report consisted of engineering details. Obviously, some of these are dated. Nevertheless, those who read the report today often remark on both the clarity of the analysis and on how much of it is still germane, nearly three decades later. In a very large measure, this is a tribute to its principal author and editor, Barney Oliver.

1.5.3   The Post-Cyclops Era

The heady Cyclops era soon gave way to a period during which money for science flowed less copiously. By the time of the 1977 SETI report [Mrp77], based upon a series of Workshops chaired by Philip Morrison, a more restrained approach to the search was being developed. That report, intending to assuage fears that

Figure 1.5: Dr. Larry Lesyna of Grumman Corporation, brings a clear Cyclops perspective to a new SETI generation.

only a full-blown Cyclops instrument could succeed, encouraged the initiation of a search using existing antennas, and "low-cost state-of-the-art receiving and data processing devices".

The 1977 SETI report once again endorsed the targeted search strategy, and envisioned an examination of nearby stars for signals of terrestrial strength, and reconnaissance of the entire Galaxy for far more powerful broadcasts. It continued to focus on the low end of the microwave window.

Four clear requirements for the search were:

1. large antennas;

2. low-noise, broadband microwave front-end amplifiers;

3. devices for resolving the spectrum into narrowband components;

4. an automated system for recognizing signals.

To get SETI underway, existing hardware could be used for items (1) and (2). The Cyclops Design Study envisioned an optical analog system for item (3), the spectral analysis. It was recognized in 1977, however, that integrated circuits (ICs) had evolved to a stage where a million-channel, all digital Fourier analyzer was feasible. Thus the report encouraged a modest search, as both proof-of-concept and proof-of-necessity, for the far more ambitious Cyclops-type instrument.

By the time the third major SETI study appeared in 1983, the *SETI Science Working Group Report* [Dra83], funding for a NASA program had been initiated (1981), terminated for a year due to Senator William Proxmire and his Golden Fleece Award, and reinstated. Battles for the initiation, defense, and survival of NASA's SETI efforts during this period were won largely due to the indefatigable efforts of John Billingham, at Ames Research Center.

The *SETI Science Working Group Report* laid out in detail what would become the NASA SETI program. It was an effort known originally as the Microwave Observing Project, and finally as the High Resolution Microwave Survey (HRMS). While much of the strategy followed the well-known path blazed by Project Ozma and the Cyclops Design Study, there was something new. The *SETI Science Working Group Report* urged a two-pronged attack on the problem of uncovering cosmic sentience – a Targeted Search and a Sky Survey. The Sky Survey would be three orders of magnitude less sensitive than the Targeted Search, but would make no assumptions about the habitat of the aliens. The rationale was straightforward: "As is true for stars, the nearest transmitters may not be the brightest." Indeed, powerful signals might originate somewhere other than near a star.

It was expected that the two-pronged effort planned for the NASA program would represent a 10 millionfold improvement over the search volume examined by all previous experiments combined. This would be accomplished without the construction of new antennas.

The *SETI Science Working Group Report* once again identified the low end of the microwave band as most auspicious for a SETI search. The targeted search would examine approximately one thousand nearby (less than 100 ly) solar-type star systems, one at a time. The proposed search band was 1 to 3 GHz, with resolutions as fine as 1 Hz and integration times of up to 1,000 seconds. The maximum number of simultaneous channels was expected to be 8 million. The Sky Survey would use 32 Hz channels, integrate for three seconds, and

examine frequencies between 1 and 10 GHz. The former would be carried out
on the world's largest radio telescopes, including the Brobdingnagian 305 meter
Arecibo antenna. The latter would be tasked to NASA's 34 meter Deep Space
Network antenna, at Goldstone, California; Tidbinbilla, Australia; and Madrid,
Spain.

The fact that observing time on large telescopes is highly prized, and therefore
difficult to get, combined with the frequency limitation of existing telescopes,
contributed to the decision to restrict the total bandwidth examined in the
Targeted Search. Similarly, the decision to use channels of 1 Hz width, rather
than something narrower, was made in the interests of covering a substantial
amount of total bandwidth with only a half-decade's worth of observations.

While leakage signals were expected to be both weak and complex, deliberate
beacons would presumably have a highly monochromatic component that could
be easily detected. A fan beam antenna could be used by ETIs to sweep targets,
for example by rotating the beam in the plane of the Galaxy. This would result
in the same average received power as an isotropic transmission, except that the
pulses would be considerably more visible. Since this is an obviously attractive
situation, the NASA program was conceived as being able to search for signals
that might be concentrated in both frequency and time.

The expected sensitivities for the NASA program were 10^{-23} W/m^2 for the Sky
Survey and several times 10^{-27} W/m^2 for the Targeted Search. Even the former
would be adequate to detect a signal from an antenna the size of the Arecibo
with its 1 MW antenna, at 50 ly.

The *SETI Science Working Group Report*, the last major report before the
NASA SETI program, provided a blueprint for that effort which was largely
followed. After seven years of research and development, the Targeted Search
began observations at Arecibo on October 12, 1992. Simultaneously, the Sky
Survey was switched on at Goldstone. One year later, with less than 0.1%
of the planned observations completed, Senator Richard Bryan introduced the
amendment that ended the NASA SETI effort. Given the limited exploration
that had been made in the first year, this was as if Isabella had called back
Columbus while his ships were still within sight of the Spanish coast.

1.5.4   Current Targeted Searches and Sky Surveys

At present, there are three SETI sky surveys in operation: The Search for
Extraterrestrial Radio Emissions from Nearby Developed Intelligent Populations (SERENDIP), Southern SERENDIP, and the Billion Channel Extraterrestrial Assay/Mega-channel Extraterrestrial Assay (BETA/META). Additionally, there is Argus, an amateur effort, and one targeted search, Project Phoenix.
The last is a direct outgrowth of the targeted search component of the NASA

SETI program, and is privately funded via the SETI Institute. The numbers would suggest that sky surveys are today's strategy of choice but such a conclusion is unwarranted, as two of the three sky surveys are commensal. They are performed concurrently with conventional radio astronomy observations, and examine patches of sky dictated by others.

The types of signals sought by targeted searches and sky surveys differ. When scanning the sky, no slowly-pulsed signals can be effectively found, for only continuous wave emissions will be certain to be 'on' when the beam passes over their source. Nonetheless, because of the inefficiency of square law detection, pulses are easier to detect than CW signals for a given average power. Thus, the NASA Targeted Search was better attuned to pulses than to CW. The pulses were assumed to be regularly spaced in time, invariant in shape and, like the CW signals also sought by the Targeted Search, possibly drifting in frequency.

1.5.5   Frequency Selection

Although hundreds of radio lines have been discovered since the publication of Cocconi and Morrison's article, the 1,420 MHz (21 cm) line of hydrogen remains the most popular frequency for SETI. Neutral hydrogen is enormously abundant, of course, constituting 1% to 10% of the observed mass of spiral galaxies. It will likely continue to be radio astronomy's spectral *prima donna*.

However, its favor among those doing searches derives from more than its obvious esthetic appeal as a logical hailing frequency. The microwave band was long ago recognized as a relatively quiet part of the spectrum. In 1959, the Galactic continuum-radiation at low frequencies was known, as was the atmospheric radiation at high frequencies. Within a dozen years, two additional components were added to the noise spectrum. The first was a 'floor' generated by the 3 K remnant radiation from the Big Bang – an inevitable price to be paid for the Universe's birth. The second, originally noted in A. G. W. Cameron's book *Interstellar Communication* [Cam63], and reiterated by Barney Oliver in the *Project Cyclops* report, is the spontaneous quantum noise, $h\nu/k$. This serves as an upper bound that defines a region of minimal noise in the radio band between 1 and 100 GHz. This relatively quiet region is called the 'free-space microwave window'. For observations made from Earth's surface, atmospheric water vapor and oxygen lines (and their re-radiation emission) clog the spectrum between 10 and 60 GHz, thereby delineating a somewhat narrower terrestrial microwave window.

One might consider extending these traditional windows towards the low end by observing at high Galactic latitude. However, there is insufficient variation of the low-frequency synchrotron radiation to make this especially attractive. The noise temperature of Galactic radiation changes by only a factor of two between a declination (b) of 5 and 90 degrees (Galactic pole).

The *Project Cyclops* report endorsed the lower end of the microwave window (in particular, the so-called 'water hole' extending just beyond the lines of H I and hydroxyl (OH), approximately 1,400 to 1,700 MHz). It did so for several good reasons, including the appeal of encountering other societies at a virtual oasis. Some of these endorsements were derived from straightforward engineering considerations, such as the less stringent frequency stability required at the low end of the band.

A more fundamental argument arises when one considers Doppler shifts due to planetary motions. The rationale is simple:

> Assume that accelerations, caused by axial and orbital spins, are of magnitude a inducing a frequency drift $d\nu/dt = \frac{a}{c}\nu$ for a signal of frequency ν.

> If the channel bandwidth is B, then the maximum integration period is $t = B/(d\nu/dt)$, because this is the time required for an extraterrestrial signal to drift to an adjacent channel.

> However, an optimal matching of bandwidth and integration period will have $t = 1/B$.

> Therefore:

$$B^2 = d\nu/dt = a\nu/c \qquad (1.2)$$

> and

$$B = (a/c)^{1/2}\nu^{1/2} \qquad (1.3)$$

In other words, the total noise, proportional to bandwidth, increases by an additional factor of $\nu^{1/2}$, simply because of the velocity drifts due to rotational and orbital motions of the ETI's planet. Once again, this favors the low end of the microwave spectrum. (Note that a de-drifted beacon does not suffer this disadvantage.)

Simple considerations of noise show that a signal will have the greatest visibility for a given transmitter power if it is confined in either frequency or time. Drake and Helou [Dra78], and Cordes and Lazio [Cor91a] have noted that even a perfectly monochromatic signal will be Doppler broadened to significant breadth when scattered by turbulent, ionized interstellar gas. The minimum width of any line will typically be 0.01 to 0.1 Hz at ∼1,400 MHz, depending on path length.

In an effort to match this natural bandwidth, Paul Horowitz has constructed a series of SETI receivers with resolutions of 0.05 Hz [Hor93]. Observations

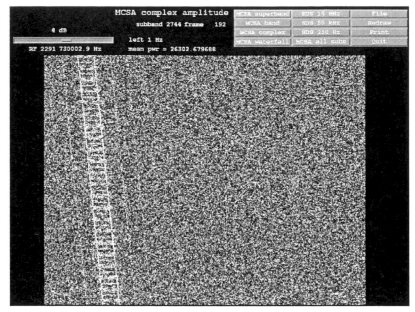

Figure 1.6: This is a portion of a modulated data stream within the lower sideband from the Pioneer 10 spacecraft.

with such spectrometers require 'de-chirping' the spin and orbital motion of the Earth, as this can be as much as 0.16 Hz s^{-1}. Over the course of 100 seconds (the required sampling time to get 0.01 Hz resolution), this would amount to a 'de-chirp' of 1,600 channels. The benefit of this procedure is that terrestrial interference is smeared out over all those channels, whereas the ETI's signal piles up in one.

In addition to removing the Earth's motions, the use of very narrow bandwidths without any means of looking for drifting signals can only be successful if the ETIs are considerate enough to adjust their transmitter to remove their own planetary motions. Indeed, unless the total bandwidth considered is large, the ETIs had better also remove the radial velocity shift between their star and ours. These narrowband spectral machines are the only devices that have tried to match their resolution to the best that interstellar space can offer.

What about the message? While the details of signal modulation have only occasionally been considered in the SETI literature, an interesting order of magnitude calculation was made in the *Project Cyclops* report. This showed that microwave arrays a few kilometers in size are capable of accommodating bit rates adequate to send conventional television over distances of hundreds of light years. However, it requires six million times as much power to fill a TV bandwidth of 6 MHz as it does to send a 1 Hz beacon of equal or greater detectability. While television may strike readers as a parochial technology, note that it meets the minimal requirements for re-creating visual reality for humans. Assuming that the ETIs are living in an environment comparable to ours, the bandwidths of their visual systems, and therefore their low-grade television systems, might be similar.

Another computation made as part of the Cyclops Design Study was to compare microwave, infrared, and optical signaling systems. The well-known conclusion was that microwave is a superior signaling mode, partly because of the far smaller energy cost per information bit. Modern high-powered, pulsed lasers, which can confine a signal in time, initiated a reconsideration of this point of view, as described elsewhere in this book.

1.5.6   Existing and Historical Target Strategies

As early as 1985, Michael Papagiannis [Pap85] wrote that SETI had already "accumulated about 120,000 hours of observations in about 50 different search projects". The majority of this telescope time was expended either scanning the sky or zeroing in on mostly solar-type stars.

Despite the popularity of searching individual stellar systems, such an approach might require the patience of Job. At a 1985 Green Bank conference, David Frisch outlined the problem, assuming (as many have done) a symmetrical approach to transmitting and receiving [Mel85]. Suppose galactic societies optimize their hailing signals so that they broadcast in a given direction for a time t, which is only as long as it takes for other civilizations to detect them. If the number of candidate star systems is N_*, and if as much total time is spent listening as transmitting, then the wait before hearing a signal will be $N_*^2 t$. This could be a very long time, unless the ETIs are broadcasting with enormous power.

Suppose our goal is to have a SETI success within 100 years, presumably a timescale also considered reasonable by the ETIs, and further suppose that $N_* = 10^6$. Then we might expect the signal duration to be distressingly short – only three milliseconds, since the two stars spend only a millionth of a millionth of the time looking at each other. Thus, we would only have three milliseconds for our detection system to identify a signal from the background noise. Either the receiving system or the transmitter must be substantial. A similar argument applies to the fraction of time two high-gain antennas pointing randomly spend looking at each other, about a millisecond per millennium.

One way to deal with such disheartening arguments is to either observe large clots of stars, or consider special locations. For example, Woody Sullivan and collaborators have made limited scans of the galactic plane [Sul97]. Nathan Cohen and others have examined galactic clusters [Coh80]. Carrying this idea to its logical extreme, some observers have searched nearby galaxies, which

Figure 1.7: The 140 foot radio telescope at the National Radio Astronomy Observatory in Green Bank, West Virginia (used for Project Phoenix observations in 1996-1997).

ensures that even small beams will encompass tens of millions of stars.

In an early Arecibo experiment, Frank Drake and Carl Sagan [Sag75] observed such nearby objects as M33 and part of M31. More recently, selected regions of the Small Magellanic Cloud were scrutinized by Seth Shostak, Ron Ekers, and Roberta Vaile [Sho96]. While the distances for even this nearby galaxy are typically a thousand times that of the solar-type stars examined in most Targeted Searches, the power demands placed on the aliens may be less an objection than their incentive for such intergalactic broadcasts. As Barney Oliver remarked, "What would be the motivation to beam Andromeda if it takes four million years for a reply, especially when, for the same power, you could flood your own galaxy with a signal?"

Special locations that have been the subject of SETI searches include the galactic center [Sho85], the north galactic rotation axis (Lord and O'Dea), and supernovas. A more recent suggestion was to observe edge-on, wide binaries during eclipse, because the axis of radio traffic internal to such a binary system would be aligned with a vector to Earth [Sho97] at such times. At optical wavelengths, a limited search for alien artifacts has been made of the Lagrangian points of the Earth-Sun-Moon system.

Such approaches are attempts to improve the quality of targets chosen for high-sensitivity SETI observations, by specifying where, and sometimes when, a beamed transmission might be observed. So far, such approaches represent only a small fraction of the search effort.

For a more complete listing of SETI searches to date, refer to Appendix L, *Archive of SETI Searches* on page 381.

1.5.7   SETI and Astronomy

SETI and astronomy, especially radio astronomy, have always been tightly linked. Historically, SETI has been viewed by astronomers as a scientifically valid search with very uncertain odds of success but a potentially huge payoff.

Every ten years, the National Research Council commissions a group of astronomers and astrophysicists to survey their field and recommend new research initiatives for the coming decade. Following are excerpts related to SETI from the last three decadal reports:

Astronomy and Astrophysics for the 1970s (Report of the Astronomy Survey Committee, Jesse L. Greenstein, Chairman), 1972:

> "Our civilization is within reach of one of the greatest steps in its evolution: knowledge of the existence, nature, and activities of independent civilizations in space. At this instant, through this very document, are perhaps passing radio waves bearing the conversations of distant creatures – conversations that we could record if we but pointed a telescope in the right direction and tuned to the proper frequency.
>
> Indeed there exist the know-how and instruments to search for ex-

Figure 1.8: The SETI Institute's Dr. Peter Backus, high above the Arecibo telescope, reviews the Project Phoenix log book.

traterrestrial civilizations. Each passing year has seen our estimates of the probability of life in space increase, along with our capabilities for detecting it. More and more scientists feel that contact with other civilizations is no longer something beyond our dreams, but a natural event in the history of mankind that will perhaps occur in the lifetime of many of us. The promise is now too great, either to turn away from it or to wait much longer before devoting major resources to a search for other intelligent beings.

In the long run, this may be one of science's most important and most profound contributions to mankind and to our civilization".

Astronomy and Astrophysics for the 1980's (Report of the Astronomy Survey Committee, George B. Field, Chairman), 1982:

> "While the Committee recognized that this endeavor has a character different from that normally associated with astronomical research, intelligent organisms are as much a part of the Universe as stars and galaxies; investigating whether some of the electromagnetic radiation now arriving at Earth was generated by intelligent beings in space may thus be considered a legitimate part of astronomy. Moreover, the techniques that can now be most effectively brought to bear on a SETI program for the 1980s are those of astronomy.
>
> It is hard to imagine a more exciting astronomical discovery or one that would have greater impact on human perceptions than the detection of extraterrestrial intelligence".

The Decade of Discovery in Astronomy and Astrophysics (Report of the Astronomy Survey Committee, John H. Bahcall, Chairman), 1991:

> "Ours is the first generation that can realistically hope to detect signals from another civilization in the Galaxy. The search for extraterrestrial intelligence (SETI) involves, in part, astronomical techniques and is endorsed by the Committee as a significant scientific enterprise. Indeed, the discovery in the last decade of planetary disks, and the continuing discovery of highly complex organic molecules in the interstellar medium, lend even greater scientific support to this enterprise.
>
> Discovery of intelligent life beyond the Earth would have profound effects for all humanity".

These historic quotes show one aspect of the strong scientific endorsement SETI has enjoyed in the past. As the discipline of SETI looks forward to the next millennium, it is clear that there are important reasons for this momentum to be sustained. Exciting new discoveries, such as planets circling other stars, of possible microfossils in a Mars meteorite, and of a probable liquid ocean beneath the frozen surface of Europa, make the SETI enterprise all the more promising. It is becoming more and more apparent that there are likely to be billions of potential sites for life in our own galaxy.

1.5.8   Conclusions

In some sense, SETI has come a long way since Drake's pioneering observations in the spring of 1960. Figures 1.9 and 1.10 show the growth in antenna collecting area and the number of simultaneously-observed channels for the principal SETI observations. The growth in antenna collecting area amounts to a factor of five per decade, and the number of simultaneously-observed channels amounts to a factor of 100. This progress can be summarized by noting that today's targeted search, Project Phoenix, is 10^{14} times more powerful, in terms of parameter space examined, than Project Ozma.

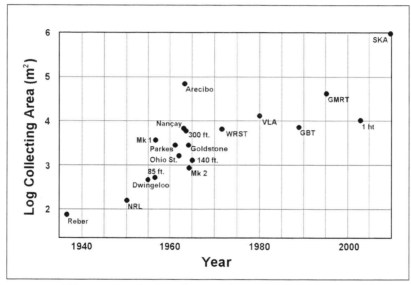

Figure 1.9: Radio Telescope Antenna Size vs. Time.

Estimates of the chance of success, based upon the Drake Equation, have become more optimistic in the last half-decade. This optimism has been encouraged by the discoveries of extrasolar planets and the growing evidence for possible life elsewhere in our solar system. The motivation to continue the search has been strengthened.

Despite the obvious progress, much of what is being done today was set in motion over twenty years ago. Project Phoenix adheres closely to the strategy and technical descriptions delineated in the 1977 *SETI Science Working Group Report*. This is not necessarily bad, and in fact, it might be considered something of which to be proud, particularly in view of the funding vagaries that have beset this enterprise.

While this chapter has necessarily dwelled upon the past, succeeding chapters address the future. In that regard, note that we have yet to complete the initial

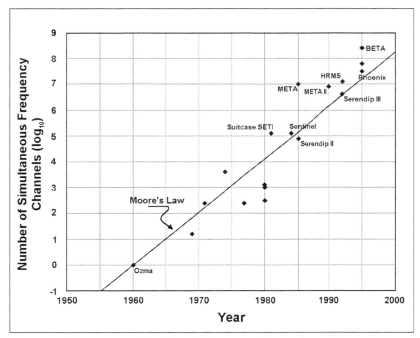

Figure 1.10: Number of Frequency Channels vs. Time.

reconnaissance envisioned two decades ago at the start of the NASA effort. So far, the Targeted Search has examined approximately 400 of its stellar candidates at a sensitivity level that is typically 10^{-25} W m^{-2}, which is two orders of magnitude poorer than first intended. (However, at Arecibo the sensitivity reached by Phoenix is about 10^{-26} W m^{-2}.) We have never really done the Sky Survey covering the 1 to 10 GHz band as originally defined. We now have a better appreciation of the debilitating effects of interstellar scintillation, brought to the SETI community's attention by Jim Cordes and others. This will affect our approach to surveys and, indeed, to any searches whose targets are farther than a few hundred light years away.

Building new hardware has traditionally received more attention from SETI researchers than considerations of strategy. As information technology revolutionizes the methods of SETI and radio astronomy, the power of equipment to detect signals will increase exponentially for many years to come. However, it must be remembered that novel search strategies and detection methods have the potential for setting us on a new road entirely. As our equipment gets better, we can expect to apply its power in entirely new ways.

Chapter 2

The Science of SETI

This chapter reviews the status of science and engineering related to SETI as of 2000, and it attempts to answer the following questions:

- What do we know about life on Earth, and what are the extreme conditions under which it can exist?
- Based upon what we know about life on Earth, where else in our solar system might life exist, now or in the past?
- What do we know about the possibility of life elsewhere in our galaxy?
- What do we know about planets around other stars?
- What do we know about the possibility of detecting terrestrial planets around other stars?
- What are the Habitable Zones around other stars, within which life as we know it can exist?
- How do the cosmic background and our environment limit our ability to detect extraterrestrial signals?
- What are some of the trade-offs between microwave and optical/infrared searches?
- What are the scientific risks – false positives, false negatives and bad science?

2.1   SETI Decision Tree

Issues covered in this chapter are highlighted in the SETI Decision Tree, Figure 2.1.

2.2   Life in our Solar System

In examining the science of SETI, it is useful to remember our not-so-distant past. The steady progress of our scientific body of knowledge produced an ever-increasing basis for the probability of life existing elsewhere than planet Earth. Five hundred years ago, Copernicus started a revolution in thought by showing that the Earth was not located at the center of the Universe, but instead revolved around the Sun. This implied that the Earth was similar to the other planets and that the Sun was like other stars. In one stroke, this discovery challenged theories about the nature of the Universe and raised the question of whether other planets like Earth might exist.

How has our scientific knowledge expanded since then? We walked on the Moon, dug in the lunar soil, and even lobbed a golf ball across that rocky surface. Most importantly, we photographed our home planet suspended in the blackness of space, protected by a fragile atmospheric cocoon. Viking missions went to Mars to sample the soil, and the Mars Pathfinder actually rolled about on the surface. Spacecraft traveled to many planets, sending back pictures and data that often challenged previously held scientific beliefs.

Here on Earth, we are gaining important knowledge on how living systems emerge. Recent studies have shown the huge range of conditions under which microbial life forms can, not only exist, but flourish. Microbes are found in places that previously were thought to be chemically toxic to all life, such as acid-rich hot springs, alkali-rich soda lakes and saturated salt beds. They are found in extremely cold places, such as in Antarctic rocks, and at the bottoms of perennially ice-covered lakes. They are found in very hot places, such as in deep-sea hydrothermal systems, at temperatures of up to 113 °C. Bacteria have been discovered in deep subsurface aquifers, possibly deriving energy from a geochemical reaction between basalt and water. Bacteria have even been found in the cooling systems of nuclear reactors. Some micro-organisms survive ultra-violet radiation, while others tolerate extreme starvation, low nutrient levels and the near lack of water. Not only can some microorganisms survive in extreme conditions, but their spores also can survive for a very long time under even more extreme conditions. Surprisingly, spore-forming bacteria are reported to have been revived from the stomachs of wasps entombed in amber for more than 25 million years. Clearly life is remarkable, diverse, tenacious and adaptable to extreme environments, and it had to be.

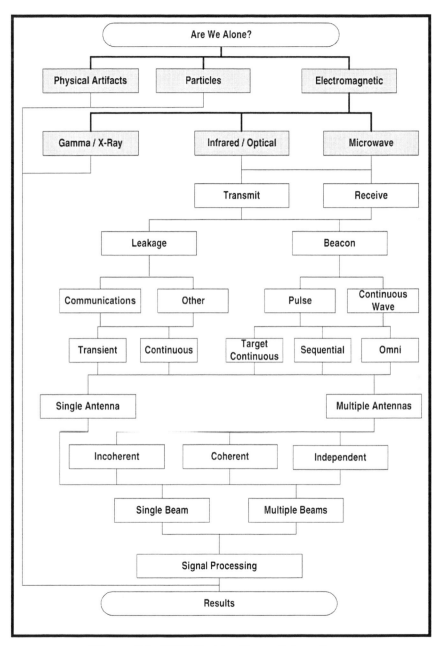

Figure 2.1: SETI Decision Tree – Science Issues

On the early Earth when life was just beginning, the conditions were extreme. The late heavy bombardment of meteors, asteroids and comets was still producing ocean-vaporizing impacts. The fainter young Sun meant that the Earth teetered on the brink of glaciation, depending upon the mix of greenhouse gases in the atmosphere. More than once it succumbed, freezing even the equatorial regions [Wil98].

Just as evidence of earliest life on Earth has been sought backwards in time in older and older rocks, the search for life elsewhere in our solar system extends outward in space. The search extends to Mars, Europa, and other planets and moons where habitable environments may have existed, and perhaps still do.

Mars once had liquid water on its surface. NASA spacecraft obtained images of features that look like dried-up riverbeds and flood channels that existed long ago when Mars was warmer and supported a thicker atmosphere. Recent images from the Mars Global Surveyor have eliminated all doubt about the accuracy of these observations, and raise the distinct possibility that life once may have gained a foothold on the red planet. The discoveries of possible microbial structures in Martian meteorites, found on the surface of the Antarctic ice sheet, increases the possibility. Proving the past or present existence of life on Mars, however, will be a major challenge. Scientists have begun to look for more detailed fossil evidence of ancient microbes in the Antarctic meteorites, and will develop methods to analyze the new samples. The question of whether the structures have a mineralogical or biological origin may not be answered any time soon.

New images taken by the Galileo spacecraft of Jupiter's Moon-sized satellite Europa show an ice crust. An enormous ocean may exist beneath this crust, heated gently by Jupiter's tides and by radioactive elements in the underlying rock. Could Europa support life forms akin to those found in the ice-covered seas of Earth's polar regions or deep in our own oceans? Throughout the Earth's deep oceans, life exists that never sees sunlight and feeds off chemicals from volcanic vents.

The atmosphere of Saturn's giant moon Titan is mostly molecular nitrogen like Earth's, but contains a wealth of organic molecules in a complex chemistry powered by sunlight. Might Titan's surface, as yet unexplored, contain the right conditions for experiments in prebiotic chemistry repeated slowly at low temperature over billions of years? Answers may come as early as the next decade.

History shows that each new major discovery about the Universe provides more insight into its uniformity. The factors that produced life here on Earth are likely to exist everywhere.

There is ample reason to believe that somewhere out there, around a star much like our own Sun, there exists a civilization that has produced technology to communicate with itself over long distances. This technology will take advantage of the laws of nature, just as does ours, and will discover where nature makes it most easy to communicate over long distances. It is the faint trace of such communications that SETI will someday discover. Is there scientific risk in such a pursuit? Of course, if you define risk in terms of obtaining a signal within a specific time frame. Is it worth the risk? Absolutely! The discovery, when it happens, will answer the oldest question of humankind: Are we alone?

Figure 2.2: Dr. Kenneth I. Kellermann, of the National Radio Astronomy Observatory, an early SETI observer who brings the radio astronomer's perspective to the SSTWG.

2.3   Our Galaxy

Life as we know it requires water in liquid form, on or near the surface of a planet in orbit around a star. There are at least 10^{22} stars in the Universe. The nearest 10^{11} stars are organized into the Milky Way Galaxy, a lens-shaped system of stars, gas, dust, and dark matter which is about 100,000 ly in diameter. The Milky Way Galaxy has a dense nucleus, about a light year in diameter, which contains millions of stars, and a central black hole. Surrounding the nucleus is a bar-shaped bulge 10,000 ly long. Further out is a thin, flat disk of relatively young stars and gas, where most active star formation takes place. In the outermost regions is an approximately spherical halo of old stars, globular clusters and dark matter. We are in the disk, about halfway out in radius and halfway through the thickness.

Suppose we need to search N stars. What volume of space will contain N stars? How far from Earth do we need to search? Figure 2.3 is a simplistic view of the number of stars (N) found within a sphere centered on us. The distance, in light years, is the radius of the sphere. It is the distance out to which we would need to search in all directions in order to sample N stars.

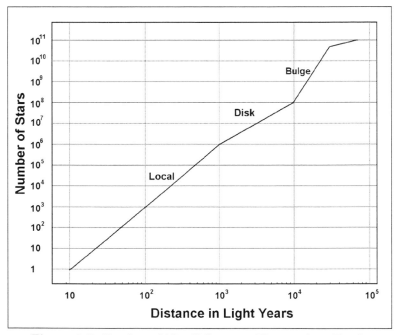

Figure 2.3: Number of 'Good' Stars vs. Distance from Earth.

Out to about 1,000 ly, we are looking locally, and the stars are distributed equally in all directions. This region is marked "Local" on the plot. This leads to an $N = O(d^3)$ relationship, where N is the number of stars within distance d of Earth. (The number of stars increases as the cube of the distance from Earth.)

However, the Galaxy is only about 2,000 ly thick at our location, so beyond 1,000 ly the increase of stars with distance slows to $N = O(d^2)$. (The number of stars increases as the square of the distance from Earth.) This region is marked "disk" on the plot.

Farther out, at about 10,000 ly, we reach the galactic bulge, and the number of stars rises rapidly, until at about 30,000 ly we have seen half the stars in the Galaxy. This region is marked "Bulge" on the plot.

Reaching all of the approximately 10^{11} stars in the Galaxy, requires searching out to a distance of about 70,000 ly. Beyond that (not shown on the graph), searching more stars requires extending the search to other galaxies, the closest of which is about 10^6 ly away.

Figure 2.4: The approximate volume of space that will be scrutinized by the 1hT (ATA) during its first 15 years of operation. Roughly one million solar-type stars are within its reach. The inner sphere represents the volume of space currently being searched by Project Phoenix. (Photo courtesy of L. Ly, SETI Institute.)

The impact of these numbers on the search strategy is discussed in Chapter 5 on page 111.

2.4   Planets

We now know that planets orbit other stars like our Sun. As of August 2000, about 50 exoplanets were associated with solar-type stars, and new exoplanets were being discovered every few months [Mar00].

We expected that, when exoplanetary systems were discovered, they would be similar to our own. To our surprise, we found that for about 3% of solar-type stars, this is not the case. They are graced by 'super Jupiters' or 'hot Jupiters', with very short orbital periods, some as short as four days! These massive planets have orbits very close to their star. Even though we did not anticipate these 'hot Jupiters', we found them first because of the observational bias inherent in the radial velocity studies used to discover those planets. The planets which will be discovered first will be those that have the most mass and the shortest orbital periods, because the data for the full cycle of the orbital period will be collected most quickly and the requirement for measurement precision is least.

Our instruments are now precise enough to detect planets with masses as small as Saturn. Unfortunately, we have not had these improved instruments long enough to be able to detect exoplanets, with masses similar to those of Jupiter and Saturn, at the same orbital distances as they have in our own solar system. There are observational residuals that look promising, but 12 to 30 years of data will be needed to observe the full orbital cycle, and provide confidence in any purported detection. Thus, if the other 97% of solar-type stars harbor planetary systems more like ours, we should begin finding them within the decade. Also, since the stellar sample sizes are large enough, a negative result would be very significant. Although theorists have speculated that large moons orbiting 'hot Jupiters' might provide habitats for life [Gek96] [Wil97], the odds for finding ETIs will improve if we begin to find other planetary systems similar to the only one we know that has produced intelligent life.

There are currently five possible methods for detecting extrasolar planets, each with its own, different, observational bias. They should soon begin to inform us about the demographics of the giant exoplanets. Finding terrestrial-sized planets is much harder and more problematic. So far, they only have been reported in orbit around a pulsar.

The histogram in Figure 2.5 shows the distribution of masses of known extrasolar planets. About 1,000 stars were surveyed, representing a nearly complete sample of solar-type stars within 30 pc [Ext00].

2.4.1   Pulsar Timing

Pulsars, rapidly spinning neutron stars, with their precision clocks, have permitted the detection of planets having terrestrial masses [Wol92]. The existence of condensed bodies in orbit around a neutron star, which is the remnant of an energetic supernova explosion, is startling and unpredicted. There is little or no inclination within the scientific community to view these bodies as likely habitats for life. They do however pose a substantial theoretical challenge, the solution of which may inform us better about the planetary formation processes within our own protosolar nebula, and those of similar main sequence stars.

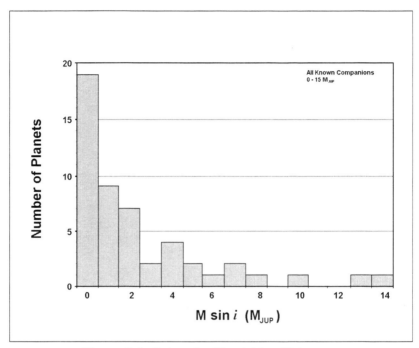

Figure 2.5: Extrasolar Planet Mass Histogram.

2.4.2 Microlensing and Nanolensing

Deviations from the expected light curves of microlensing events can also reveal the existence of terrestrial-sized planets [Ben96]. A microlensing event occurs when the motions of two stars cause them to be precisely aligned along the line of sight to the Earth. Then, the gravitational field of the foreground star acts as a lens for the background star and causes a significant brightening of the background star over a period of weeks. Unlike other stellar brightening events such as novas, the light curve of a microlensing event has a predictable symmetric shape, which is the same for all frequencies.

	Star Name	Msini (M_{jup})	Period (d)	Semimajor Axis (AU)	Eccentricity	K (m/s)
1	HD83443	0.35	2.986	0.038	0.08	56.0
2	HD46375	0.25	3.024	0.041	0.02	35.2
3	HD187123	0.54	3.097	0.042	0.01	72.0
4	TauBoo	4.14	3.313	0.047	0.02	474.0
5	BD103166	0.48	3.487	0.046	0.05	60.6
6	HD75289	0.46	3.508	0.048	0.00	54.0
7	HD209458	0.63	3.524	0.046	0.02	82.0
8	51Peg	0.46	4.231	0.052	0.01	55.2
9	UpsAndb	0.68	4.617	0.059	0.02	70.2
10	HD168746	0.24	6.400	0.066	0.00	28.0
11	HD217107	1.29	7.130	0.072	0.14	139.7
12	HD162020	13.73	8.420	0.072	0.28	1,813.0
13	HD130322	1.15	10.72	0.092	0.05	115.0
14	HD108147	0.35	10.88	0.098	0.56	37.0
15	HD38529	0.77	14.31	0.129	0.27	53.6
16	55Cnc	0.93	14.66	0.118	0.03	75.8
17	GJ86	4.23	15.8	0.117	0.04	379.0
18	HD195019	3.55	18.20	0.136	0.01	271.0
19	HD6434	0.48	22.0	0.15	0.3	37.0
20	HD192263	0.81	24.35	0.152	0.22	68.2
21	HD83443c	0.16	29.83	0.17	0.42	14.0
22	RhoCrB	0.99	39.81	0.224	0.07	61.3
23	HD168443	7.18	58.10	0.29	0.53	470.0
24	GJ876	2.07	60.90	0.207	0.24	235.0
25	HD121504	0.89	64.	0.32	0.13	45.0
26	HD16141	0.22	75.80	0.351	0.28	10.8
27	HD114762	10.96	84.03	0.351	0.33	615.0
28	70Vir	7.42	116.7	0.482	0.40	316.2
29	HD52265	1.14	119.0	0.493	0.29	45.4
30	HD1237	3.45	133.8	0.505	0.51	164.0
31	HD37124	1.13	154.8	0.547	0.31	48.0
32	HD169830	2.95	230.4	0.823	0.34	83.0
33	UpsAndc	2.05	241.3	0.828	0.24	58.0
34	HD89744	7.17	256.0	0.883	0.70	257.0
35	HD202206	14.68	258.9	0.768	0.42	554.0
36	HD134987	1.58	260.0	0.810	0.24	50.2
37	HD12661	2.83	250.2	0.799	0.20	89.3
38	IotaHor	2.98	320.0	0.970	0.16	80.0
39	HD92788	3.86	337.7	0.97	0.27	113.1
40	HD177830	1.24	391.0	1.10	0.40	34.0
41	HD210277	1.29	436.6	1.12	0.45	39.1
42	HD82943	2.3	442.6	1.2	0.6	73.
43	HD222582	5.18	576.0	1.35	0.71	179.6
44	16CygB	1.68	796.7	1.69	0.68	50.0
45	HD10697	6.08	1,074.0	2.12	0.11	114.0
46	47UMa	2.60	1,084.0	2.09	0.13	50.9
47	HD190228	5.0	1,127.0	2.3	0.43	91.0
48	UpsAndd	4.29	1,308.5	2.56	0.31	70.4
49	14 Her	5.55	2,380.0	3.5	0.45	98.5
50	Epsilon Eridani	0.8	2,518.0	3.4	0.6	19.0

NOTE: Stellar masses are derived from Hipparcos data, stellar composition, and stellar evolution theory.

Table 2.1: Properties of Detected Exoplanets as of 13 August 2000.

If the lensing star has a planetary system, the planets have a well-defined statistical probability of acting as secondary lenses that generate 'nanolensing' events. These are superimposed on the microlens light curve. This enables us to detect planets smaller than Mars, located tens of thousands of light years away. Microlensing searches have observed several microlensing events each year for the past several years, with planets beginning to be reported. These studies will eventually lead to a large sample of planets both large and small. The statistical bias in this sample should be correctable, giving us a complete picture of the distribution of planetary masses and circumstellar distances, for stars throughout the Milky Way.

One problem with this technique is that of follow-up. The lensing event itself will not repeat, and the lensing systems are too far away for study by any other method.

An international network of seven telescopes, named PLANET, was established to follow any "lensing event alert" announced by either of the large dark matter surveys, Massive Compact Halo Objects (MACHO) and Optical Gravitational Lensing Experiment (OGLE). PLANET provides the high sampling rate needed to find the short anomalies in the lensing light curve that could indicate the presence of a planet in orbit around the lensing star.

2.4.3   Radial Velocity Surveys

The current precision of instruments used in radial velocity studies (about 3 m/s) is sufficient to permit the detection of planets having the mass of Saturn [Mar00]. The instrumentation, however, has not been operational for sufficient time to allow for the detection of such bodies at the orbital distance of Saturn.

The planetary bodies found thus far have masses greater than or comparable to Jupiter. They also are closer to their host stars than Jupiter is from the Sun, and many of them have highly eccentric orbits. Contrary to the suggestion that planets with the mass of Jupiter might be difficult to form in the available time, planets with masses from 1 to 10 times that of Jupiter appear to be present in orbits of less than a few Astronomical Units (AU) around 3% of solar-type stars. Selection effects have led to the early discovery of these planetary systems, which are quite different from our own [But97], [Mar00] , [May97]. See Table 2.1 on page 46 and Figure 2.5 on page 45.

Theorists have adroitly explained how these giant planets could have formed at large distances from their parent star, early in the development of the planetary system. Subsequently, they migrated inward due to tidal drag and angular momentum transfer. They stopped at the inner edge of the protoplanetary disk, just short of falling into the star [Bos96], [Lin96]. The presence of such giant planets may indicate planetary systems in which any terrestrial planets have already either been drawn into the central star, or thrown out of the system by dynamical interactions with the eccentric giants. Still, they remain of interest to SETI.

While the "super-Jupiters" will not provide a likely habitat for life as we know it, any large moons orbiting them may provide conditions suitable to the origin of life [Wil97]. It is not clear how these "super-Jupiters" moons would fare from impactors. In our outer solar system, Jupiter deflects potential cometary impactors away from the inner solar system. This may have been important for the evolution of intelligent life on Earth[Wet94]. The "super-Jupiters"' moons, however, may not have any such protection.

Figure 2.6: Dr. Jill Cornell Tarter, the holder of the SETI Institute's Bernard M. Oliver Chair for SETI.

2.4.4 Transits

Photometric observations from space can detect planets as small as the Earth by measuring the drop in stellar brightness as a planet transits in front of its star. Spectroscopic binaries provide a special case in which it may be possible to detect terrestrial planets with groundbased observations [Doy00]. However, a systematic search for Earth-like planets can be made only from space [Koc98] since such a search requires a precision of $\approx 10^{-5}$.

Figure 2.7: Dr. Sandra Faber of the University of California at Santa Cruz, brings an optical perspective to the deliberations of the SSTWG.

2.4.5   Astrometry

Space-based astrometry, such as the Space Interferometer Mission (SIM) [Sha95], may well be able to detect terrestrial-sized planets around nearby stars by measuring the reflex motion of the star. The Jet Propulsion Laboratory (JPL) is actively working on the technology for this mission and has proposed it for launch in 2005.

2.4.6   Detecting Terrestrial Planets

Because larger planets are easier to detect than smaller ones, most methods only found planets much larger than Earth, located in planetary systems very different from ours. Our planetary system has a mixture of large and small planets in relatively circular orbits. Do we still expect such systems to be common, and can we hope to detect Earth-sized planets in such systems?

It is clear that 97% of solar-type stars do not have "super-Jupiters", with highly eccentric orbits and short periods. This is good news for the search for potential habitable sites. As the observational timeline increases, radial velocity and astrometric studies should be able to tell us about giant planets in large orbits, which is what we expect to find in the majority of cases.

Detecting planets with terrestrial masses in Earth-size orbits around stars similar to the Sun, remains a difficult technical challenge. Radial velocity observations are limited by photon statistics to larger planets, and since observers are already using the largest telescopes and extremely good detectors, they will be hard pressed to improve enough to find terrestrial planets. Similarly, ground-based astrometry will probably not reach the precision of about $10^{-6} arcsecond$ that is required for detecting terrestrial planets.

Other techniques look more promising, and this is an area where we expect considerable progress in the next few decades. Pulsar observations have already detected Earth-size planets. Gravitational microlensing surveys from the ground [Ben96] are improving, and more intensive follow-up allows us to observe the short deviations from the expected light curve that characterize lensing by planets. Transit experiments should be able to detect Earth-size planets, and these experiments may be launched in the next few years.

On a longer timescale, both NASA and the European Space Agency (ESA) have outlined a series of space-based missions that will include astrometric precursors. These precursors will lead to large, imaging interferometers that can make single-pixel images of any orbiting planets around nearby stars, and can conduct crude chemical assays of their atmospheres [Leg00]. Within the first few decades of the twenty-first century, we may have inferential evidence that some form of life exists on exoplanets.

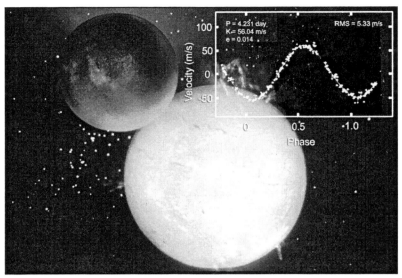

Figure 2.8: Artist's conception of the star 51 Pegasus, its first detected planet, and its radial velocity curve.

2.5   Habitable Zones

In 1961, when the National Academy of Sciences convened a small meeting in Green Bank, West Virginia, and discussed Frank Drake's agenda in the form of the Drake equation, the Habitable Zone was a simple and static concept. It described the distance from a star at which any planet might maintain liquid water on its surface. This single-variable definition emphasized the equilibrium temperature of the planet provided by stellar insolation. As subsequent observations within our own solar system and the known exoplanetary systems have shown, the situation is far more complex, and it changes over the life of the star. The Venus-Mars-Earth "Goldilocks" story (too hot, too cold, just right) is very complicated and not yet fully understood.

While the young Mars was probably wet and warm, it had too little mass to sustain plate tectonics. Plate tectonics, and the resulting plate subduction and volcanism, recycle gases from weathered rocks to the atmosphere (the rock cycle). The CO_2 that initially provided much of the insulating blanket needed to warm the surface, reacted with water vapor to form a weak carbonic acid rain. The carbonic acid weathered the surface rock to form carbonates. Without plate subduction, the carbonates remained on the surface, and the CO_2 could not be released back to the atmosphere by sustained volcanism. As the atmosphere was depleted by reacting with the surface rock, the surface water escaped and Mars became colder and drier.

Figure 2.9: Dr. Christopher Chyba, holder of the SETI Institute's Carl Sagan Chair
for the Study of Life in the Universe.

Young Venus had some liquid water, as evidenced by the deuterium/hydrogen
ratio in its atmosphere today. Apparently, this was not enough to lubricate the
geologic plates and allow them to move. Therefore, there was no plate tectonics
and no rock cycle. However, there was also not enough water vapor to generate
the carbonic acid needed to fuel the rock weathering cycle and remove CO_2 from
the atmosphere. The hot radioactive core of the newly formed Venus caused
volcanic action that released CO_2 into the atmosphere with no relief valve, thus
creating a runaway greenhouse effect. This was unlike Mars, where volcanism
caused by its radioactive core died out early, due to its much smaller mass.

Even the Earth itself was not always "just right". The young Sun was 30%
fainter than it is today. No matter what the exact composition of the early
atmosphere of the Earth was, it had to contain large amounts of greenhouse gases
to keep the Earth warm enough for liquid water. There is evidence in the rock
record going back 3.9 billion years that this blanket was not always sufficient.
The Earth was frozen solid when South Africa lay at the equator about 2.2
billion years ago, which was when the partial pressure of oxygen spiked and
upset the delicate greenhouse balance. This appears to have happened again
about 600 million years ago, just prior to the Cambrian explosion of species
[Ker99]. We do not know if this history represents a common and generally
unfavorable environment for terrestrial planets, which Earth just barely escaped.
Alternatively, there may be numerous, as yet unknown, negative feedback loops
that can rectify most extreme fluctuations. Perhaps nature is more imaginative
and resourceful than we now appreciate. Instead of two out of three potentially

habitable planets failing to provide long-term habitats for life, elsewhere it may be zero out of three or three out of three.

Much real estate in the solar system is not planetary. Because of our experience with life as a planetary phenomenon, we have not thought much about the potential of other sites.

The recently discovered violent activity of Io broadened our thinking. Tidal heating of the satellites of the massive planet Jupiter reminded us that the Sun and radioactive decay are not the only sources of heat in our solar system. As we plan for missions that will explore the possibility of a liquid ocean and life beneath the ice of Europa [Chy98] and Callisto, we are challenged to think about how habitable environments might be established on satellites of 'hot Jupiters' snuggled up to their stars.

In his concluding remarks at the 1993 Third Decennial US-USSR Conference on SETI [Dys93], Freeman Dyson suggested:

> "In spite of the reduction in intensity by a factor of 10^{10} from sunlight to Sirius-light, the space-habitat is still a thousand times better than the Earth. Dilution of starlight by distance does not diminish its quality as an energy-source. An ETI species living on a small comet with 10^9 tons of available mass anywhere in the galaxy can collect about 100 megawatts of starlight, enough for a self-contained ecology to live on."

Today Earth is wet and warm, but it is going to get much hotter, and not just from global warming. In another five billion years, when our Sun becomes a red giant, the Earth will be enveloped within the expanding atmosphere of the Sun. If we are still around, we will be strongly motivated to move farther out in the solar system. The location of the Habitable Zone around any star evolves over time [Kas97].

Figure 2.10[1] on page 54 shows the Zero Age Main Sequence (ZAMS) Habitable Zone as a function of stellar mass. The long-dash lines delineate the most probable formation zone for terrestrial planets. The short-dash line is the "tidal lock radius", for which an Earth-like planet in a circular orbit would be synchronously or slowly rotating as a result of tidal damping. Note that all such Earth-like planets in the Habitable Zone around M stars are within this tidal lock radius.

---

[1]This figure is from Daniel P. Whitmire and Ray T. Reynolds, "Circumstellar Habitable Zones: Astronomical Considerations", *Circumstellar Habitable Zones–Proceedings of the First International Conference*, Lawrence R. Doyle, ed., Travis House Publications, Menlo Park, California, 1996, page 123. Reprinted with permission of Travis House Publications.

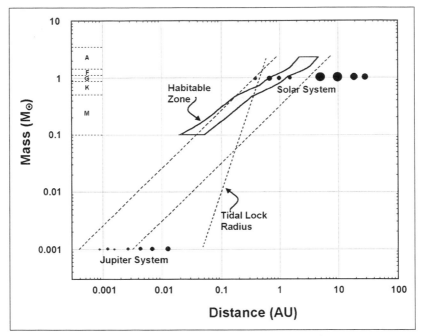

Figure 2.10: The Habitable Zone as a Function of Stellar Type.

2.6 Cosmic Background

For this discussion, it is useful to subdivide the electromagnetic spectrum into three broad wavelength bands:

- radio/microwave
- infrared/optical
- X-ray/gamma ray

Except for the human eye, sensors for all these wavebands were developed in the twentieth century. All are worthy of consideration for SETI searches, and all will appear in any general assessment of relative payoff. Figure 2.11 represents the power flux emitted, per solid angle on the sky, by the astrophysical sources in our Universe. SETI search sensitivity depends in part on the sky brightness, whose components are defined as follows and plotted in Figure 2.11.

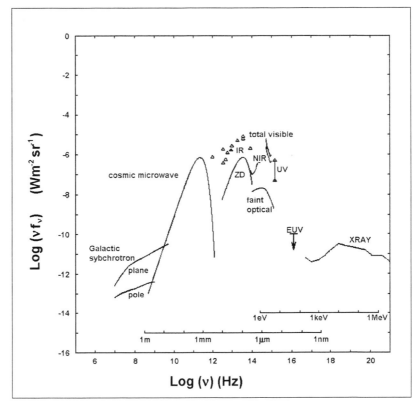

Figure 2.11: Components of Sky Brightness (Average Power Flux Received).

Following is a list of the component labels used in the plot and their descriptions:

plane   Galactic synchrotron – average radio continuum to-
                     wards the galactic plane (0°) [All63], 258);

pole   Galactic synchrotron – average radio continuum to-
                     wards the galactic poles (90°), parallel to the axis of
                     rotation of the Galaxy ([All63], 258);

Cosmic   Cosmic Microwave Background (CMB, from
Microwave   Planck's formula for 2.7 K blackbody radiation);

IR   Infrared contributions ([Tow97] Fig. 1);

ZD   Zodiacal dust contribution ([Tow97] Fig. 1);

total visible   Visible sky brightness in the galactic plane (upper
                     curve) and the galactic pole (lower curve), excluding
                     airglow lines ([All63], 258);

UV   Ultraviolet all-sky average including stars (upper)
                      and excluding stars (lower) [Gon80];

EUV   Extreme ultraviolet. Upper limit from Extreme Ul-
                      traviolet Emission (EUVE) satellite [Par80];

XRAY   Diffuse X-ray and gamma ray components ([Zom82]
                      113, 128, 143);

faint optical   Faint light from distant background stars, nebulae
                      and galaxies.

2.6.1   X-ray and Gamma Ray Background

Figure 2.11 shows that the celestial background radiation in the X-ray and
gamma ray bands is comparable to the background levels from galactic radio
noise and considerably less than the power in the millimeter-band cosmic microwave background (CMB). This suggests that X-rays and gamma rays could
be equally effective for communications purposes. Indeed, if extremely large
information bandwidth were a requirement, the X-ray and gamma ray bands
would be the only possible electromagnetic communications regime. However,
in practice, these very short wavelengths are difficult to generate. They are also
very difficult to focus with known technology, and demand space-based communications facilities. A more fundamental drawback is the very high energy per
photon that these wavelengths exhibit. This leads to a small amount of signal
complexity or information per unit of transmitted energy. For these reasons, we
do not consider these short wavebands further in this book.

2.6.2   Radio/Microwave Background

The background in the microwave bands is well known and ubiquitous – any
antenna pointed anywhere in the sky will see the noise remaining from the
earliest moments of our Universe. This noise is characterized by a temperature
of about 2.7 K.

The major background sources are the CMB, noise from our Galaxy, and noise
from our atmosphere. Of course, the atmospheric noise only applies to observations from the Earth's surface. These background sources are shown in Figure
2.12. From it, we observe that:

- At low microwave frequencies (below 1 GHz), the background is dominated by synchrotron emission from relativistic electrons spiraling in the
  magnetic field of our Galaxy.

- At higher microwave frequencies (above 10 GHz), the background for
  Earth-based observation is increasingly affected by atmospheric thermal

noise.

- At microwave frequencies, unlike optical, the noise from a solar-type star in our field of view contributes nothing of significance. As shown below, most SETI radio signals are likely to be much brighter than the adjacent star.

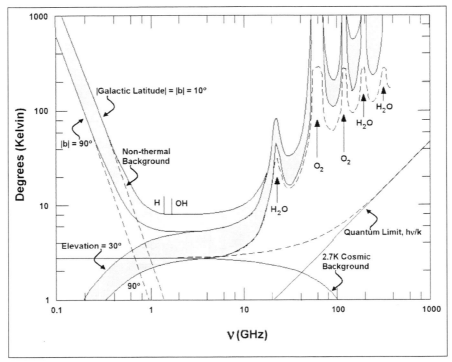

Figure 2.12: Background Noise Sources in the Microwave.

2.6.3 Infrared/Optical Background

For SETI, we also need to consider the background on very short timescales. Optical SETI assumes that short optical pulses will not occur naturally. Therefore, short optical pulses would make a good beacon. What are the backgrounds – both instrumental and natural – that interfere with the search for short optical pulses? The nanosecond timescale is a natural one for this purpose at our current level of technology. It approximates the speed of our high-speed photon counting detectors and high-power pulsed lasers, and it represents easily achievable performance for pulse electronics.

There are four aspects to the infrared/optical background noise:

- night sky background
- stellar background
- other astrophysical background
- terrestrial background

These are discussed in the following four subsections.

Night Sky Background

The night sky is dim and very non-thermal. The background arises chiefly from terrestrial airglow lines plus broadband zodiacal light. Its intensity at dark mountaintop observatories is about 0.01 photon m^{-2} s^{-1} sr^{-1} Hz^{-1} which is about 13 orders of magnitude more dilute than a 6,000 K blackbody (star).

This is so faint that the photon arrival rate R in one broadband Keck pixel ($A = 100 \text{ m}^2, \Omega = 3 \times 10^{-11}$ sr, $B = 0.3$ PHz) is only on the order of 10^4 photons/s. In a more modest one meter telescope this rate R drops further, to about 100 photons/s. Applying narrowband wavelength filters would reduce these count rates further.

On timescales shorter than $1/R$, a detector pixel receives fewer than one photon. The Poisson probability of finding a pixel accumulating two, three, or four events is a rapidly decreasing function of the count. This fact can be used to design pulse search strategies that discover dark sky locations from which the photon arrival statistics are unusual. If more than a few photons per second are detected in a dark sky location, then perhaps a laser or other artificial source is transmitting the signal.

Stellar Background

For sky pixels containing a visible star, the photon statistics are governed by photon-arrival shot noise. This is the same as for the dark sky, but the count rate is higher. A new factor is twinkling (atmospheric scintillation), which causes the incoming photon flux to vary on a timescale of about 0.01 second. This twinkling complicates the statistical detection of unusual photon count accumulation. The twinkling factor can be severe for small telescopes, but large telescopes gain some advantage due to averaging over multiple atmospheric paths. It may be possible to characterize the statistical behavior by comparing nearby stars.

A solar-type star at 1,000 ly distance ($m_\nu = 12$) yields a steady flux of about 1×10^6 photons per second per square meter over the visible band ($B = 0.3$ PHz). Therefore, a pulse containing 30 photons would be detected as "highly unusual" if it were shorter than about a microsecond. A longer duration pulse could also be detected as "highly unusual", but it would have to be more intense to overcome the increased starlight accumulation associated with longer time intervals.

Slow variations in optical power caused by a beacon can of course also be detected, down to a small fraction of the star's average brightness. However, all stars except white dwarfs are believed to be naturally variable. For example, the Sun varies by 0.1% over a timescale of a few days, owing to chromospheric activity and sunspots. Flare stars of class dMe show optical flares amounting to 10% of the luminosity of the star lasting about 1,000 s. To be of interest to SETI, broadband stellar variability would have to exhibit some sort of remarkable pattern unexplainable by variable star models.

Other Astrophysical Background

Atmospheric Cerenkov telescopes, such as the gamma ray telescope on Mt. Hopkins in Arizona, detect gamma rays and cosmic rays by the atmospheric Cerenkov flashes they produce in the night sky. However, data on Cerenkov flashes in the atmosphere [Har90], produced by cosmic rays, suggest that these are innocuous. A 10^{12} eV primary cosmic ray produces a flash of about 5 ns duration, with the light falling on a 'footprint' on the ground with a radius of about 150 m. From within the footprint, the source 'image' looks like a diffuse blob in the sky, about 2 degrees (Full Width of Half Maximum (FWHM)) in size. It produces about 30 visible light photons per square meter during the pulse. The primary flux varies approximately as $E_{\mathrm{pri}}^{-1.7}$ for energies between 10^{12} and 10^{15} eV, and as $E_{\mathrm{pri}}^{-2.3}$ at higher energies. The photon fluence per flash varies approximately as E_{pri}.

The rate of Cerenkov events, as seen from an arbitrary point on the ground, is given by the product of the flux, the area of the footprint and Ω_{image}. Thus, for 10^{12} eV primaries, the rate of Cerenkov events is about 15 per second. However, because the Cerenkov image is diffuse on the sky, the fluence for each event, as seen by a focal-plane aperture corresponding to a 10 arc second field of view, is only about 6×10^{-5} photons per flash. Therefore, a narrow field detector rarely detects even a single photon, and never two or more.

It would be extremely interesting if any other astrophysical phenomenon could produce nanosecond pulses. Such a source would have to be coherent on a distance scale of tens of centimeters. Discovery of such sources in the course of a SETI search would be an important serendipitous result.

Terrestrial Background

We do not know of any atmospheric phenomena that are capable of producing nanosecond light flashes at a level that could interfere with the detection of multiphoton nanosecond pulses. For example, lightning is a pulsed terrestrial source. However, with electrical current paths of a few kilometers, the prompt optical emission lasts for a few microseconds. This signature distinguishes lightning from faster phenomena.

We can plausibly imagine the production of pulses shorter than a microsecond for artificial terrestrial events such as electrical sparks. It just requires some creativity to imagine how to direct such pulses into the telescope without broadening the pulses.

2.7   Microwave Propagation and Coherence

Density irregularities in the interstellar medium cause variations in the refractive index, which can bend the signal paths. This multipath phenomenon causes signal fading just as in terrestrial communications. Cordes and Lazio ([Cor91a], [Cor91b]) discuss this in considerable detail. The characteristics of the interstellar medium have been studied extensively in connection with pulsars. Cordes and Lazio use these results to predict the propagation of narrowband signals. They find that the interstellar medium is fairly friendly to these signals, with 100 second coherence times possible in most directions. They also find that dispersion decreases as $f^{-6/5}$. This could push interstellar communication to higher frequencies than those envisioned by the *Project Cyclops* report.

To examine other factors, let us assume that the beacon is transmitting directly at us, with perfect equipment and using complete Doppler compensation at the transmitter end. Many of the signal detection algorithms currently in use by SETI projects exploit the presumed coherent nature of the signal being sought. However, interplanetary scintillation (IPS) and interstellar scintillation (ISS) can destroy that coherence, and the Earth's atmosphere imposes limitations. These factors are discussed in the following two sections.

2.7.1   Scintillation

ISS is caused by gas, dust and plasma. At frequencies below 2 GHz, we will be in the strong scintillation case even for some targets within 1,000 ly. However, the timescale for these strong scintillations is likely to be approximately 10^4 s, except along lines of sight with especially high dispersion measures. At the other extreme, transmitters in the galactic plane, at typical galactic distances, can have ISS time variations over a period of just 10 s.

IPS , which is caused by plasma irregularities in the solar wind, may be more of a problem. For Project Phoenix, coherent integrations up to 300 s are allowed. To ensure that solar IPS does not significantly degrade the coherence, Project Phoenix limits observations to directions greater than 60 degrees away from the Sun. To estimate the avoidance even at 300 s was an extrapolation from data measured by spacecraft near the Sun. We may not know enough to estimate the IPS on 1,000 second timescales.

IPS on the transmitting end may be even worse for SETI [Cor91b]. Solar avoidance removes our own contribution to the problem, but the situation with respect to the distant host star is less clear. Underlying many of the search strategies in subsequent chapters are assumptions about the physical location of the transmitter: planetary surface, low orbit, synchronous orbit, gravitational minima, removed from local dust, or in thermally benign regions. Each of these assumptions implies a different projected line of sight to the host star. If that distance is too small, the effects of their solar wind will have to be added to those of our own.

2.7.2   Atmosphere

The Earth's atmosphere is also a problem. At wavelengths longer than 20 cm, the ionosphere is the biggest problem. Experience with Very Long Baseline Interferometry (VLBI) has shown that 1,000 s coherent integrations can only be achieved under especially good conditions at 20 cm wavelengths. During solar maxima, or whenever the ionosphere is disturbed, the longest coherence times drop to a few hundred seconds. However, we can almost always count on coherence times of 100 s. Hence, we need sensitivities over a short period, which can only be obtained with a large collecting area.

2.8   Optical Propagation –
Extinction and Smearing

Simple calculations show that extinction and scattering may limit the detection of optical pulses to distances within a few thousand light years, while infrared pulses may be detected all the way to the galactic center.

At optical wavelengths, the interstellar medium both scatters and absorbs. There are 25 magnitudes of optical extinction between us and the galactic center. That is why we cannot see our own galactic center, or even much of the galactic plane. That is also why maps of galactic structure must be made from

Hydrogen I (H I) studies using observations at longer wavelengths, primarily from radio astronomy.

Unlike radio frequencies, where plasma scattering is the dominant effect, optical scattering is caused by interstellar dust grains. Away from the plane of our Milky Way Galaxy, intergalactic space is transparent. Within the plane, the effect of scattering is severe and limits optical SETI to distances on the order of 1 kpc, at which the visual extinction is about two magnitudes. The effect of scattering on a laser pulse is to reduce the 'prompt' pulse height. Simultaneously scattering produces delayed tails on two timescales – a close-in tail (seconds later) from forward scattering by large grains, and a much longer tail from diffuse scattering. The prompt pulse is not scattered and therefore is not widened. (The term "ballistic photon" has been used in a similar context.) The amplitude of the prompt pulse is reduced by a factor of about $e^{-\tau_{\text{scatt}}}$, where τ_{scatt} is the scattering optical depth. Absorption also reduces the prompt pulse, so the surviving fraction is about $e^{-\tau}$, where τ is the total optical depth.

The energy in the unscattered pulse from optical beacons, out to about 1,000 ly (300 pc), is reduced by a maximum of 40%. If optical signaling is a local activity over distances of less than 1 kpc, then extinction and smearing do not seem to be serious problems. For greater distances, it is necessary to use longer wavelengths. At $\lambda = 10\mu$m, optical beacons would be effective over distances of 10 kpc or more. The extinction by clouds in our own atmosphere, however, is another matter.

2.9   Assessment of Scientific Risks

Are the scientific underpinnings of this book sound? Do we understand the technology correctly? What if technology changes faster? Without question, it is a difficult task to plan twenty years into the future. However, a solid base of trends gives us reason to be optimistic that the premises of this book will remain valid. Technology risk is addressed separately, in Section 6.12 on page 222.

Three components of scientific risk are considered in the following sections:

- False Positives
- False Negatives
- Bad Science

2.9.1   False Positives

A credible detection of ETI life would be astounding news, and one of the top scientific discoveries of all time. It is crucial to avoid false positives because

Figure 2.13: Dr. Leonard S. Cutler, of Hewlett-Packard Labs, a leading expert on atomic frequency standards, enjoys an interesting discussion with a fellow SSTWG member.

of the high impact but low probability of success. A false report of such a discovery would reflect badly on the investigators who performed the work and, by implication, the organizations that supported it. It would waste an enormous amount of scientific time and talent on futile follow-up. It could easily affect funding and set the entire field back many years.

Over the past three decades, we have considered how to proceed with microwave searches. We reduce the chance of a false positive by requiring each candidate signal to pass several confirmation steps. First, at even a single telescope, the source must behave as expected for a source at a great distance. It should disappear when the telescope is pointed off target, for example, and have a nonzero Doppler drift in the reference frame of the telescope.

Second, the signal must be confirmed at a second site. It should appear at the same spot in the sky, as is proper for sources at interstellar distances. The relative Doppler shifts and drifts should be correct for such a source. These tests rule out any sources on the Earth's surface or in near-Earth orbit.

These procedures have been used by the Phoenix Project since 1995. Through the end of the March 1999 observing run at the Arecibo Observatory (AO), a total of 483,197 signals have been detected. Of these signals, 327 required additional follow-up activity; all eventually turned out to be false positives. This rate of about one per day of telescope observation will increase as the speed of searching increases, but should remain manageable.

Deep space probes are the only known sources that can pass these tests. They are few in number, are publicly known, and can easily be accounted for. In fact, they provide us with excellent calibration sources.

Finally, once a signal is discovered and confirmed, the last filter will be provided by the scientific community and careful description of what was actually observed. Before any signal can be accepted as a detection of ETI, the entire scientific community must be given a chance to provide other explanations for the observed signal. The discovery of pulsars provides a good example of this. When these unanticipated regular pulses were observed, one possibility left open was that they were from another civilization. They were immediately confirmed by other observatories, indicating the discovery was real. Within a few weeks, however, theorists had realized that dense spinning astrophysical objects such as neutron stars were capable of generating such signals without the aid of intelligence.

All new searches need to incorporate these same principles.

Artifacts of Human Origin

Specifically, one possible class of false positives is that which arises from intelligence, but of the human kind. Three possibilities that must be guarded against are hoaxes, clandestine activities, and unknown deep space scientific probes. All should be caught by the protocols described above.

A hoax would have to be extremely elaborate to survive the confirmation tests. Complex calculations would be needed to generate the right Doppler shifts. Access would be needed to the internal designs of multiple telescopes to generate the right on/off source behavior. Models of the sidelobes of telescopes would be needed to calculate the correct signal levels, and these models do not even exist at this time. The hoax would need to be duplicated at each site used for follow-up, with a detailed design suited for the telescope at that site. Generating a hoax signal that would be correctly reconstructed by a long baseline interferometer would be particularly challenging. It would require state-of-the-art Earth motion monitoring, time and frequency standards, and submillimeter positional accuracy. While not categorically impossible, such a hoax would require substantial effort, and need to be conducted in total secrecy.

More than a million people have downloaded the SETI@home screensaver to help with SETI data processing. The SETI@home site attracted nuisance cyberattacks. However, the data processing complexities have protected the integrity of the search so far. While the heightened public interest warrants more security and vigilance than perhaps is needed for the average research program, it should continue to be feasible to provide it.

Unknown deep space hardware is a realistic possibility. Many interplanetary probes have fallen silent, and their locations are currently unknown, since they are too small to track without their cooperation. One of these probes could resume transmitting. There is also the possibility of secret government programs, which might have launched deep space craft for some unknown objective. Such sources would be correctly identified as man-made during the verification process, because they would move slowly with respect to the distant stars. If they were located within our solar system, their parallax could be determined fairly easily. As infrared and optical searches begin, there is less experience with what may have been launched by governments or other agencies for clandestine reasons, so extra care is needed.

2.9.2 False Negatives

False negatives are a more insidious problem. Here, the risk is that a signal is actually present, but we do not see it. A published record of non-detection may discourage anyone else from making the same measurement, thus delaying the eventual discovery for an arbitrary length of time.

This problem is addressed by regularly performing end-to-end tests of the entire system. For microwave searches, deep space probes provide a known signal source of a specific type that the system should detect if working properly. Periodic observation of these known sources provides confidence that a truly ETI signal would be detected if present.

This is also a strong argument for cooperation with the radio astronomy community. Their routine use of astrophysical calibration sources serves as a continual check on system performance.

2.9.3   Bad Science

Bad science occurs when neither the positive nor the negative result can be believed, or when confounding effects are not properly taken into account. This risk is minimized by several factors. First, the proposals and experimental protocols are peer-reviewed to ensure that the basic ideas are sound. This is another good side effect of sharing facilities and people with the radio astronomy community. Next, the potential problems of false positives and false negatives are carefully addressed. Finally, detailed and accurate records of the observations are essential, as in any scientific endeavor.

In all its endeavors, the SETI Institute should arrange for ongoing neutral, critical reviews by external experts. The SSTWG meetings leading to this book are a good example of the many benefits of this approach.

Chapter 3

Interstellar Communication Engineering

This chapter discusses the mechanics of communicating across interstellar distances, particularly in the microwave and optical bands, using pulses or continuous carriers. The objective is to consider the engineering aspects of interstellar communication, which are appropriate to any ETI intending to advertise its presence. The main problem is to decide which are the optimal wavelengths to monitor within the range of 30 cm to 300 nm.

Focusing problems affect both the transmitter and the receiver, and the consequences place optical and microwave searches approximately on a par with each other. This suggests that we should search both bands, particularly since recent experience suggests that lasers would make very effective beacons. Aiming accuracy seems achievable within an assumed Habitable Zone. Revisit times are considered, and methods for collecting, detecting, and transmitting both pulsed and continuous signals are compared.

The microwave band is reviewed as a mature and well understood technology. The mathematics of antennas and signal-to-noise ratios are demonstrated and identify the most appropriate frequencies to watch. Receiver sensitivity is discussed and corresponding transmitter characteristics are identified.

The infrared and optical bands are examined in terms of current and foreseeable laser technology, including appropriate wavelengths, power requirements, and signal-to-noise calculations. Pulsed beacons are recognized as the easiest to transmit and to detect.

3.1   Setting the Scene

Radio astronomers view the Galaxy in all its complexity and richness. They see star-forming regions, supernovas, pulsars, the rich chemistry and exotic physics of element formation, and molecular evolution. All of these far exceed our predictive capability.

In contrast, a communications engineer takes one look at the Galaxy, and as a first approximation, sees 'empty' space and smiles at the simplicity of it all – there is no loss due to terrain! Engineers who have to deal with those of us demanding that our cellular phones work from inside our car, across hills and through city buildings, must allow huge link margins to handle the terrain problem and its associated background noise. Both cellular phones and microwave links running parallel to the Earth pick up about 200 K in ground noise, straight into the antenna. At decimeter wavelengths, pointing the antennas into space drops the noise by a factor of 10 to 20.

There is, of course, a small price to pay when communicating at interstellar distances – the power of the signal decreases as distance squared. Just what would it take to make the local calling area of your cellular phone the nearest million solar-type stars? It requires going out to about 1,000 ly in distance, instead of the normal 1,000 km, a factor of 10^{13} in distance.

How do we solve the corresponding spreading loss ratio of 10^{26}? A larger antenna could provide a gain of about 10^7 compared to a cellular phone antenna. If we could use one on each end (analogous to two Arecibo-type systems talking to each other), it would increase to about 10^{14}. Boosting the transmitting power from one watt to a million watts would yield another factor of 10^6. That leaves a factor of a million, which we take care of simply by talking very slowly. Instead of the 10 thousand bits per second of a satellite phone, we send only one bit each 100 seconds. Factor in the nice quiet background of space, and we have solved the problem with plenty of margin, and without uncertain and time-variable attenuation.

Then choose a convenient transmitting frequency and a suitable direction to point the antenna, and hope that those on the other end are paying attention!

Figure 3.1 highlights the interstellar communication issues that are discussed in the remainder of this chapter.

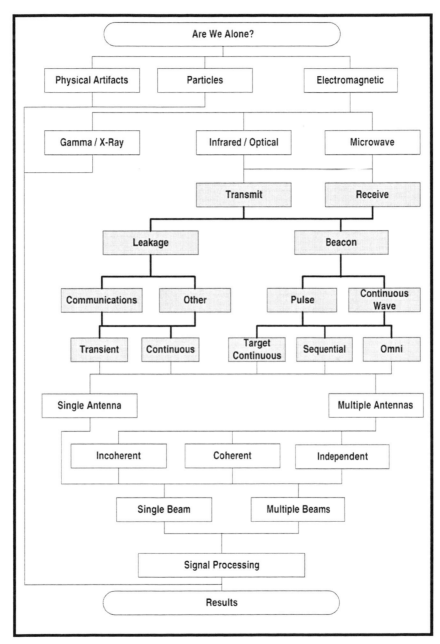

Figure 3.1: SETI Decision Tree – Interstellar Communication Issues.

3.2   Spectrum Choices

As discussed previously, propagation of interstellar signals is relatively effective from the low end of the microwave band through the optical. At the low end, the limit is set by the rapidly increasing galactic noise. At the high end by absorption of ultraviolet light by the interstellar medium. This still leaves a rather wide range of wavelengths, from about 30 cm to about 300 nm, a range of about 10^6. Where in this range should we concentrate our efforts?

There has always been, throughout the development of SETI, a relationship between the state of our technology and the signals for which we search. In essence, we attempt to find probable signals for which we can design ideal detectors. Clearly, what we can detect cost-effectively depends on the state of our own development, but so does our view concerning probable signals.

When the *Project Cyclops* report was written, the focus was on searching for a close twin to our technology. The transmission of continuous sinusoidal signals, in the form of TV and radio carriers, was common. For such signals, we knew how to design a matched filter with a sensitivity that was limited only by the time available to search a particular frequency. The idea of a continuous signal was particularly appealing because it eliminated the need to consider when and how long to search. Search any time, and for as long as possible.

Signals were assumed to be omnidirectional, because our own transmissions were not directed relative to the stars, and because we had no inexpensive way to beam individual stars. Such signals were easiest to detect in the microwave region, where photon energies were low, coherent signals were common, and omnidirectional transmitters were easily built.

Since the *Project Cyclops* report was published, our technology dramatically improved. As analog and digital signal processing evolved, the kinds of signals that we could easily detect changed, as did the signals we transmitted from Earth. The signals became more complex and more directed. These changes in technology impacted our approach to SETI. The first impact of this transition was the search for pulsed signals in Project Phoenix. This added the dimensions of both time and pulse width to the search. Yet, within the search domain, the 'probable' signals would be highly detectable and analogous to a beamed radar, which was a common terrestrial transmission.

In the microwave, extremely short pulses are dispersed by the interstellar medium, and therefore, are not good candidates for detection. In the optical, however, such signals propagate through interstellar space unaltered, and they can be detected by photon counting. While heterodyne detectors are ideal for continuous microwave signals, they are noisy for continuous laser signals. For optical laser pulses, they can be replaced by photon detectors that can count even a few optical photons. This significantly improves the signal-to-noise ratio.

Even at the current state of our technology, 30 years after the *Project Cyclops* report, continuous signals are difficult to detect in the optical. The Cyclops Design Study assumed that only heterodyne techniques would be effective in detecting the presumed narrow bandwidth continuous wave signals. Such detection requires simultaneous measurement of photon energy and phase, introducing quantum noise, which increases with photon energy. Detection of continuous optical signals may be easier with future narrower band detector systems that perform direct photon detection.

Because our knowledge of the stars has continued to improve through the years, beaming of individual stars seems ever more plausible. Lasers are getting more powerful and pulses ever more detectable. Thus, today, although we do not transmit strong pulsed lasers to the stars, such signals may be transmitted in the future. Since our technology is changing ever faster, this book considers not only plausible technologies identical to ours, but also those that might evolve from a similar technology. Thus, pulsed laser signals are discussed at considerable length as a promising candidate. It is not that other signal types have become less plausible, rather, as our technical powers increase, it becomes more reasonable to search for other, probable, detectable signals.

This book is somewhat less constrained than was the *Project Cyclops* report. We consider innovation in multibeam and multielement receiving and transmitting arrays. We consider not only signals that we currently transmit, but also signals that we are likely to transmit. We even plan with innovation in mind, as we design projects that can only be built economically if computing costs keep dropping.

The ideas of beaming and searching for very short optical pulses are new consensus favorites of the SETI community. The idea of continuing innovation in computing is crucial to the projects proposed in this book. This is not only the basis of what goals we set, but also of what we expect from more advanced technologies – the kind we are likely to detect.

During the last three decades, most proponents of SETI have held that microwaves are superior to optical waves for detection, because microwaves require less energy per bit. The argument, reduced to its basics, is that below about 50 GHz the required energy per bit is determined by the need for a signal to exceed the 3 K background radiation. This leads to a requirement of kT per bit, or about 4×10^{-23} J. For higher frequencies over 50 GHz (6 mm), a single photon has more energy than the cosmic background, and provides an adequate signal-to-noise ratio. However, one photon per bit is still necessary. At optical frequencies, this is much more energy than kT. For example, at 1 μm a single photon has about 2×10^{-19} J, so sending a single bit requires about 5×10^3 times more energy than at microwave frequencies. Thus, looking only at the energy of information carriers, microwave transmission seems significantly more effective than optical.

However, we must consider not only the energy required per bit but also the ability to send bits to a desired receiver. For a given amount of money, we can afford very different apertures at different frequencies. The aperture and wavelength affect the ability of a transmitter to focus, and of a receiver to collect information carriers. Such technological considerations can be factored into the calculations, with significant consequences. Consider the following examples from our current technology:

- Today, two advanced antenna designs are the Keck Telescope for optical studies and the Green Bank Telescope (GBT) for microwave studies. Both cost approximately the same amount but operate in different wavelength regimes. The Keck and GBT both use corrective technology because pure structural stiffness was not practical, and both push the state of the art for a single aperture. The Keck works down to about 300 nm wavelength, and has a 10 m diameter. The GBT works down to about 3 mm and has a 100 m diameter.

- As another example, the Very Large Array telescope (VLA) in New Mexico covers the same frequency range at the same cost as the GBT. It has a 130 m equivalent diameter, but is composed of 27 dishes or apertures, each 25 m in diameter.

At first, the difference between microwave and optical collecting areas seems to magnify the asymmetry introduced by the unequal energies per bit. A microwave transmitter must achieve an energy of kT across the 100 m aperture of the GBT, whereas an optical transmitter must put its single photon into an aperture 100 times smaller. Thus, the optical transmission must generate an energy per square meter that is 5×10^5 times greater than the microwave transmission.

However, we must also consider antenna gain. Substituting the collecting areas into the formula for parabolas shows that the gains of our two example antennas are very different. Compare the microwave transmitter (the GBT working at $\lambda = 1$ cm) with the optical transmitter (the diffraction limited, Keck telescope working at 1 μm). We see that the GBT has a transmitting gain of 10^9, but the Keck has a gain of 10^{15}. In this case of directed transmission, the optical transmitter requires an *equal* amount of energy per bit received, as that of the microwave transmitter.

There is some justification for caution when using very high gain antennas. This is because:

- We may not be able to effectively use extremely high gain, because we do not know enough to deliver the signal to inhabited planets. It is difficult to point at distant stars, because they will move in unknown directions while our signal travels to them. However, the pointing problems appear to be solvable. Nearby stars have their potential Habitable Zones only partially illuminated by very high gain beams, but this is only a problem for the closest nearby stars.

- The difference in energy caused by the physics is not alterable, whereas the offsetting higher gain of the transmitter is just an artifact of our current technology. Certainly, an advanced technology could build transmitters that would selectively beam either microwave or optical signals to just completely cover the Habitable Zone. Then, microwaves regain their power advantage over optical.

- The counter argument to this holds that while the aliens *could* build any transmitter they want, they will choose instead to build the most cost-effective one. If one assumes that ETIs have *any* required technology, then there is nothing in the physics that prevents them from building an extraordinary beacon that screams 'Here we are!' so loudly we cannot possibly miss it. This has not happened, so presumably if there are civilizations interested in communicating, they are facing budget or technology limits, or both. For them, as for us, it may be cost-effective to build surfaces with high relative precision at smaller physical sizes, if for no other reason than pure material stiffness, which improves at smaller scales. This makes their choice of wavelength very hard to predict.

- We could use a higher microwave frequency, because at 1 cm, the GBT and Keck are comparable. This implies that we should be searching higher in the microwave band, but, at least for the example above, the optical signals have lower energy requirements than the signals we are currently seeking at the lower end of the microwave band.

- Most SETI researchers have worked in the centimeter microwave portion of the electromagnetic spectrum. This choice was based upon favorable considerations of energy efficiency, atmospheric transparency, available technology, and the existence of a unique wavelength marker at H I (21 cm). Of course, since our search has not succeeded yet it is hard to argue that this choice of search strategy is optimum – but it has motivated searches, of increasing sophistication, sensitivity, and thoroughness.

In conclusion, microwaves are still a very good choice from a purely physical perspective, because they provide the lowest required energy per bit. When additional engineering issues are considered, however, a broad minimum appears. This minimum extends from the microwave region through the optical region, limited by galactic noise on the low end and Habitable Zone illumination on the high end.

Although technological considerations seem more ephemeral than those based on physics, we should carefully analyze the case where ETI technology is broadly similar to ours. In such a situation, our ETI counterparts will be able to build higher gain antennas at shorter wavelengths, and frequencies up through the optical region may be their choice.

This is not an argument against microwaves, which still represent an excellent, and largely unexplored choice for interstellar communication. Instead, it is a recognition that other civilizations, possessing technologies similar to ours, might choose a higher frequency. We simply fail to find the arguments in favor of microwaves so compelling that we should limit our search to *only* this band. Instead, we now recommend investigations into a wider range of wavelengths, from the microwave region down to and including the optical region.

3.3   Microwave and Optical Bands Compared

The suggestion by Cocconi and Morrison [Coc59], that the search be carried out at the 21 cm wavelength of neutral hydrogen, came at a time in our technological development when no other astronomical lines were known in the microwave, and there were no lasers. Since then, laser technology [1] has developed rapidly, a Moore's Law doubling of capability approximately every two years, and many microwave lines of astronomical interest have been discovered. This has somewhat lessened the allure of searching near the hydrogen-line.

Additionally, the elucidation of the consequences to SETI of interstellar dispersion, which was first observed in pulsar observations, broadened the thinking about optimum wavelengths [Dra78], [Cor91a], [Cor91b]. Even operating under the prevailing criterion of minimum energy per bit transmitted, one is driven upward towards millimeter wavelengths.

Further considerations encourage even shorter wavelengths. For example, a transmitting civilization might wish to minimize transmitter size or weight, or use a system capable of great bandwidth, or perhaps design a beacon that is very easy to detect.

In comparing the relative merits of radio versus optical, it has sometimes been incorrectly assumed that in detecting optical signals one would always prefer coherent (heterodyne) detection, for which the noise background is given by an effective temperature $T_n = h\nu/k$. For very high resolution spectroscopy one must use such a system, heterodyning the optical frequency down to microwave frequencies, where radio detection techniques can be used. If one is interested in the detection of short pulses, however, then it is far better to use photon-counting detectors (e.g., photomultipliers). That is because the process of heterodyning and linear detection is *intrinsically* noisy due to additional uncertainty in the amplitude, because it entails a measurement of phase. The added noise is immaterial in the radio region where there are many photons per mode, but it is serious in the optical where the photon field is dilute.

---

[1]The term "laser" here means an intense collimated light source; there could be other technologies that are functionally interchangeable.

Unlike microwave technology, which is relatively mature, optical technology is still rapidly improving. Many of the problems noted by the *Project Cyclops* report, which largely rejected optical techniques, have now been resolved.

The basic argument, in Section 3.2 on page 70, is derived from a more detailed analysis by Townes [Tow97]. He compared received S/N versus wavelength, making reasonable assumptions about antenna apertures and accuracies, detection methods, transmitter power, and so on. The conclusion is that optical methods are comparable to radio methods with regard to the single figure-of-merit of S/N *delivered for a given transmitter power*. Other factors are obviously important and could easily tip the balance. For example, penetration of an atmosphere favors microwaves, while pulsing and high data rates favor optical.

3.3.1 Pulses and Carriers

What is natural at radio frequencies may not be so at optical. At *radio* frequencies it is easy to use coherent detection, using the ordinary heterodyne techniques of mixing with a local oscillator (LO) to a lower frequency. With classical filter techniques, or with contemporary digital processing using discrete Fourier Transforms, one can achieve extremely narrow bandwidths. The bandwidth is limited only by oscillator stability (one part in 10^9 is easy and inexpensive, one part in 10^{12} is routine but expensive), and patience (the resolution is the inverse of the coherent integration time). Furthermore, the interstellar medium is kind to radio carriers. At gigahertz frequencies, a carrier is broadened only by *milli*hertz in its passage through the interstellar medium, as long as one avoids the most congested region of the galactic center. Even towards the Galactic center, the carrier is broadened by only a few hertz. In other words, a signal that is a spike in the *frequency* domain is a natural candidate for interstellar signaling at microwave frequencies.

Interstellar dispersion, natural impulsive phenomena (lightning, etc.), and artificial impulsive interference (transients, spark plugs, etc.), all make pulses in time less distinctive. Finally, the relatively low carrier frequency, along with dispersion, prevents high bandwidth communications over the same channel. However, dispersion does have a characteristic signal that enables its separation from extraterrestrial signals, as was done by Hankins et. al. [Han96], when looking for short radio pulses from the Moon caused by high-energy neutrinos.

At *optical* wavelengths, the situation is reversed. One cannot implement extremely narrowband systems with optical filters or gratings. One is forced to use optical heterodyne techniques, ultimately applying precise radio frequency (RF) spectroscopic methods to the resulting microwave intermediate frequency (IF). This results in added noise, as mentioned above and described by Townes [Tow97]. However, the exercise is futile, because at optical wavelengths, the higher carrier frequencies ($\sim 10^{14}$ Hz) result in much larger absolute Doppler shifts. For example, 1 km/s corresponds to 5 kHz at 1.4 GHz, whereas 1 km/s corresponds to 1 GHz at 1 μm.

Extreme bandwidths are available because the infrared/optical frequencies are so high (0.1 to 1 PHz). In principle, pulses can be quite brief (picoseconds to femtoseconds). Unlike the radio band, optical pulse broadening due to interstellar plasma dispersion does not appear to be a problem. Again, unlike radio, absorption by interstellar dust can be a problem in some parts of the sky, especially in the direction of the galactic center.

Furthermore, natural and artificial sources of nanosecond flashes are probably absent (see Section 2.6.3 on page 57). In other words, a signal that is a spike in the *time* domain becomes a natural candidate for interstellar signaling at optical wavelengths. An added bonus is that the stellar background becomes negligible at nanosecond timescales.

These considerations suggest that a pulsed laser beacon is an attractive scenario. Indeed, in the next section we will sketch plausible parameters for such a beacon that appear to be entirely reasonable and can accomplish the task of making contact with a civilization out to $\sim 1,000$ ly or more. However, it would be wrong to ignore completely the possibility of an optical spectral laser line as a means of contact. Section 3.5.1 on page 85 compares pulsed and continuous laser beacons and the systems required to detect them.

3.4   The Microwave Band

This section explores the technology available for interstellar communication in the microwaves. This is largely a review since microwaves are a mature and well understood technology. The combination of good technology, good interstellar propagation, and low energy per bit make microwaves attractive for interstellar communication.

3.4.1   Signals, Noise, and Antennas

An omnidirectional or *isotropic* transmitter of power P_t produces a flux density at radius R of:

$$f = \frac{P_t}{4\pi R^2} \tag{3.1}$$

since the power is spread equally in all directions.

We receive this with an antenna of area A, leading to a received power P_r of:

$$P_r = \frac{AP_t}{4\pi R^2} \tag{3.2}$$

Figure 3.2: Nobel laureate Dr. Charles H. Townes, of the University of California at Berkeley, is a compelling proponent of optical SETI.

The receiver receives noise in addition to the desired signal. The nature and strength of the noise depends on the portion of the electromagnetic spectrum we are using. In the microwave bands, the main noise sources are galactic radio emissions, the Earth's atmosphere, ground radiation, the cosmic background radiation, and the telescope electronics. In the optical bands, the noise consists of photons from the night sky, which is not completely dark, and from other astronomical objects in the field of view, such as the parent star. The optical case is considered further in Section 3.5 on page 82.

For the case of microwaves, Figure 2.11 on page 55 shows the external sources of

Figure 3.3: Dr. John W. Dreher of the SETI Institute is the leader of the microwave antenna group.

this noise; in practice we need to add the internal contribution of the instrument. A good approximation is to express the sum of the powers from all sources of noise Ψ as an effective system temperature T. This gives a noise level of:

$$\Psi = \frac{h\nu}{e^{h\nu/kT} - 1} \tag{3.3}$$

where k is Boltzmann's constant (1.38×10^{-23} J/K), T is the system temperature, ν is the frequency, and h is Planck's constant. An additional quantum contribution of $h\nu$ must be added if coherent detection is used:

$$\Psi = h\nu + \left(\frac{h\nu}{e^{h\nu/kT} - 1}\right) \tag{3.4}$$

For achievable system temperatures (> 3 K) and lower microwave frequencies (less than about 50 GHz) equation 3.3 simplifies to $\Psi \approx kT$. This is white noise, equally spread across all frequencies, at a level of kT per unit bandwidth. Therefore, within any particular bandwidth B, the noise power is given by kTB.

For reliable detection, the signal must exceed the noise by a margin M, where M depends on the number of observations and our tolerance for false alarms. So we have:

$$kTBM = \frac{AP_t}{4\pi R^2} \tag{3.5}$$

We solve for R to get:

$$R = \sqrt{\frac{AP_t}{4\pi kTBM}} \tag{3.6}$$

Of course, the antennas used for radio astronomy are not usually isotropic. In general, they achieve a very high gain in some particular direction, with consequent loss of sensitivity in all others. For single elements, this is accomplished by focusing. If single elements are inadequate, this is accomplished by combining signals of many elements in phase. Classically, because electromagnetism is time reversible, we can calculate the antenna properties for receiving or transmitting, whichever is most convenient mathematically. The result applies to both, since one is a time-reversed version of the other.

The best known means of focusing is the parabolic dish. A uniformly illuminated parabolic dish of diameter d, operated at a wavelength λ, gives an on-axis gain G_0 of:

$$G_0 = \left(\frac{\pi d}{\lambda}\right)^2 \tag{3.7}$$

At an angle Θ from the axis, the gain is:

$$G(\Theta) = G_0 \left(\frac{2J_1(\Theta\sqrt{G_0})}{\Theta\sqrt{G_0}}\right)^2 \tag{3.8}$$

where J_1 is a Bessel function of the first kind. This implies a half-power beam width in radians of:

$$2\Theta_{1/2} = (1.029...)\lambda/d \tag{3.9}$$

Since a 3% error is small compared to all the other uncertainties in our calculations, we often simplify the beam width to λ/d.

If we array N transmitters, each with effective power P, forming a single beam so that the voltages all add, the effective power P_{eff} of the combination is:

$$P_{eff} = PN^2 \qquad (3.10)$$

One factor of N reflects the ratio of power in the array to that of a single element, while the other factor reflects the area ratio. That is, array power and array gain are both proportional to the number of elements.

3.4.2 Receiver Sensitivity

In the microwave region, the theoretical limit to sensitivity is set by the cosmic background radiation. This radiation, which comes from all directions and hence cannot be avoided, has the spectrum of a black body at 2.76 K. Existing receivers on real radio telescopes come within a factor of three or four of this theoretical optimum, even from the Earth's surface. At first glance, this is very surprising! Clearly, the huge metal structure of the radio telescope will be at the ambient temperature of about 300 K. Furthermore, it is immersed in an atmosphere of approximately the same temperature. How then can we get a noise level that is about two orders of magnitude less?

Fortunately, physics comes to our rescue. From reciprocity, the thermal radiation of an object is reciprocal to its absorption. If it does not absorb, it does not radiate. In the microwave range, metal surfaces reflect and the atmosphere is transparent, so neither contributes much to the system temperature.

Of course the metal surfaces of radio telescopes are not perfect reflectors, but they are much better than 99% in the microwave range. This means that the approximately 300 K temperature of the reflector, when multiplied by the absorption of much less than 1%, contributes much less than the unavoidable 3 K background radiation. Moreover, the reflector shields the antenna feed from ground radiation.

Likewise, the atmosphere is quite transparent in this range. This helps in exactly the same way. Since the atmosphere does not absorb, it does not re-radiate, and we can see through it to the quiet sky beyond.

The net result is that we can approach the theoretical limit of the 3 K background, even from the ground, at the lower end of the microwave window (up to about 10 GHz). Above this frequency, the atmosphere is no longer transparent, and we are forced to move to high elevations or into space for optimally sensitive observations, but we can still use ambient temperature mirrors. At still higher frequencies such as the infrared bands, the reflectivity of mirrored surfaces drops dramatically, and the telescopes and mirrors start to glow in their own thermal radiation. Then both space-based operation and cooled mirrors are required.

It is, of course, true that ground scattering from structures other than the mirror contribute significantly to antenna temperature even below 10 GHz, but this does not mitigate the surprising fact that the contribution from the warm mirror is small. Diffraction and scattering of ground radiation into the telescope adds 2% or 3% of this warm value to the overall system temperature.

Figure 3.4: Dr. Sander Weinreb, a world renowned radio telescope electronics wizard, adjusts his plate voltage.

3.4.3   Transmitters

Good microwave transmitters are easy to build. Large amounts (megawatts) of continuous power can be produced, with excellent spectral purity. When combined with existing large dish antennas, this results in a very powerful beam, easily visible across interstellar distances. The technology can be purchased today and is used in the JPL and Arecibo planetary radars. However, these systems, while powerful, have significant limitations. They can only point in one direction at once, and take considerable time to move from target to target.

Multiple beams can be formed with phased array technology, where the required electromagnetic field is calculated and then generated with a large number of small transmitters whose waveforms can be fully controlled. Such a transmission system is capable of phasing together a large number of elements to form many arbitrary spot beams on the sky. This phasing technology is the ultimate generalization of concepts now being considered for the next generation of radio astronomy receiving antennas, and is described in more detail in later chapters.

The transmission system for a flexible beacon is even likely to have good efficiency since the individual transmitters could be derived from personal, portable communications electronics (such as cellular phones). Since maximum battery life is crucial in such applications, a great deal of work has gone into making these transmitters efficient.

Phased array beacons become more attractive as computing improves and personal communication technology is produced in higher volume. Unfortunately, this same proliferation of handheld transmitting technology makes the radio frequency interference (RFI) problem worse, threatening the very search for ETIs that it makes more feasible.

3.5   The Infrared and Optical Bands

This section covers the science of optical SETI as we understand it today. The starting point is the argument by Townes [Tow83], [Tow97] and others, as summarized in Section 3.3 on page 74, pointing out that both infrared and optical wavelengths are very reasonable for SETI.

3.5.1   Producing Signals

Laser technology is in a phase of rapid expansion relative to the mature technology at radio frequencies, as shown in Figure 3.5 and Table 3.1.

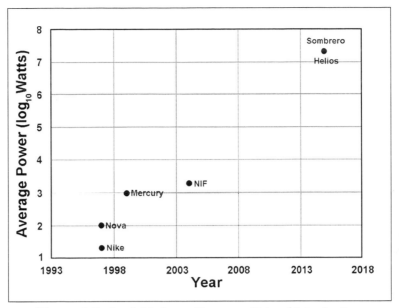

Figure 3.5: Available Laser Power vs. Date of Introduction.

Characteristic	System					
Laser Name Date	Nova 1997	NIF (2004)	Mercury 1999	Helios (2015)	Nike 1997	Sombrero (2015)
Type	SSL	SSL	SSL	SSL	Gas	Gas
Gain Medium	Nd:glass	Nd:glass	Yb:S-FAP	Yb:S-FAP	KrF	KrF
Pump	Lamp	Lamp	Diode	Diode	E-beam	E-beam
Pulse Energy	~ 0.1 MJ	~ 2 MJ	~ 0.1 kJ	~ 2 MJ	~ 2 kJ	~ 2 MJ
Pulse Duration	~ 1 ns	~ 1 ns	~ 1 ns	~ 1 ns	~ 1 ns	~ 1 ns
Pulse Rep. Rate	0.001 Hz	0.001 Hz	10 Hz	10 Hz	0.01 Hz	10 Hz
Wavelength	353 nm	353 nm	1047 nm	349 nm	248 nm	248 nm
Efficiency	0.1%	0.5%	10%	10%	$\sim$1.5%	$\sim$7%
Average power	100 W	2,000 W	1,000 W	20 MW	20 W	20 MW

NOTE: The Nova and NIF are flashlamp pumped solid-state lasers (SSLs); the Mercury laser is a diode pumped SSL.
NOTE: Dates in parentheses are projected dates for the laser to be operational

Table 3.1: Characteristics of Some Current and Projected Lasers.

A review of the recent advances in optical laser power levels has been presented by Mourou et al. [Mou98]. To summarize their paper, laser power and pulse energy have been growing by astonishing factors over the past 40 years. Kilojoule and petawatt levels have both been achieved in the past 10 years, and the growth rates have approximately doubled every two years.

At the Lawrence Livermore National Laboratory (LLNL) a pulsed laser with near diffraction-limited beam quality has achieved 1 kJ of pulse energy in a 1 ps pulse (i.e., 10^{15} W or about 1 PW) [Per96b]. This is a flashlamp-pumped Nd-glass laser amplifier ("Nova"), operating in a chirp-pulse amplification mode for extremely high peak power, but it cannot be cycled faster than once per hour.

Recently, however, the inertial fusion program has been working on prototypes of highly efficient diode-pumped solid-state lasers (DPSSL), using Yb-doped strontium fluorapatite (Yb:S-FAP). This is an elegant design [Kru96] for a scalable architecture of beamlines that is intended to be combined to produce a single high quality beam. The full combination – "Helios" – would produce $\sim$3 ns pulses at $\lambda = 349$ nm (tripled from 1047 nm), with a pulse energy $E_p = 3.7$ MJ and a repetition rate of 10 pulses per second. However, it is not our mission to develop a *transmitting* system, only to think out what might be plausibly achievable so that we can plan appropriate *receiving* strategies. Although not terribly important for optical SETI, it is worth noting that these lasers, intended for fusion power plants, deliver very high efficiency – about 10% from wall plug to photons.

This rapid change in available laser power makes it easy to imagine that an extremely high-energy level might be achieved by a motivated ETI. Therefore, when thinking about optical searches one should not be unduly influenced by the particular parameter set of a laser such as Helios. Rather, one should assume that a 'designer' laser can be tailored (in terms of pulse width, power, repetition rate, and wavelength) to the optical requirements.

Are there motives driving optical power and energy levels upward? One obvious use of high-energy levels is information transmission. JPL, for example, is investigating optical means for communicating with deep space probes, driven by the desire to minimize the size and mass of the probe. Sending to such a probe requires high-power ground stations with accurate aiming.

Preferred Wavelength

The microwave search for ETIs was energized by Cocconi and Morrison's 1959 suggestion [Coc59] to use the 21 cm hydrogen hyperfine line as a probable beacon frequency. Since then, many other special microwave frequencies have been proposed.

No compelling special wavelengths have been suggested for the optical searches, although Charles H. Townes pointed out that extinction in the visible makes infrared a preferred band. There have been suggestions to exploit the Fraunhofer stellar dark lines to reduce stellar background; moreover, a unique dark line might serve as a probable wavelength. Possible choices include the calcium H and K lines, and the iron lines, all at the blue end of the spectrum, as well as the yellow sodium D lines.

A dark line at a longer wavelength would be better, because it would present fewer problems of extinction and smearing. Perhaps, an ideal line would be a dark line in the near-infrared that coincides with the wavelength produced by a particularly favorable lasing material (e.g., the 10.6 μm $^{12}CO_2$ line [Bet93]).

If it proves feasible to construct pulse detectors that can search over a wide range of wavelengths (e.g., using gratings and array detectors), the idea of a 'magic wavelength' may lose some of its luster, just as it has with the microwave band.

However, as calculated above, stellar background is not a problem for the pulse energies that are needed to produce detectable count rates. There seems to be little reason to consider narrowband filters, or multichannel spectroscopy.

Required Power for Pulses and Spectral Lines

What level of transmitter power and energy is needed to yield a detectable signal for pulsed or narrowband optical beacons? The transmitter calculation begins with estimates of the received flux that would be detectable by present-day technology. Of course, whether the needed transmitter is in use by the ETIs is completely unknown. But a quantitative comparison of transmitters does serve to distinguish which options are easier than others, and therefore may suggest which search strategies might be more rewarding.

Signal-to-Noise Calculations

Optical S/N calculations differ in several ways from the calculations usual in radio work. There are several reasons for this difference. A radio receiver processes a single electromagnetic mode. In contrast, an optical image sensor consists of many pixels, each of which is a square law photodetector fed by hundreds of independent electromagnetic modes. A radio receiver nearly always works at its diffraction limit. In contrast, a ground-based optical observatory cannot approach diffraction limited resolution due to telescope imperfections and atmospheric seeing. Further, the noise seen by a radio receiver is thermal. In contrast, the noise seen by an optical receiver is highly nonthermal; it is the shot noise of photon arrivals from faint foreground and background processes whose total phase space intensity is vastly smaller than optical blackbody radiation. Finally, optical detectors have improved to the point that their quantum efficiencies approach 100%, and their noise levels are dominated by photon arrival statistics. These differences oblige us to present the optical S/N considerations in photon language.

Figure 3.6: The Harvard optical SETI project is being carried out using a 61 inch telescope. On the right is Paul Horowitz, project leader.(Photo courtesy of P. Horowitz.)

Consider a telescope whose light-gathering area is A. The telescope feeds one or more photodetectors. The area of each photodetector determines the solid angle Ω that it views on the sky. Each detector senses the arrival of every photon it receives. The instrument can also include a filter or disperser to provide spectral selectivity. Aimed at the night sky, each pixel receives photons at a rate determined by the diffuse night sky brightness, additional counts from any astronomical object that lies within its field of view, and from the thermal emission of the telescope. The mean count accumulation for time interval τ is:

$$< N >= A\Omega\beta\tau + A \int_0^\tau F dt \qquad (3.11)$$

where β is the diffuse photon flux in the system's wavelength range and F is the photon flux from the object, if present.

Measurements of F and its possible time variations cannot be exact because of the Poisson statistical nature of the accumulation of independent random events. Equation 3.11 will be a real number, but the observed count N will always be an integer. The Poisson distribution has a rather different character for small mean values versus large mean values. For mean values < 1, the Poisson probability of a random outcome N falls off approximately exponentially with N. For mean values > 1, the Poisson probability peaks near the mean and acquires a more nearly symmetrical Gaussian character. Its root mean square (RMS) width is given by the square root of the mean.

These two extreme limits are of interest in connection with optical photometry for SETI. Short timescale measurements, in which a brief pulse of photons might be detected in the gaps between the photometric events of the steady components, and longer timescale variations for which the sky or star background would be significant but subtractable. This distinction leads to two varieties of optical searches.

Preliminary experiments attempted to detect short optical pulses from a few target stars. They demonstrated the need for a pair of detectors in coincidence, rather than a single detector with its pulse height threshold set to reject events of few photoelectrons. This particular experiment used photomultiplier tubes (PMTs), but the same would likely hold true for any other equally sensitive detector.

The main instrumental background noise in these experiments arose from occasional large pulses in the dark current of the photodetectors. These probably arose from radioactive decay events in the photodetector glass, from ion feedback, from scintillation in the glass caused by electron impacts from within, and from muons and electrons induced by cosmic rays. A beamsplitter and coincidence circuit effectively eliminates all but muons and electrons (a muon can pass through both photomultiplier tubes). If necessary, the latter can be eliminated by an anti-coincidence arrangement with an external muon scintillator.

A second optical diagnostic is provided by spectroscopy. Line features abound in astronomical targets at infrared/optical frequencies. Stellar atmospheres impose absorption and emission features on the underlying continuous spectrum, while extended stellar envelopes produce emission lines. Ground-based observations of the diffuse night sky reveal emission lines of atmospheric origin plus faint diffuse emission lines characteristic of the local interstellar medium. Narrow absorption lines are seen in nearly all astronomical spectra and are known to be due to resonance absorption by interstellar atoms and ions.

Extraterrestrial civilizations could employ narrowband optical transmitters for communications purposes. To be discovered in a spectroscopic survey, such emissions would have to be unlike the various natural lines and bands. Characteristics that could lead to a detection might involve an unusual polarization, time modulation, Doppler variation, a wavelength that is difficult to explain on the basis of known emitters, or lines much narrower than thermally broadened natural lines.

If the transmitter were located within about an arcsecond of a star, the signal would be mixed with the starlight by atmospheric seeing at our end. The transmitter would have to be powerful enough to make the combination of signal and starlight remarkable in some way. On the other hand, if the transmitter were spatially separated from the star by a larger angle, the background to be overcome would be far lower and the detection threshold correspondingly reduced.

The detection threshold and false alarm rates depend very much on the details of how such a survey is conducted. The received line power that would be regarded as significant for SETI depends upon which observed characteristics are used to govern the detection.

Two Examples

Consider a model with a spatially coincident star and transmitter. Suppose that the line is narrower than 1Å (0.1 nm), and suppose that the receiver is equipped with an ideal, photon noise limited, spectrometer with correspondingly matched resolution. Then, in a 1 meter telescope, the continuum of a solar-type star at 10^3 ly distance produces 100 photons/s in each 1Å interval. If the line flux is 1 photon/s, a one hour accumulation would show a 6σ excess at the line's wavelength. This is statistically interesting, but potentially is fraught with confusion from natural stellar lines. A detected line would be very much more interesting if it exhibited some other unusual characteristics.

Consider the case where the transmitter is spatially isolated and does not lie in the seeing patch of a star. At 1 photon/m^2/s, it would appear only as bright as $m_V = 27$ in continuum photographs. Yet, if the proper 1Å filter were chosen it would be far brighter than the 0.01 counts per second background and could be detected in less than a minute of observing time. A more practical search would cover all optical wavelengths at once by means of an objective grating. The sky background would rise to 100 cps/pixel, and the beacon could be detected anywhere in the field as a 'monochromatic star', at any optical wavelength, in a few hours of observing.

Present-day charge-coupled device (CCD) detectors appear well suited for line search work in the visible, and there are also several good infrared array detectors. Examples of extremely sensitive profiles of stellar spectra are found in the contemporary astronomical literature.

Can We Avoid Spectroscopy?

It would be very helpful to find a simple alternative to high-resolution optical spectroscopy. It has been suggested that we might detect a weak laser line embedded in an intense broadband stellar spectrum by exploiting the different photon statistics of laser light versus that of blackbody radiation. This might be done by accumulating statistics of time intervals between detected photons, or by taking moments of the distribution of photon counts accumulated in equal time intervals.

One can think of a laser as a classical oscillator of constant amplitude, which produces photon detections with Poisson statistics. In contrast, blackbody radiation, whether broadband or filtered, produces non-Poisson arrivals with greater fluctuations. The statistics for blackbody radiation depend in detail upon the spectrum.

Unfortunately, the timescale of the correlations is of the order of the inverse bandwidth of the accepted light; thus one is forced to use extremely narrowband filters in order to see the non-Poisson nature of blackbody light (of the order of $\Delta\lambda/\lambda \sim 10^{-6}$ for $\tau \sim 1\,\mathrm{ns}$), which defeats the power of the technique to detect a line of unknown wavelength. Moreover, even when so filtered, the statistics reverts rapidly to blackbody when the detected photon rate is dominated by starlight. Consequently, this technique is not useful for weak laser lines, or for lines of unknown wavelength.

Pulsed Beacons

Given modern optical technology, a simple thought experiment shows that optical pulsed beacons can be detected using a very simple apparatus. Imagine looking at a solar-type star 1,000 ly away with a telescope of 10 m^2 collecting area. You would expect to see about 10^7 visible light photons per second, or 1 photon per 100 ns. Existing instrumentation can easily count the incoming photons at this rate. Now suppose you observed 10 photons in 1 ns. This is statistically unlikely at the 10^{-20} level, so the signal is clearly artificial or from some new astrophysical source.

Thus, a 1 photon/m^2 pulse is enough to be seen as artificial. How hard is this to generate? As we discuss below, the transmitted beam probably needs to cover the whole Habitable Zone of the target star, and the optics and pointing mechanism can easily support this required spot size. If we assume a 10 AU diameter Habitable Zone covered at 1 photon/m^2, we need 1.5×10^{24} photons total. At a wavelength of 1 μm, each photon carries approximately 2×10^{-19} J, so the total pulse energy needed is about 3×10^5 J. This is about a factor of 10 larger than existing lasers, but less by a factor of 10 than lasers now being designed for fusion power plants. Generating a 10 AU spot size at a range of 1,000 ly requires a telescope with a diffraction-limited seven meter aperture, probably located in space. This is very close to what is planned for the Next Generation Space Telescope.

So a very plausible laser, aimed through a soon-to-be-built telescope at a star 1,000 ly away, will generate a pulse strong enough so that a medium sized telescope with simple instrumentation could clearly recognize the signal as artificial. This shows that optical pulses are clearly feasible as interstellar beacons.

As we have seen, very plausible optical systems can communicate across interstellar distances. But what about the case where the civilizations are deliberately trying to establish contact? How might a sending civilization approach this task using optical technology? Our views about which beacons are most plausible will determine our detection strategy.

The most obvious approach is a targeted multiplexed optical pulse beacon. That is, we assume that the sending civilization wishes to irradiate, in sequence, the planetary zones of the nearest N solar-type stars (or a selected subset, if they know more), going out to a range R_{max} comparable to the mean distance between advanced civilizations. Within a factor of 10, we might assume that N is in the range of 10^3 to 10^6, corresponding to distances R_{max} of 100 to 1,000 ly. We do our part by aiming our telescopes at nearby stars and looking for pulses.

Given this overall strategy, the usual questions remain. How often will they send a pulse? This determines how long we should observe each star. Can they be expected to aim the beacon accurately enough? Is there a preferred wavelength regime, or perhaps a single 'magic' wavelength? Is there a compelling parameter set (analogous to the use of an RF carrier at the hydrogen hyperfine wavelength of 21 cm) that suggests an obvious optical search strategy?

Aiming Accuracy

Let us assume that the transmitting civilization has a catalog of target stars, with positions, proper motions, and ranges known with sufficient accuracy. If we are trying to obtain about a 10 AU diameter beam at the target, a reasonable aiming requirement is about 1 AU accuracy. At a range of about 100 ly, this corresponds to a positional accuracy of 0.03 arcsecond and a proper motion

Figure 3.7: Dr. Alan Roy from Australia, currently a post-doc at the NRAO in Socorro, NM, brings a combination of engineering and astronomy. (Photo courtesy of A. Roy.)

uncertainty of 150 μarc second/year; at about 1,000 ly the corresponding figures are 3 marcsecond and 15 μarc second/year.

The required range accuracy depends on the star's proper motion but not its range. For example, to keep the aiming error down to about 1 AU for a star whose transverse proper motion is 10 km/s, the range uncertainty cannot exceed about 0.25 ly. These accuracies are relaxed if the transmitted beam is broadened to illuminate a larger zone at the expense of received signal strength. These figures suggest pretty accurate shooting, although they certainly seem possible;

perhaps the advanced civilization doing the transmitting would choose to spread the beam a bit, raising the laser power to compensate.

The positional accuracy is presumably attainable, because if the sending civilization can transmit a beam of the required width they can certainly observe the star's apparent position to the same precision. The required proper motion and range accuracies also appear to be completely feasible, certainly for a civilization with good optical technology in orbit. The JPL SIM [Sha95], currently in development for a 2005 launch, is aiming for better accuracy than is required here.

Beam Size Versus Planetary Zone

As we scale up the transmitting aperture, it is necessary to check that the resulting diffraction-limited beam size does not become smaller than the planetary zone we wish to illuminate. This, rather than the size of the aperture, limits the gain for nearby stars. For example, for a Habitable Zone diameter of 10 AU, a distance of 100 ly, and a wavelength of 1 μm, the maximum usable gain is reached at a 0.7 meter aperture. At 1,000 ly a 7 m aperture is the maximum needed. For most practical ranges, the beam size remains the same in absolute units.

Revisit Times

How often might we see a pulse? The pulse repetition rate towards a given target might be limited by one of two factors – the total power the ETs are willing to spend, or the time to re-aim their beacon.

The total power looks quite promising. Assuming they use 10^4 J per pulse, and are willing to radiate 1 GW total, they can hit 10^5 targets per second. Even if their list is very large, perhaps 10^8 stars, then each target is visited every 1,000 seconds.

Another possible limit is the beam steering rate. First, we look at what we can do. To send a pulse to each of $N = 10^3$ stars (approximately the number of solar-type stars within about 100 ly) with a single laser system, the sender could use an assembly of fast beam steering mirrors of relatively small size and weight, in combination with a large objective mirror that is steered slowly. We could build such a system today that would settle in ~0.1 s to diffraction-limited pointing. Such a device would let them send an optical pulse to 10 stars per second, thus 100 seconds to illuminate 1,000 stars. (Of course, to illuminate stars anywhere in the sky, they would need either several laser stations on their planet or an orbiting station; the latter is probably better.) It is interesting to note that pulsed laser systems on the drawing board today, in particular the

LLNL 'Helios' laser, have a design repetition rate of 10 pps.

It is not difficult to imagine an optical search program that spends a few minutes observing each of a thousand stars. In fact, the job could be done in just a few clear nights, perhaps even by amateur astronomers.

For the more difficult scenario of $N = 10^6$ stars, the revisit time stretches to 10^5 seconds, about a day. The task of the recipient is to observe 10^6 stars, each for a day – no longer a reasonable amateur search program. The task of the searcher scales as N^2, with N itself scaling as R_{max}^3. Thus, the effort scales as R_{max}^6. The sender can use a network of transmitters; the target can mobilize amateur astronomers, who have long since become victims of 'variable-star ennui' and desperately need a new challenge.

Our current technology can reach about 10^6 stars. Most likely, the sending civilization has more advanced technology, and the revisit time would be limited only by the total power the ETI wished to spend. If they are willing to emit 10^9 watts total in 10^4 J pulses, then they can target each of 10^8 stars every 10^3 seconds. Thus, we cannot be certain about what pulse rate to expect.

3.5.2   Collecting and Detecting Optical Signals

Collecting and detecting optical signals has a long history and it uses a mature technology. As in the microwaves, physics is kind – the atmosphere is sometimes transparent and the mirrors do not need to be cooled, so observations can be done from the ground with physically large mirrors.

Once the optical signal is collected, it needs to be analyzed. Modern technologies come close to detecting each photon that arrives.

Looking for narrow spectral lines is a common task for astronomers, and existing technology divides the optical spectrum into about 5×10^4 bins. Within each bin, the SNR is near the theoretical limits set by aperture and photon flux.

Looking for very short pulses is not a common task for optical astronomers, but it is for particle physicists. The technology is readily available and does not appear to be very difficult. It is relatively easy to build a system to detect at least 10 photons, coming from an identifiable nearby star, within a 1 ns period. Assuming that no great wavelength specificity is needed, then the pulse detection system could be very simple. It could consist of a reflecting telescope of modest aperture (e.g., 1 m), an optional multilayer filter, followed by a beamsplitter and photomultiplier tubes. The electronics could consist of a pair of pulse height discriminators, to reject dark counts and single photoelectron events, driving a coincidence circuit. The observer points the telescope at each star in turn, guides on the photomultiplier 'singles' rates (upstream of the pulse height discrimination), and waits for the unique coincidence signature generated by a burst of at least 10 photons within the 1 ns resolving time of the photomultiplier tubes.

Of course, more complicated systems are possible. One could record and time-tag all large multiphoton events, perhaps in a wavelength-dispersive array detector, and look for coincidences later. In this way, one is not committed to any particular threshold, pulse width, or wavelength. The downside is a glut of data, but there is surely a happy middle ground.

Pulse Detection

Receiver Area	1 m^2	10 m^2	100 m^2	1,000 m^2
Photons Needed	10/m^2	1/m^2	0.1/m^2	0.01/m^2
Fluence/Pulse (J/m^2)	2×10^{-18}	2×10^{-19}	2×10^{-20}	2×10^{-21}
Isotropic Transmitter (J)	2×10^{21}	2×10^{20}	2×10^{19}	2×10^{18}
Diffraction Limited Optics (10 m, 1 μm) (J/Pulse)	2×10^6	2×10^5	2×10^4	2×10^3
Pulse Width, dark sky (s)	$< 10^{-2}$	$< 10^{-2}$	$< 10^{-2}$	$< 10^{-2}$
Pulse Width, on star (s)	$< 10^{-6}$	$< 10^{-7}$	$< 10^{-8}$	$< 10^{-9}$

Table 3.2: Looking for Broadband Optical Pulses.

Table 3.2 shows the pulse energy required for an optical beacon to be reliably detected by telescopes of various sizes. For good discrimination against the statistical pileup of single photon counts, about 10 photons must be detected in a time interval that is brief compared with either the mean time between counts coming from the night sky, or from the target star. The table lists the required pulse energy at the transmitter for two cases: an isotropic transmitter, and one beamed towards Earth using diffraction-limited optics with a 10 m aperture and a wavelength of the order of 1 μm. The extreme directivity of this second possibility explains the remarkable reduction in required pulse energy (See page 92).

Also shown in the table is the count rate on a one arcsecond patch of dark sky, integrated over the entire visible band (B = 0.3 PHz). The table shows the longest feasible pulse widths that are detectable for this same band when light from a solar-type star at 1,000 ly distance is also in the beam. We conclude that optical pulses in the microsecond to nanosecond regime are highly visible because, being so brief, they stand out significantly from the steady starlight. The transmitter energy levels are, however, very demanding unless the energy is beamed our way.

Continuous Line Detection

Table 3.3 shows what is needed to detect a continuous line assuming:

- The spectrum can be resolved to one part in 5×10^4.
- We integrate for one hour.
- We require a 6σ detection.

Receiver Area	1 m²	10 m²	100 m²
Photons per second	20	200	2,000
Photons per hour	7.2×10^4	7.2×10^5	7.2×10^6
Standard Deviation (J)	2.7×10^2	8.4×10^2	2.7×10^3
Flux Needed (ph/s/m²)	0.45	0.14	0.045
Watts (10 m aperture)	9×10^4	2.7×10^4	9×10^3

Table 3.3: Looking for Narrow Emission Lines.

Another factor of five should be obtainable by working within a dark line of the spectrum. Even so, it can be seen that the power required for detection of continuous lines is much higher than for pulse detection.

From these two tables, we can see that using narrow filters is unnecessary when searching for pulses from stars beyond about 100 ly, but may still be desirable. This is because any pulsed beacon has to put a few photons into the aperture to be reliably distinguished from the expected single photon events. For any star, except those very nearby, any reasonably short pulse ($\approx$1 ns or less) is already in the regime where at most one photon from the entire broadband emission of the star is expected. Therefore, dividing the light from the star by wavelength does not help pulse searches.

For nearby stars and big telescopes this argument fails. For example, a solar-type star at 30 ly radiates 10^{10} photons per second into a 10 meter telescope – enough to overwhelm existing, broadband, pulse counting. However, there are few nearby stars, and they can be observed with smaller telescopes or filters.

Today, we could build both pulsed and line beacons. The pulsed beacons seem to offer an advantage of about 1,000 in required energy per bit. A single pulse suffices for pulsed beacons, whereas continuous beacons must be kept on a single target for an hour to get a large enough S/N.

3.5.3   Sending Signals

In the optical region, as in the microwave region, once a beacon is found we can use the same technology for communication. The JPL is working on deep space optical communication. Their goal is to reduce the size and mass of the communication portion of deep space probes by utilizing the very high gains of optical transmitters for a given physical size. This effort also includes ground stations with large collecting areas and high-power transmitters.

For interstellar communication, consider the pulsed case discussed earlier using a 100 m^2 telescope as a receiver, with a comparable 10 meter telescope being used as a highly directional transmitter. Because its diffraction-limited beam solid angle is 1×10^{-14} sr, its gain with respect to an isotropic emitter is about 1×10^{15}. To obtain 2×10^{19} J/pulse isotropic energy would require 2×10^4 J/pulse – an energy level that is already obtainable today on Earth. Therefore, the same system used for contact will work well for communication.

If narrow-beam optical links between ETIs are feasible today using technology comparable to our own, would we be able to eavesdrop on such communications using the same technology? Probably not – the volume of space occupied by a collection of tight beams between ETIs would be very small. A random location, such as that of the Earth, would be very unlikely to be inside such a beam. However, if the technological growth that yielded narrow-beam optical links were to continue, then broader beams and indeed broadcasting could become feasible. Such developments are impossible for us to determine except through observation.

3.6   Summary

In spite of suggestions, going back nearly 40 years [Sch61], that optical and infrared methods permit efficient interstellar communications, most searches to date have been done at microwave frequencies. Historically this made sense because Project Ozma was carried out in the year the laser was invented. At that time, microwave technology was much more advanced than the laser. However, recent advances in laser technology make it clear that optical methods are altogether practical, both for establishing contact and for carrying out directed communication. For example, a present-day technology pulsed laser attached to the Keck telescope would outshine our Sun by more than three orders of magnitude (for brief intervals), viewed in broadband visible light by a distant civilization. This is most remarkable, in just the same way as it is that our television carriers, in their narrow bandwidths, outshine our Sun.

There is no clear-cut choice of the most desirable wavelength for the search for ETIs. Microwave, infrared, and visible all have advantages and disadvantages, depending upon what factors (power, size, bandwidth, simplicity of detection, etc.) one wishes to optimize. Although optical photons individually require more energy than microwave photons, optical beam sizes are typically much smaller, and directed communication can make up the disadvantage in energy per bit. A diverse strategy is best for the search for ETIs, and searches in the microwave, infrared, and visible are all worthwhile.

This diversity is relatively inexpensive. Infrared/optical search experiments can be simple and inexpensive. They can be performed concurrently with existing survey activities, use existing spectrometers, and even mine existing databases.

We have identified no compelling 'magic' wavelengths in the visible or infrared bands. None may be needed, however, if short pulses are used, because stellar backgrounds are negligible on nanosecond timescales. Because of the effects of extinction and time smearing, infrared wavelengths (of a few μm or greater) are favored over visible wavelengths for ranges greater than a few hundred parsecs.

In the infrared and optical, pulsed-laser searches at the nanosecond scale appear to be considerably more sensitive than continuous-wave searches. This is unlike the situation at microwave frequencies, where interstellar dispersion, natural and artificial pulsed backgrounds, and lower carrier frequencies all favor narrowband continuous beacons. Nevertheless, both pulse and continuous searches are worthwhile.

Calculations suggest that a multiplexed optical pulsed beacon is reasonable for an ETI to use for establishing contact out to about 100 ly. The equivalent of a Helios-class laser and a 5 m transmitting aperture, along with appropriate beam-steering optics, would suffice to contact a reasonably sophisticated amateur astronomer on Earth. For sources out to about 1,000 ly, however, amateur astronomers would have to defer to the professionals in a serious optical search for ETIs. There is no reason to assume, however, that the LLNL laser plan for the year 2030 is an upper bound on what an advanced ETI civilization might use to establish contact via targeted optical pulsed transmissions.

The SETI community must consider alternative optical search strategies in an attempt to identify a particularly compelling strategy which could be the basis for Earth-based optical search efforts in the near term. Different choices of wavelengths, pulse widths, repetition rates, revisit times, etc., must be considered for both sender and receiver.

Our discussion of optical search techniques has been more indicative than definitive. This is nothing more than a reflection upon the relative youth of efforts to search at optical wavelengths. The invention of the maser, dating to a *Physical*

Review article by Arthur Schawlow and Charles Townes [Sca58], trailed Heinrich Hertz's production of radio waves by nearly seven decades. While optical telescope technology is an old art, the ability to generate and detect strong, clearly artificial, optical signals is young. Consequently, strategies for optical searches are less advanced than those for microwave searches.

However, some trends are indisputable. The existence of highly directional optical instruments (the gain of the Hubble Telescope is approximately 10^{13}, whereas that of Arecibo is about 10^7) has inclined optical search proponents to search for deliberate signals that advanced societies may beam to the inner realms of solar systems.

Without doubt, the first optical SETI experiments will be targeted searches, and the difficulty of high-resolution spectral dissection has persuaded researchers to look for short pulses. Nanosecond pulses are a preferred choice only because of contemporary technology – pulses that would represent 1% of the optical bandwidth could be a thousand times shorter. Consequently, the experiments described in this chapter are, at best, a short extrapolation of today's capabilities, and do not begin to exploit the full potential of optical communication.

That will change. For forty years, the attraction of searching for cosmic company in the radio regime has been compelling. That compulsion is now yielding to a more eclectic view, as SETI research becomes multispectral. The 'holy grail' – all-the-sky, at all-frequencies, all-the-time – can already be discerned on the horizon of radio technology. Its optical counterpart cannot be far behind.

Chapter 4

Statistics, Computers, and Signal Processing

This chapter makes it clear that computing costs will not dominate the total costs of developing SETI systems over the next twenty years. Computer processing costs are now about 1/10,000 of their equivalent 20 years ago, and are continuing to become less expensive. For example, hardware or software to rapidly carry out Fast Fourier Transform (FFT) calculations was virtually unknown 40 years ago, but is now very fast and inexpensive.

Radio telescope antennas and optical telescopes are essentially mechanical devices, and thus tend not to follow the rapid, exponentially decreasing, cost curve that computing devices follow (Moore's Law). Therefore, over time, the cost of computing becomes smaller with respect to the cost of the mechanical components of the telescopes.

This chapter discusses the essentially mathematical aspects of signal processing and information theory that are both relevant to, and independent of, any assumed ETI civilization. These methods are first applied to the process of identifying an extraterrestrial signal (microwave or infrared/optical). This requires separating the signal from the noise, and correcting for distortions due to the differential rotations and orbits of transmitting and receiving bodies. Then they are applied to the evaluation of the signal to decide whether or not it is actually from an ETI.

Specific topics discussed in this chapter include:

- the detection of pulsed and continuous signals in the presence of noise;
- the mathematics of signal detection in the microwave and optical bands;
- the implications of signal coding.

4.1 Signal Processing Stage

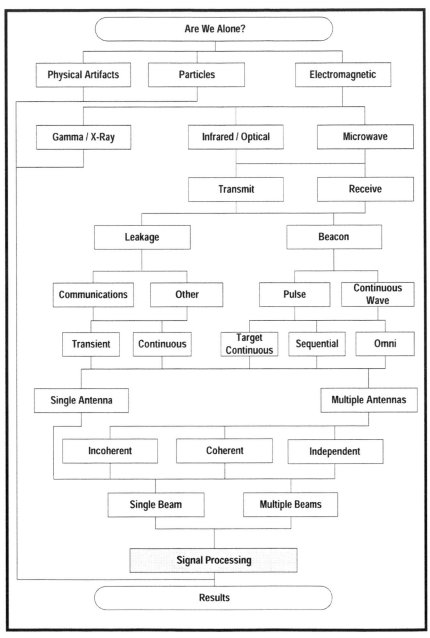

Figure 4.1: SETI Decision Tree – Signal Processing Issues.

Figure 4.2: Dr. Greg Papadopoulos, Chief Technology Officer of Sun Microsystems, introduces computational innovations to SETI.

4.2 SETI Algorithms

Today's SETI detectors are fundamentally software algorithms that simulate various summing and filtering operations [Cul85a]. Years ago, such operations were performed using circuit elements [Oli71]. Such an approach, however, would be vastly expensive compared to modern computing. In fact, it has been the coming of age of digital technology that makes SETI possible on its current growing scale.

SETI algorithms currently rely on the equivalent of filter banks, which slice frequencies into narrow bands. The hope is that one band will contain an interesting signal, and that the noise from the receiver and sky in this band will be smaller than the signal. The development of an efficient, $n \log n$, algorithm to perform such frequency segmentation has led to the almost universal use of the FFT to perform this operation. This means that even very wide frequency regions can be divided into rather narrow bands. Any of these bands could contain a signal at a good signal-to-noise ratio (S/N).

In 1980, dividing 1 MHz of spectrum into a million channels, sampled at the Nyquist rate, cost a million dollars. This high cost was due to the fact that general-purpose computers were not fast enough to do the job. Therefore, special purpose components had to be used, and the labor to build these components was very expensive. Today, the same operations can be performed in a general-purpose computer for a cost of about a thousand dollars. Furthermore, it is encouraging that FFTs continue to become less expensive.

Very narrow carriers are a significant component of the Earth's transmissions. They pass over the horizon and leak into space. Since FFTs easily detect narrow carriers, they seem ideal for detecting technologies like ours.

SETI currently seeks signals with concentrated components, either in frequency or time. Components concentrated in frequency are similar to continuous sinusoidal carriers. Components concentrated in time are most simply carriers pulsed in time, which have a width determined by the uncertainty principle. Such signals are ideal for detection using FFTs with the correct resolution. Furthermore, algorithms can run in parallel because of three reasons:

- Any region of frequency is much like any other.
- Simple signals tend to be stable in frequency.
- Results from one frequency do not involve information from another.

There is a caveat. If constant-frequency signals are transmitted from the surface of a rotating, orbiting body like the Earth without correction, then at typical wavelengths of interest, they can drift many hundreds of Hz during an observation. In one sense, this is a trivial problem. If processing domains are many megahertz wide, the drifting of a signal from the domain handled by one processor into that handled by another is rare. The problem can be neglected, viewed as a small amount of missing frequency space, or the domains can be overlapped slightly. The larger problem is that signals do not remain in the very narrow bands one wishes to use to obtain an optimal S/N ratio.

Note that an optimal S/N ratio is obtained using a matched filter. Since the FFT decomposes data into sinusoids, it is a matched filter for pure, continuous wave signals that do not drift. Of course, the match is only as good as the integration time allows.

The very narrow FFT bands, which are desired for obtaining the best S/N ratio, result from long integration times. During integration, the drifting signal from a planet changes frequency, depositing minuscule amounts of power in each of many narrow, analysis bands. Integrations, then, must be short enough to contain a drifting signal within a band during analysis.

Typically, the analysis band is chosen to be approximately 1 Hz in width, which is about 100 times broader than that which could be used if the transmitter were in an inertial reference frame. In order to recover the detectability of a signal drifting across many 1 Hz analysis bands during an observation, the power in each band and spectrum must be computed and summed along all possible signal paths. This summation technique loses sensitivity by factors from 5 to 10, compared with the ideal case of narrowband, matched filter analysis.

A bin is a particular band and spectrum. Assume a signal can begin in any of m bins, and drift for as much as one band per spectrum during n spectra per observation, while still yielding a strong response in the bins along its path. This means that approximately $2mn$ accumulations must be updated every spectrum, as opposed to the m required for power accumulation of nondrifting signals. Fortunately, it is possible to create good approximations to drifting paths by taking two paths of equal length and summing them in two nearly identical ways. Thus, longer paths require not $2n$ accumulations per starting channel but $2\log_2 n$. Thus, at least, for power accumulation, line detection algorithms are in the same pleasant regime as spectrum analysis [SetXX].

Detection of Pulses

Assuming for the moment that pulses of interest are matched to the same 1 Hz, one second, spectrometer mentioned earlier, their detection is more complex than carriers. Pulses drift in the same way that carriers do, but they have unknown spacing and length. If their duty cycle is high, they can be detected, more or less efficiently, using continuous wave detection techniques.

The Fourier Transforms of signals that are periodic in frequency and in time are the same type of signal with reciprocal dimensions. A constant signal in frequency of infinitely narrow width is analogous to a pulse in time of infinitely narrow duration. We can conceive of detecting a constant, narrow carrier because detection conditions are controlled by the receiving civilization. Just turn on the detector for as long as necessary to collect enough energy for a detection, or until boredom brings an end to the experiment.

Pulses are more difficult. However long we may listen, assuming the bandwidth is known, we can never be certain that a signal has fallen into the pulse gate. Maybe the pulse has not yet been transmitted. Maybe, in the extreme analogy of one pulse being analogous to one carrier, the pulse has already happened. We can build ever bigger spectrometers to cover more frequencies at a given sensitivity trying to capture the elusive carrier of unknown frequency. We cannot, however, build along the temporal dimension. We must wait, and hope we have not missed the boat.

Still, the interstellar medium limits ideal single pulse and carrier experiments. We transmit carriers whose coherence is limited by transmission through the interstellar medium. Similarly, very narrow pulses are dispersed and therefore broadened in their interstellar travels. For continuous waves, we get around this problem with either multiple looks for carriers, or by averaging small coherent transforms to obtain incoherent statistics. In the same way, it is not necessary to examine all possible pulse durations to get ideal matches to pulses. Instead, we can average pulse bins together to incoherently add pulse energies. Since frequencies can be analyzed simultaneously, limited only by the computing resources available, and since antennas are typically band-limited in their characteristics allowing propagation of monochromatic[1] rather than monotemporal[2] signals, carriers tend to be favored for detection. The analogies for the reciprocal pulse case, however, should not be overlooked.

In order to see how sparse arrays of pulses are detected, such as events above a threshold power in a spectrometer, consider an incoming spectrum of m bins. If the probability of an above-threshold event is p, then there are an average of mp pulses in the current spectrum. Assume that n spectra have already come in since the observation began. Thus, second pulses must lie in a drift-limited triangle with the apex on a current pulse and with area $\frac{(2n)(n)}{2} = n^2$. The number of such pulses in the triangle is then pn^2, and the total number of pulse pairs per observation is mn^2p^2 [Cul86]. In any pulse detector, there is a trade-off between sensitivity, wanting to make p large for signal and noise events alike, and the desire to winnow data so that less searching is necessary.

A good compromise is to require another pulse, predicted from any two pulses found in the current and earlier spectra. This lowers the false alarm rate by another factor of p. Integrating all the predictions that must be made during an entire observation, we find a requirement for order mn^3p^3 statistical tests of pulse triplets. This can be a very large number indeed, if p is too big. Fortunately, the probability distribution of the power for Gaussian noise, such as that typically found in receiving systems, decreases exponentially. It is easy to set a threshold that will make the noise field sparse, while simultaneously allowing three regularly spaced low duty cycle pulses of modest total energy to stand out against the random background. Typically, with only a few pulses in an observation, pulses are three or four times more visible than carriers with the same average power. If pulses are strong, then their energy is concentrated. Therefore, the noise is separated from the signal by the thresholding operation,

---

[1]A monochromatic signal is one that has a single frequency and therefore wavelength, and is continuous.

[2]A monotemporal signal is one that is a single pulse in time, but contains components at all frequencies and wavelengths.

rather than being averaged with the signal samples as is done with drifting CW detection using power spectra [Cul86], and [Cul85b]. Thus, the result of using simple algorithms is that low duty cycle pulses originating from planets become almost as detectable as non-drifting (i.e., inertial) beacon signals, while planetary CW signals are not.

There are two very difficult computational problems that arise if we wish to detect pulses and carriers with a matched filter system. Both signals have ideal, exponential statistics, if they are entirely captured in an FFT bin. However, to do this requires a great amount of Fourier transforming. For pulses, one must match the transform length to the pulse, and even overlap transforms to ensure that pulses are not split between analysis intervals. We want to detect carriers in the narrowest possible bandwidth, excluding most noise. However, as the channel narrows, the potentially drifting carrier may not remain within the FFT channel during the entire observation. As the time doubles for a spectral analysis and the channel narrows by half, the allowable drift rate to keep a signal in a channel decreases by a factor of four. Thus, the maximum allowable drift rate for optimum detection decreases as the square of the spectrum length. Therefore, many FFTs must be performed at the correct drift rates to optimally capture narrow carriers and achieve ideal detection of planetary leakage. Additionally, many FFT lengths must be tried for pulses, since no particular length is compelling.

To ideally capture all pulses between a second and a microsecond, about twenty different FFT resolutions are required. To perform ideal CW detection, using a single long spectrum for 100 second observations, would require ten thousand 100 second transforms all at slightly different microdrifts. Fortunately, using one second transforms costs only about a factor of 10 in sensitivity, and only one set of these is required. As FFTs become less expensive, they will undoubtedly be used to search more sensitively for drifting signals. However, increases in FFT computation by factors of ten thousand will gain only a factor of 10 in sensitivity for CW signals. On the other hand, each new octave of FFT length detects pulses almost independently of those having different lengths. Thus, systematic searches of the pulse length domain are potentially rewarding, because, in addition to exploring new territory, they are almost as sensitive as the ideal matched search.

4.3 Complicated Signals

Transmissions from the Earth are becoming increasingly complex. Signals using Frequency Shift Keying (FSK), Phase Modulation, and Spread Spectrum, for example, are common. However, matched filters to all conceivable signal types have two problems. First, such a scheme would be so computationally intensive that our searching of stars and frequencies would fall far off the exponential growth curve. Second, the number of statistical tests would raise the false alarm rate so much that the sensitivity of an individual test would need to be dramatically dropped.

To give an extreme example of this, consider the following. If we tested an observation for all combinations of bits possible with our levels of digitization, an immediate difficulty would arise. Since, given arbitrary encoding, any configuration could be a signal, we would always get a false alarm in an observation. How we could investigate such reports is difficult to imagine. Thus, it seems best to look for simple signals, hoping for some cooperation from our alien colleagues. Although the Earth is sending fewer leakage signals that are both simple and strong, later sections will show how technologies like ours could create very powerful receiving antennas that could also be used as attention-getting beacons. Broadband searches for strong, stochastic signals will also be discussed.

4.4   Application of Statistics to SETI

In this section, we apply the statistics of Appendix P on page 455 to typical SETI observations, and show which rules apply in each case. The numerical examples are chosen to be typical of SETI technology on Earth in 1999.

4.4.1   Microwave

In the microwave, the background consists of Gaussian white noise, and a signal stands out by having a higher than expected power in a single channel. The main problem comes from the enormous number of channels that are searched. For example, if we search a 1 GHz bandwidth with 0.01 Hz resolution, we are looking at 10^{11} observations every 100 seconds, or 10^9 observations per second. A false alarm percentage of 10^{-12}, which would be considered extraordinarily low by most standards, still gives one false alarm every 20 minutes!

To prevent this we need relatively high thresholds. The signal power observed in a channel exceeds the average power by a factor of p with a probability of e^{-p}. This is shown in Table 4.1, where the *Signal Power / Average Power* column shows the S/N ratio. The *Probability* column shows the odds that an observation exceeds the given threshold by chance alone, and the *False Alarm Rate* column shows the rate of observations that exceed the threshold by chance alone, assuming 10^9 observations per second. For a project that cannot easily repeat observations, a signal about 40 times the noise is required to be statistically significant. An example of such a project is the Search for Extraterrestrial Radio Emissions from Nearby Developed Intelligent Populations (SERENDIP). It was one of the first SETI projects, a commensal search that does not have control of the telescope.

Signal Power / Average Power	Probability	False Alarm Rate
1	0.37	3.7×10^8 per second
5	6.7×10^{-3}	6.7×10^6 per second
10	4.5×10^{-5}	4.5×10^4 per second
15	3.1×10^{-7}	310 per second
20	2.1×10^{-9}	2.1 per second
25	1.4×10^{11}	0.83 per minute
30	9.4×10^{-14}	8.1 per day
35	6.3×10^{-16}	20 per year
40	4.3×10^{-18}	0.13 per year

Table 4.1: Microwave Statistical Example.

Re-observation of the same source allows weaker signals to be detected reliably. If you observe the same target twice (as does BETA), and demand that a signal shows up in both observations, then a threshold of 20 times the average power gives the same false alarm rate as a single observation at 40 times the average power. It also provides some defense against soft errors and terrestrial interference.

Phoenix takes this strategy further – it makes hundreds of short (about one second) observations. This generates a very high alarm rate per observation, which is reduced to practical levels by accumulating the probabilities over many observations. Although this approach is not as sensitive to continuous wave signals as a single long observation, it also allows for the detection of pulses or drifting continuous wave signals.

4.4.2 Optical Pulses

We use direct photon detection for optical pulses. The background is the Poisson distributed arrival of individual photons from the star, which is also in our field of view. The photons are coming in at a rate r, and for any time interval τ, this leads to an expected number of photons $x = r\tau$. Then the probability of getting exactly N photons in this interval is:

$$P(N) = \frac{x^N e^{-x}}{N!} \tag{4.1}$$

Figure 4.3: Dr. Richard Stauduhar, a SETI Institute mathematician, contemplates the infinite.

For example, suppose we expect to receive one photon every 100 ns from a star. Then, for any nanosecond interval, $x = 0.01$ and the odds of receiving N photons are shown in Table 4.2. During most nanosecond intervals, we receive zero photons. In about $1/100$ of the nanosecond intervals, or 10^{-2} of the time, we receive a single photon. This corresponds to the incoming rate of 10^{7} photons per second. We get two photons in a given nanosecond interval about 5×10^{-5} of the time (or 5×10^{4} times per second overall), three photons about 170 times per second, and four photons about once every four seconds.

Continuing this trend, the rate of natural occurrences drops dramatically with increasing photon count, until seven photons in a nanosecond will only occur naturally about 2×10^{-18} of the time, or about once every 16 years. Therefore, from this star, if we observe pulses of seven or more photons more often than once every 16 years, we can safely state that this is either a new astrophysical phenomenon or an artificial signal. By using still shorter time periods, we can increase the sensitivity somewhat – the arrival of four photons in the same picosecond is about as unlikely as seven photons in a given nanosecond.

Photons	Probability	How Often By Chance
0	0.99	n/a
1	10^{-2}	10^7 per second
2	5×10^{-5}	5×10^4 per second
3	1.7×10^{-7}	1.7×10^2 per second
4	4×10^{-10}	24 per minute
5	8.3×10^{-13}	3 per hour
6	1.4×10^{-15}	3 per month
7	2×10^{-18}	0.06 per year

Table 4.2: Pulsed Optical Statistical Example.

4.4.3 Optical Lines

For this search, we would use an integrating photon detector. Here the background is the expected number of photons in a spectral bin. How many extra photons do we need to stand out? In theory, once again we have Poisson arrival statistics.

Because of the large number of photons though, the number arriving in a given time will have very nearly Gaussian statistics. If we expect N photons, for $N \gg 1$, then the standard deviation of the number of photons is $\sqrt{N}$. For example, if we expect 1,000 photons, then we expect the standard deviation to be about ± 30 photons per observation. The statistics for this case are shown in Table 4.3 on page 110.

Extra Photons	Probability	False Alarm Rate
30	0.32	4.4 per second
60	4.6×10^{-2}	0.63 per second
90	2.7×10^{-3}	2.25 per minute
120	6.3×10^{-5}	3.15 per hour
150	5.7×10^{-7}	0.68 per day
180	2.0×10^{-9}	0.87 per year
210	2.6×10^{-12}	0.11 per century

Table 4.3: Continuous Line Optical Statistical Example.

In the table, the "False Alarm Rate" numbers are calculated by assuming 5.0×10^4 discrete frequencies and one hour integrations. These are reasonable numbers for an optical search for continuous lines, and they are used elsewhere in this book.

From Table 4.3, we need about a 5σ variation to keep the false alarm rate down to about 10^{-6}, so about 155 additional photons are needed for a reliable detection. Note that the false alarm percentage is much less stringent than in the microwave case, because far fewer channels are observed (perhaps 5×10^4 as opposed to 5×10^{11}) and each observation takes much longer.

Chapter 5

Extraterrestrial Sources and Search Strategies

This chapter discusses where we should look for signals from ETIs, and for what evidence we should search. Intelligence cannot be detected, but it can be inferred by the manifestation of a technology. If we find an ETI civilization among our nearby stellar neighbors, it will probably be older than ours. Therefore, their technology will presumably be more advanced than ours.

Leakage – Photons are the most easily detectable of all potential information carriers, and also are the easiest to produce. Currently, our own radio technology creates spectral features that are completely uncharacteristic of stars. Similarly, other comparable civilizations should be highly detectable in the microwave band over great distances. This is unintentional leakage. Deliberate beacons would be even more easily detected, if we happened to be looking at the right place at the right time. As our use of radio develops, leakage is likely to be reduced and the signals will appear more and more broadband like background radiation, becoming much harder to identify. Presumably, the same applies to our neighbors, although they may also be communicating with each other between stars.

Survey Type – We need to choose whether to survey the whole sky or to search specific targets. The two strategies are capable of detecting different kinds of signals. A whole sky survey can detect a continuous wave and a slowly pulsing signal, but it will find a beacon only if it happens to be on when the search beam passes over it. Targeted searches, while being able to detect both beacons and continuous waves, are restricted by our choice of targets, which may not be the best. There are, however, targets of opportunity and regions dense with stars worthy of special scrutiny.

Microwave vs. Infrared/Optical – A crucial question, of whether to look for beacons or leakage in the microwave, is discussed by analogy with the reduction of our leakage, and the likelihood that leakage from ETIs is even less detectable. Beyond about 1,000 light years, absorption by interstellar dust favors the microwave and far-infrared over near-infrared/optical wavelengths. The likely luminosity functions of extraterrestrial transmitters are discussed for non-targeted and targeted beacons.

Beacons – Beacons are considered on the basis of the kinds we would use if we were trying to attract attention, then assuming that ETIs are doing the same and discussing the ways we could detect them. We now know how to make high-power optical laser pulses of very short duration, and we could detect them at interstellar distances. It is therefore reasonable to expect the same abilities of any well-developed extraterrestrial civilization. Various beacon configurations are compared. Other possibilities are examined, particularly those of detecting microwave power transmissions or leakage from interplanetary navigation beacons.

Ideal Detector – The ideal detector would cover the whole sky, all the time, at all frequencies with very good sensitivity. This requires massive computational capacity. Possibilities do exist for hybrid strategies using targets of opportunity in space and time.

Transmission and Reception Strategies – Likely transmission strategies that may be used by ETIs are discussed, together with receiving strategies that would match them. Targets of opportunity are explored for both beacons and corresponding receivers in frequency, space, and time. Opportunities for serendipity are considered in historical perspective, as are commensal opportunities for receivers. The application of game theory to the selection of a best possible strategy is mentioned.

5.1   Searching for an ETI Civilization

How does one search for an ETI civilization? Given our own current state of development, our only possibility is to search for some manifestation of another technology that can be detected across the vast distances between the stars. Finding such, we will conclude that an intelligent civilization was responsible. This may sound overly abstract, but intelligence as such is undetectable; only the consequences of its actions can be observed. Water-covered worlds, inhabited by extraordinarily intelligent analogues of our own whales and dolphins, remain beyond our grasp.

Within the horizons of this book, inferential evidence of life on a nearby planet may be achieved. However, it will not be possible to distinguish between algae and alumnae. Technology is our only discriminator. Where and how should we look, and for what? What manifestations of their technology might we expect to see against the natural astrophysical background? What is the best strategy for detecting such manifestations? Answering such questions inevitably requires assumptions about the nature of ETI civilizations, technologies and motivations.

First, we need at least a provisional answer to two very closely related questions:

- How long do we expect technical civilizations to last?
- How far away is the nearest technical civilization?

The answers to these questions will determine how many stars we believe need to be searched to make a detection. This, in turn determines the distance out to which we must look.

Secondly, we need to consider what signs of an ETI civilization we might observe. There are two main possibilities:

- We could observe some physical artifact of their civilization.
- We could detect a signal emitted by their civilization.

In this chapter, we argue that looking for signals is still a better choice than looking for artifacts.

Once we decide to concentrate on signals, there are again two main possibilities:

- They are trying to contact us using some kind of beacon.
- We look for some unintentional signal (leakage) from their technology.

Addressing these points tells us the required sensitivity and duty cycle that we need.

Finally, given all these assumptions, we must decide what frequencies to look at, where in the sky to look, and at what times we should look. We use game theory to help with this analysis. The result is our search strategy.

Figure 5.2 on page 114 highlights the portions of the SETI Decision Tree that we examine in this chapter.

Figure 5.1: Section of the VLA telescope with the Moon in the background. (Photo courtesy of VLA, NRAO.)

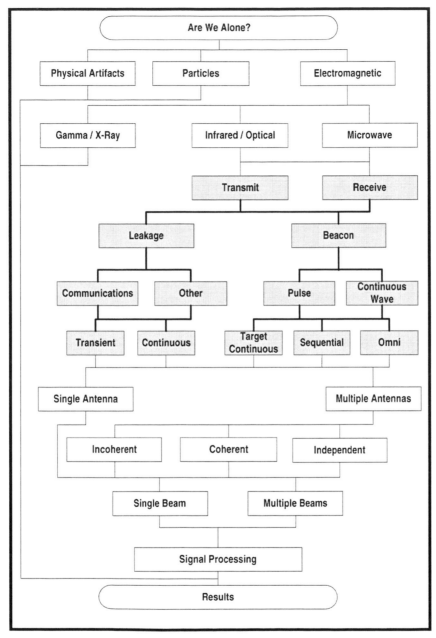

Figure 5.2: SETI Decision Tree – Sources and Strategies.

5.2   ETI Lifetimes and Distances

How long can we expect an ET civilization to last? The answer to this question
is crucial in determining how far away the nearest communicative civilization is
likely to be. We can write:

$$N = R_{com}L \qquad\qquad (5.1)$$

where N is the number of co-existing communicating civilizations in the Galaxy,
R_{com} is the rate at which communicative civilizations arise in the Galaxy, and L
is their expected lifetime. The value of R_{com} is uncertain for reasons discussed
in Chapter 2 on page 37. Optimists predict that the value of R_{com} is about one
per year, and pessimists predict many orders of magnitude less.

The number of stars we could expect to examine before finding a communicative
civilization is simply the number of suitable stars in the Galaxy divided by N.
Given the finite sensitivities of our searches, and in the absence of any other
data, we would choose to search the nearest stars that meet our criteria. This
leads to Figure 5.3, which was derived from Figure 2.3 on page 42. The later
figure shows how far we must look for an assumed number of civilizations in the
Galaxy.

There are about 10^{11} suitable stars. If, for example, $N = 10^4$ we would have
to search about 10^7 stars on the average to find one communicative civilization.
Thus, we would need to look out to a distance of about 2,500 ly.

The combination of R_{com}, the rate at which communicative civilizations arise,
and L, their lifetime, determines how far we must search. Both of these are very
uncertain. Table 5.1 shows the result of different combinations, ranging from
the optimistic to the pessimistic. Since the results depend upon the product of
two very uncertain numbers, the results have a very broad spread. If we assume
that communicative civilizations arise frequently, and they are long lived, then
the nearest one might be quite close (astronomically speaking). If, however, we
assume that communicative civilizations are unlikely to arise, and that they do
not last very long, then there will be very few per galaxy at any given time. In
this latter case, we might have to search hundreds of galaxies to find even one
communicative civilization. It is the enormous uncertainty in these predictions
that lead some to question the value of searching, but with such uncertainty the
only way to find out is to do the search.

Table 5.1 highlights the fact that our immediate neighbors, if any, will be far
older than we are. Spatial and temporal coincidence are unlikely. Only a large
value of L leads to civilizations around nearby stars.

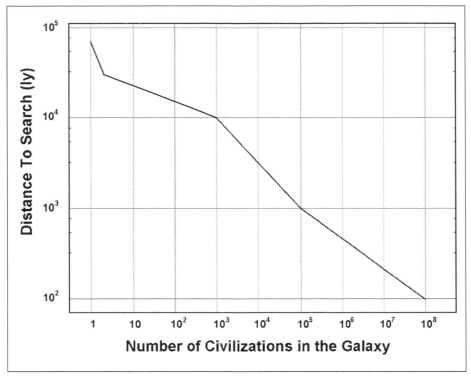

Figure 5.3: Distance to Search as a Function of Number of Civilizations.

Scenario	R_{com} (per year)	L (years)	Number of Civilizations in Galaxy	Number of Stars to Search	Distance (ly)
Optimist	1	10^8	10^8	10^3	≈ 100
⋮	1	10^4	10^4	10^7	$\approx 2,500$
⋮	10^{-2}	10^4	10^2	10^9	$\approx 10,000$
⋮	10^{-2}	10^6	10^4	10^7	$\approx 2,500$
⋮	10^{-4}	10^6	10^2	10^9	$\approx 10,000$
Pessimist	10^{-4}	10^2	10^{-2}	10^{13}	$\approx 10^7$

Table 5.1: Rates, Lifetimes, Stars, and Distances.

5.3 For what should we be Looking?

Virtually all our knowledge about the Universe beyond Earth is derived from information carried by the electromagnetic spectrum. Our capacity for detailed analysis of gravitational or neutrino spectra, for example, pales by comparison to our abilities in photon detection. In particular, the optical band is what most people think of when they imagine learning about natural extraterrestrial objects. The difficulty is that the natural Universe produces so much energy at optical wavelengths that it dwarfs the contributions of our technology.

The Universe abounds with energy, emitting vast numbers of massless particles as well as a variety of energetic particles with mass. In most cases, cosmic energies entirely overwhelm the output of our young civilization. How can we expect to find distant signs of intelligence amid the cosmic tumult?

Of all the information carriers we know, photons are the easiest to detect. By comparison with carriers like neutrinos and gravitons, they are also the easiest to produce. Hence, we concentrate on analysis of the electromagnetic spectrum, because transmission and detection of potential information is efficient.

Our total electromagnetic energy expenditure equals about 1 TW, or about 0.001% of the power falling on the Earth from the Sun. The Earth itself captures slightly less than a billionth of the 4×10^{26} W solar flux. It seems astonishing that our technology could be detected against the spectrum of the Sun.

Surprisingly, our own radio technology, intended purely for terrestrial uses, modifies the electromagnetic spectrum of our solar system. This is because microwave transmissions using simple oscillators create spectral features completely uncharacteristic of stars. Civilizations with technologies similar to ours, using just a small fraction of their available power in coherent signaling, may be highly detectable in the microwave region of the spectrum. Furthermore, these anomalous spectral signatures propagate through space almost unaltered by the tenuous interstellar medium, contending only with the rather weak, background cosmic radiation from the Big Bang.

Consequently, we devote significant resources to a search for civilizations that emit radiation with unusual spectral features. This is very much in the tradition of radio astronomy, but we are looking for technology instead of unusual natural objects.

It would be a trivial extension of our present technology to broadcast a strong, omnidirectional radio signal to attract the attention of other interstellar neighbors. In its simplest form, such a beacon would be just a stronger version of signals which the Earth already transmits. Perhaps, some nearby ETI civilizations have already taken this incremental step toward making themselves detectable.

5.4   The Signature of Our Technology

Our current electromagnetic leakage within narrow portions of the radio bands, outshines the Sun by a factor of a million. By leakage, we mean incidental signals radiating from Earth such as our TV carriers which, though intended only for nearby receivers, travel over the horizon into space. Of course, this extraordinary signature must be visible not just against the Sun, but also against the background of the Universe. If observed with sufficiently sophisticated receiving equipment, our leakage will stand out, even over interstellar distances. It seems clear, therefore, that searching for extraterrestrial technology should entail investigation of the radio spectrum. Current investigations, such as Project Phoenix, perform such searches.

There are three concerns when modeling a search for ETI technology upon the characteristics of the contemporary Earth. First, ETI leakage, as received, would not be constant in frequency. Even when a signal is emitted from a stable transmitter, motions of the planet of origin relative to the planet of reception cause the apparent received frequency to change with time, making sensitive detection difficult. Second, the narrowband character of our leakage may be transient, symptomatic of our technological youth. It may not be typical of civilizations generally. The bright lines of simple transmission schemes may evolve to spectrally diffuse encoded transmissions. Third, over time, most emissions may be sent through cables or beamed to targets, rather than transmitted indiscriminately over the horizon and into space. Nonetheless, at present, our microwave leakage creates a significant spectral anomaly for this solar system.

Even if civilizations become radio quiet and do not leak incidental signals into space, some may design ideal signaling systems to facilitate their detection. Such signals would be even more detectable than those from a present-day terrestrial analogue, and would greatly increase the detectability of a civilization in almost any region of the electromagnetic spectrum in which transmission occurs.

It is clearly technologically possible to form communication links between the stars. While we have no evidence that such links exist, strong signals intended for interstellar transmission are becoming more technically feasible as phased information systems with precise synchronization become common. Even an uncorrected omnidirectional radio beacon could easily outshine the Sun by a factor of a billion in its band, using only a typical commercial power station and no sophisticated transmitting technology. Phased arrays allow beamed transmissions to target stars in radio wavelengths, corrected for frequency shifts due to planetary motion in the directions of transmission.

If one assumes the cooperation of advanced civilizations, this makes the case for detection of radio signals better than ever. Not only can beacons be strong, but because they are beamed instead of leaked, they can be simplified for easy reception by novice listeners. Radio transmissions can routinely outshine the

Figure 5.4: Dr. Michael Lampton of the University of California at Berkeley adds an astronaut's perspective to the discussion.

Sun with power levels that do not significantly add to the Earth's energy expenditure. Radio is special because it is a region where both the background Universe and stars are quiet. Even our current radio technologies can make a big difference in the radio signature of our solar system.

All this being said, there is virtually no physical limitation on the effective power of beamed phased arrays in the electromagnetic spectrum ranging from microwave to optical. Power levels visible against astrophysical backgrounds vary greatly over the spectrum in this range. Presumptive devices of great

Figure 5.5: Dr. Zoya Popovic, of the University of Colorado, supplies the dual perspective of an engineer and a teacher. (Photo courtesy of Z. Popovic.)

Omnidirectional microwave transmitters can continuously outshine the Sun, using megawatts of power and narrow bandwidths of approximately 1 Hz. In
contrast, for optical transmitters to outshine the Sun, they must develop very
high Effective Isotropic Radiated Power (EIRP) by concentrating their signals
into very small solid angles or into very short time periods.

Assuming a blackbody radiator at the temperature of the Sun, one can compare
the required power levels to outshine the Sun at either the microwave hydrogen
line or at the peak of the solar brightness spectrum. In the microwave, a beacon
must emit about 1 kW/Hz to outshine the Sun. In the optical, it must emit
about 1 TW/Hz.

It is certainly possible to construct beacons that use nanosecond pulses to momentarily outshine the Sun in the visible. However, our technology is only on the verge of being able to do this, and such beacons would probably be very directional, yielding low duty cycles for a given target. Although the concept of phased array beacons would broaden the interesting region of the electromagnetic spectrum, radio benefits from this as much as do other spectral regions. Certainly, as one assumes more powerful signals, the details of the cosmic background become less important. Thus, though radio searches for ETI technology seem quite attractive, searches in other parts of the spectrum are highly plausible and also deserve resources.

5.4.1 Beyond the Solar System

This study did not seriously consider the possibility of interstellar travel as a means of exploring beyond the bounds of our solar system. NASA is tentatively planning to launch a small interstellar probe two decades hence, but it will require technologies not yet developed. Searching for ETI civilizations by physically visiting them would require technological solutions that are not yet understood, and may not be possible. For at least the next twenty years, our searches will rely on remote sensing, using photons as our probes.

Consequently, we need to decide what frequency of photons to capture, and whether to survey the entire sky or limit our inspection to certain places. We must decide what types of signal to look for, and when to look for them.

5.5 Luminosity Function for ETIs

In order to optimize any search strategy it is necessary to estimate the relative numbers of ETIs transmitting at different luminosities, i.e., the ETI luminosity function. We can assign each ETI a luminosity that varies with time. Some may light up briefly and die. Others may grow forever. Others may appear and disappear many times. Averaging over cosmic time scales, how would the number of brighter ones compare with the number of less luminous ones?

The luminosity function of ETI transmitters is unknown. However, the detection count versus observing sensitivity, for a given luminosity function and spatial distribution is obtained through the following argument:

> Use the inverse square law to establish a minimum detectable luminosity at each distance, and use an assumed spatial source density function to integrate the number of detections. For uniformly distributed objects in non-curved space, the expected number of detections above a given

threshold flux F varies as $F^{-1.5}$. This is independent of the luminosity function, but dependent on the uniformity assumption. Minimizing the threshold flux maximizes the sensitivity and pays big dividends.

Sensitivity can be improved by using a bigger telescope, by integrating for a longer time, or both. For a detailed discussion about calculating luminosity functions for ETIs, see Appendix O, starting on page 443.

5.6   What to Look for in the Microwave

To decide what kind of microwave search we should be pursuing in the next decades, we must ask "for what kinds of microwave signals are we searching?" The fundamental choice is to look for signals intended to be found, e.g., beacons, or for ETI signals generated for other purposes, e.g., leakage. A search for powerful beacons is relatively easy, whereas looking for leakage at the levels manifested by our own civilization is difficult.

In this section we discuss these issues and derive some plausible conclusions about the type of signals we might try to detect. We try to gain some insight into the search for ETI signals by thinking about the reverse problem, where we are doing the transmitting. We briefly consider what kinds of beacons we could build with current technology and a large, but not unlimited, budget. We also consider the incidental radio emissions to be expected from our broadcasting activities, radars, and some specialized services, such as power transmission and navigation. Finally, we summarize and discuss the results.

5.6.1   Beacons

We can gain some insight into the kinds of beacon signals we should be looking for by considering how to build one with technologies not too different from our own. Today, the best approach would be to build a phased array transmitter that could target a large number of stars in our vicinity, either simultaneously or sequentially.

An attractive concept with current technology would consist of a very large number, such as $N \approx 10^8$, low-gain elements, each of modest power, e.g., $P_x \approx 1$ W. For example, using technologies employed in cellular phones, such an array might cost less than a billion dollars. It could illuminate one star with an EIRP of $(N^2)P_x \approx 10^{16}$ W, or M stars with simultaneous beams, each with $1/M$ times this EIRP.

There are only about 10^7 stars within 1,000 ly, a good fraction of which can probably be eliminated as suitable hosts for civilizations, so it is likely that we would choose $M \ll 10^7$ for our beacon. Another advantage of this array concept is that it lends itself to being self-powered from solar energy. A more detailed discussion of the feasibility of constructing such a phased array beacon using current technology is given in Appendix B, on page 303.

The only apparent limitation on the size of the beacon we could build, using this technology, is the amount of money we choose to spend. The realization that beacons efficiently targeting many stars could be built, combined with the expectation that advanced technologies may not produce leakage radiation of an easily identifiable form, has shifted the emphasis away from strategies detecting leakage and toward those looking for beacons. See the SETI Decision Tree path in Figure 5.6 on 124.

5.6.2   TV Broadcasting

Ultra-high frequency (UHF) television broadcasting currently uses high-power transmitters with antennas of modest gain. Stations that service large markets often have an EIRP between 1 and 5 MW, and with National Television Standards Committee (NTSC) modulation, about 10% to 20% of the total signal power goes into the carrier. Typical antenna gains are 12 to 15 dBi, corresponding to 1/15 to 1/30 of the sky. Since there are thousands of these transmitters, an ETI detector that is looking at the Earth will always be in the radiation pattern of several of them.

Thus, TV broadcasting signals are one of the most readily detectable signatures of our technology. Broadcast television was expected to be replaced by cable TV (Community Antenna Television (CATV)), but that has not happened. The reasoning still seems sound because increased channel coverage, improved quality, and better reliability are offered by hard-wired cable connections. For the short term, we can probably anticipate an increase in UHF TV broadcasting as markets open in developing countries that have minimal wiring infrastructure. Indeed, in some of these areas, cellular telephones are more common than directly-wired units, just the reverse of what one might have expected!

Nonetheless, simple engineering considerations would seem to ensure that we will eventually phase out such broadcast services. Even sooner, we can expect conversion to digital TV broadcasting, with modulation schemes that do not lend themselves to easy detection at low S/N ratios. In the United States, analog TV will be phased out as soon as a high percentage of homes have the capability to receive digital TV. Crude estimates place this transition at about the year 2006, very soon by astronomical standards.

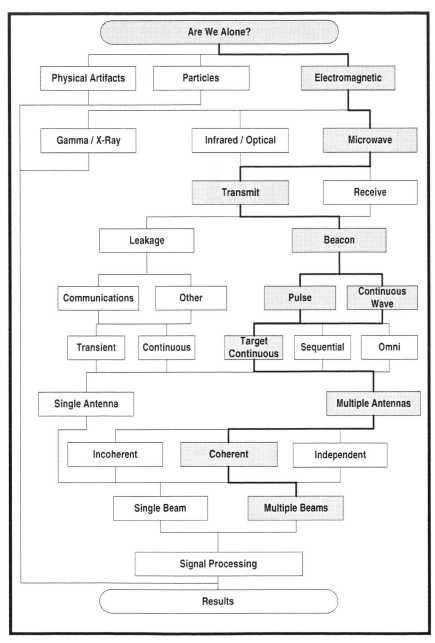

Figure 5.6: SETI Decision Tree – Beacons Targeting Many Stars.

Most new broadcast services require transmission EIRPs of 1 kW or less. Efficient modulation methods, such as Code Division Multiple Access (CDMA), produce wideband signals with gross properties similar to noise. For many satellites, especially those in Low Earth Orbits (LEO), the Doppler drift rates will be rather high ($\approx 10^{-7}$), compared to planet-based transmitters ($\approx 10^{-9}$). Combined, these properties result in signals that will be extremely hard to detect over interstellar distances. This is discussed in Section 4.3 on page 105.

Figure 5.7: Doug Thornton of the University of California at Berkeley, lends his microwave engineering expertise to the SSTWG. (Photo courtesy of D. Thornton.)

5.6.3   Radars

The required transmitter powers for a radar system scale as the range to the fourth power (R^4). Thus, we find that radars are the highest EIRP transmitters on Earth, because some radars need very long range. There are hundreds of radars with peak EIRP > 1 GW currently in continuous operation. Research instruments, in intermittent operation, have peak EIRPs all the way up to the 22 TW of the Arecibo 2.3 GHz radar. These high EIRP values reflect the very high directivities of the antennas. This implies that the probability that a given star will be in the beam is quite small – in the range of 10^{-4} to 10^{-8}. During the next few years, more powerful radars may be built for monitoring orbital debris, and for determining the orbits of Near-Earth Asteroids (NEA).

The pulsed nature of such signals may pose a challenge to SETI-type detectors, because the most economical way to detect pulses is to search only down to an S/N ratio $\gg 1$ on individual pulses. In this case, the SETI detectability range of the radar will be determined by the Effective Isotropic Radiated Energy (EIRE) per pulse. At the upper end, the Arecibo or Goldstone transmitters could be used to create one second pulses with EIRE $\approx 10^{13}$ J, while the widely deployed Next Generation Weather Radars (NEXRAD) produce EIRE $\approx 10^5$ J.

Figure 5.8: Control console for the 430 MHz pulsed radar transmitter at the Arecibo radio telescope. The Arecido 2380 MHz cw radar was used for a three-minute transmission to the star cluster M13 in 1974.

Orbital Debris Monitoring

The growing use of Low Earth Orbits for a multitude of applications is threatened by the presence of an ever-increasing amount of orbital debris at altitudes from 750 to 1,000 km. Debris as small as 10 cm in diameter can destroy a spacecraft, while 1 cm objects can do severe damage. Just how to abate this hazard is still under vigorous discussion, but at a minimum, it will probably be necessary to keep track of LEO debris. (Debris in higher orbits is not a near-term problem.) Both optical and radar methods are currently in use, and provide complementary features. However, current radar monitors, such as the systems deployed by the militaries of the USA and Russia, are only sensitive to objects down to ≈ 10 cm, primarily because of their operating wavelengths. To be able to track 1 cm objects, a system operating above 8 GHz must be used.

Currently, radars at Millstone and Goldstone are used to conduct statistical studies of LEO debris. It is plausible that in the next few decades a much more capable and dedicated LEO debris monitoring radar system will be needed. For a slant range of 2,000 km, and a receiving area of 100 m^2, the required pulse EIRE would be $\approx 10^9$ J. It is also plausible that we will have substantial satellite and other assets from LEO to beyond geosynchronous orbits, and will wish to protect them from orbital debris using even more powerful radar monitors. For slant ranges $> 4 \times 10^4$ km, pulse energies of approximately 10^{13} J would be required.

The current trend in radar is toward phased array systems with electronic beam steering. The rapid scan capability of such a system is both a help and a hindrance to SETI studies. While a given star may not remain in the beam for very long, the beam may emit pulses into a larger number of directions per unit time, increasing the odds of detection in a random direction.

NEA Trajectory Determination

The cataclysmic consequences of impacts of asteroids on the Earth have become clear since the elucidation of the KT boundary event, and it is plausible that all species will need systems to detect and monitor asteroids in dangerous orbits. At our current technological level, the detection task seems best accomplished by optical means, but radar observations play an important role in making accurate orbit calculations. Unfortunately, our current generation of planetary radars have a range limited to about 0.25 AU for asteroids 1 km in diameter with 10% albedo. We would benefit by increasing this range to ≈ 2 AU, requiring an EIRE of approximately 10^{19} J for an S/N ratio of 10 per pulse. Unfortunately, the sky coverage of such radars is extremely small. Goldstone and Arecibo, for example, transmit only about 1% of the time, and their beams are quite small. Since these antennas are about 1,000 λ (wavelengths) across, their beams cover about 10^{-7} of the sky.

5.6.4   Microwave Power Transmission

A civilization may want to transmit electrical power without using wires. This might be useful for delivering power to mobile receivers such as spacecraft or planes, or for transmitting power from mobile sources, such as 'solar-power' satellites, to fixed locations on a planetary surface.

Such an application might generate considerable leakage if the power levels are substantial, e.g., several times 10^9 W for 'solar-power' satellites. Unlike the case of communications, there is little incentive to use fancy coding or spread-spectrum techniques, so the structure of the signal should be simple. The signal would not be expected to exhibit Doppler compensation or extreme stability.

5.6.5 Navigational Beacons

Global Navigation Satellite System (GLONASS) and Global Positioning System (GPS) transmitters allow small, lightweight receivers to determine their location very easily. If a civilization has many small spacecraft traveling throughout their planetary system, they might place similar navigational beacons throughout their system. It makes sense to make the transmitters powerful so the receivers can be small and light. The receiver might consist of a single chip and a few centimeters of wire, which would be smaller and lighter than passive optical navigation systems.

What characteristics might be typical of an interplanetary navigational beacon? The use of higher frequencies would reduce the influence of the stellar corona, and would allow the use of physically smaller antennas. The signals are exactly known, so a small (information) bandwidth would be expected, although the modulation could be complex. The signals should be extremely stable (or at least predictable) since their characteristics will be used for navigation.

How much power might be used? Since the signals are known, an S/N ratio on the order of one would be enough at the intended receivers. Since the receivers are intended to be small and lightweight, they would not be actively cooled, so the system temperature might be on the order of 60 K. If we assume a 3 cm wavelength, then a small antenna might have an effective area of about 10^{-4} m^2. Combining these assumptions, and requiring a range of 10 AU, implies that about a 200 MW transmitter would be required.

5.6.6 Signals Summary and Discussion

Table 5.2 summarizes the expected performance of the five types of beacons discussed in this chapter and in Appendix B, on page 303. The first three rows of the table are for *Small*, *Medium* and *Planetary* type beacons, each with a single beam (*No.* column). (Trying to simultaneously illuminate large numbers of stars with these beacons would not really make sense.) The EIRP is given in the third column, and the probability that the beacon will be aimed at us if we look once is given in the fourth column. For each case, the flux produced at the target is calculated for ranges of 43, 200, and 927 ly, which correspond to volumes containing 10^2, 10^4, and 10^6 'good' stars, where 50% of main sequence F, G and K stars are assumed to meet the standard.

For the *SKA Type* beacon, the table illustrates the options of illuminating a target list of 10^6 stars sequentially, with 1,000 beams at 10^{-3} duty cycle, and with 10^6 simultaneous beams with 100% duty cycle. Similar options are shown for the *Advanced* beacon, extending up to 10^9 stars. The *Prob.* column is the probability that the beacon will be aimed at us if we look once, assuming that the ETIs have chosen a list of stars to illuminate (10^6 of them for the *SKA Type* beacon, and 10^9 of them for the *Advanced* beacon), and assuming that we are on that list. In the case of the *Advanced* beacon, the EIRP is not applicable because it does not vary by distance; the beam is made narrower if the star is further away. This keeps the flux constant at the target.

Beacon	Beams			Flux (Wm^{-2})		
Type	No.	EIRP (W)	Prob.	43 ly	200 ly	927 ly
Small	1	5×10^8	10^{-6}	2.4×10^{-28}	1.0×10^{-29}	5.0×10^{-31}
Medium	1	2×10^{11}	10^{-6}	9.6×10^{-26}	4.5×10^{-27}	2.1×10^{-28}
Planetary	1	2×10^{13}	10^{-10}	9.6×10^{-24}	4.5×10^{-25}	2.1×10^{-26}
SKA Type	1	10^{16}	10^{-6}	4.8×10^{-21}	2.2×10^{-22}	1.0×10^{-23}
	10^3	10^{13}	10^{-3}	4.8×10^{-24}	2.2×10^{-25}	1.0×10^{-26}
	10^6	10^{10}	1	4.8×10^{-27}	2.2×10^{-28}	1.0×10^{-29}
Advanced (Future)	1	n/a	10^{-9}	1.4×10^{-16}	1.4×10^{-16}	1.4×10^{-16}
	10^3	n/a	10^{-6}	1.4×10^{-19}	1.4×10^{-19}	1.4×10^{-19}
	10^6	n/a	10^{-3}	1.4×10^{-22}	1.4×10^{-22}	1.4×10^{-22}
	10^9	n/a	1	1.4×10^{-25}	1.4×10^{-25}	1.4×10^{-25}

Table 5.2: Beacons.

Even if a beam is aimed at us, we still need the flux sensitivity to detect it. Table 5.3 on page 131 gives the CW flux detection limits for several current SETI programs. At first, it appears that many of the current surveys are sensitive enough to detect the planetary radar, for example. The problem is that an essentially 100% duty cycle beacon would be required.

Unless the ETIs have some special way of picking us out of an ocean of stars, they are not likely to be pointing something like a planetary radar at us when we happen to take a look at them. Two beacons (transmitter and receiver), using CW with multiple beams, get around this problem when they are employed in the simultaneous illumination mode.

The predicted fluxes for an *SKA Type* beacon in the simultaneous illumination mode, however, are too low for any current survey. The *Advanced* beacon might be detected by the most sensitive of the Phoenix observations if it happens to be included on the relatively small Phoenix search list. If Project Argus [Shu97], operated by the SETI League, reaches its goal of 5,000 systems, it will provide more looks. Unfortunately, the 'looks' are not very sensitive, and they require the almost exclusive use of a beam from one of the big beacons.

CW leakage emissions from UHF TV transmitters currently have EIRPs of approximately 1 MW, and there are enough of them so that they cover most of the sky. At the distance of the nearest star, about 4 ly, the flux produced will be approximately 6×10^{-29} W/m^2, well below the detection limit of current surveys. We should expect that older civilizations will produce CW leakage

Survey	Threshold W/m^2
Phoenix	
Arecibo	7.7×10^{-27}
Nançay	8.7×10^{-26}
Parkes (long)	1.5×10^{-25}
Parkes (short)	2.1×10^{-25}
NRAO	3.0×10^{-25}
SERENDIP IV	1.0×10^{-24}
META	1.7×10^{-23}
BETA	1.3×10^{-22}
SETI League	1.1×10^{-21}

Table 5.3: Sensitivities of Current Searches.

emissions that are orders of magnitude less powerful than this level.

Finally, Table 5.4 gives some parameters for the most powerful current radars, plus estimates for the proposed LEO Debris and NEA radars discussed previously. The most important quantity for detection is the EIRE per pulse. The EIRE for the Arecibo and Goldstone radars assumes a 1 s pulse. Unfortunately, no existing surveys have any significant sensitivity to short pulsed radars.

Name	How Many	Peak Power (MW)	Average Power (kW)	Gain (dBi)	Pulse Width (μs)	Pulse Rate (pps)	EIRE (J)
NEXRAD	160	1.0	1.4	45	4.5	318	1.4×10^5
Air Traffic	153	1.1	1.1	33	1.0	980	2.2×10^3
Arecibo PR	1	1.0	1.0	73	10^6	10^{-3}	2.2×10^{13}
Goldstone PR	1	0.5	0.5	64	10^6	10^{-3}	1.3×10^{12}
LEO Debris	proposed	*	*	*	*	*	1.0×10^9
NEA Radar	proposed	*	*	*	*	*	1.0×10^{19}
*No information is available.							

Table 5.4: Parameters of Some Terrestrial Radars.

5.7   Strategies for Transmitters and Receivers

No systematic observing programs are currently possible for some potentially observable manifestations of advanced technologies (e.g., spaceships, probes, slow spaceships, energy production, Dyson spheres, and astroengineering). Observers and researchers with access to large quantities of raw data from observations at all frequencies should be encouraged to consider an ETI civilization as a potential explanation for anomalous results whenever the quality of the anomalous data demands further investigation. Although pulsars did not turn out to be "little green men", that hypothesis was at least considered. The "little green men" hypothesis was replaced by another when new data, and the rediscovery of an old theoretical model, offered a more robust explanation. Such considerations should be emulated in the future, as appropriate.

In the rest of this section, we shall consider only electromagnetic signals; either deliberate beacons or leakage radiation. Beacons can be expected to have significantly greater power than leakage radiation. Deliberate beacons may well contain embedded information that is intended to be understood by the receiver. Beacons can be omnidirectional or narrowly beamed. They may be continuous or temporally sparse; they may be narrowband or broadband. Beacons may exist at any frequency, limited only by the resources that are available to the transmitter.

Figure 5.9: Dr. Louis K. Scheffer of Cadence Design Systems, offers new ideas and important editorial contributions.

Leakage radiation is expected to be lower in power because its primary purpose involves the routine information transfer activities of the transmitting technology. Bracewell [Bra74] suggested that a network of tightly beamed communication channels might connect the advanced technologies living near many stars in our Galaxy. The probability that we would come within the beams of such transmissions is very small, although it can be maximized by targeting eclipsing binaries. The broadcast leakage we are more likely to encounter will have power levels consistent with its usage over interplanetary, rather than interstellar, distances.

There is only one electromagnetic spectrum, and it must be shared. Competition for bandwidth and the economics of spectrum usage suggest that, before long, our own leakage radiation will be complex, broadband, and almost noise-like, or *pink*. While our leakage radiation will be temporally continuous in the aggregate, any particular component may have a low duty cycle. Projections of our own technological future suggest that broadcast leakage will be confined to longer wavelengths. While forward projection of terrestrial technology is illuminating, we should be careful to avoid excluding too many possibilities. We should admit that we do not know what an advanced technological civilization might choose to do.

Now we can ask, what are the best strategies for transmitters of beacons, and for receivers of beacons and leakage. We begin with a classification of frequency space and the natural astrophysical background.

As discussed in Chapter 3 on page 67, it is possible to divide the electromagnetic spectrum into three well-understood bands: the radio/microwave, the infrared/optical, and the X-ray/gamma ray bands. Figure 2.11 on page 55, shows the power flux emitted per solid angle on the sky, by the astrophysical sources in our Universe.

Since X-ray and gamma ray observations must take place above the atmosphere, such observations undertaken just for SETI will remain too costly, at least within the horizon of this book. Astronomical observations will provide a potential for serendipitous detection. Although a consensus appears to be emerging about the nature of the fast gamma ray burster sources; in the spirit of "little green men" as an explanation for pulsars, the manifestations of advanced astroengineering should be considered for explaining anomalies in the high-energy data.

While the *Project Cyclops* report strongly favored the radio region over the optical region, several arguments have been made to soften that conclusion. It seems prudent over the next years to support those relatively inexpensive infrared/optical searches that can set the first limits on the possible population of very strong transmitters. Phoenix and other microwave searches have already begun to constrain the population of very strong radio sources. Therefore, improvements in detection probability at these wavelengths will come at a higher cost, and improved sensitivity is crucial.

We next discuss targets of opportunity for both the transmitter and the receiver. These are ideas that have been suggested in the literature over the years, as ways to optimize searches that are limited in their capability. Some have more appeal than others, and any search program should set aside a small percentage of time to explore these novel strategies. However, since none of them is extraordinarily compelling, the conclusions of this study will be focused on how to make the most effective systematic explorations that can be afforded.

5.7.1 Targets of Opportunity for the Beacon

Modern technology allows many possibilities for building beacons. Lasers with large pulsed energy, combined with accurate stellar positions from space-based astrometry, now allow optical beacons that are almost as easy to detect as radio beacons. Radio beacons can now be built with phased array technology, which allows the choice of strong sequential pulses, continuous illumination at a lower level, or any combination in between. These are discussed in considerable detail in the sections on microwave and optical search strategy, and in Appendix B, on page 303.

The *Project Cyclops* report assumed that a directed beacon would be built by aiming a dish/transmitter combination at each target star. In this case, as the range R increases the number of stars increases as R^3 and the power of each transmitter as R^2. This leads to a total increased power requirement on the order of R^5, which soon exceeds the power required by an omnidirectional transmitter. This occurs when the relatively low gain beacons point at enough stars to cover more than 4π in solid angle. For these reasons, the *Project Cyclops* report advocated omnidirectional transmitters. Of course, if we can create narrower beams, this problem vanishes, provided that we possess accurate stellar proper motion data to ensure that the beam coincides with the star after the signal propagation delay.

To avoid this problem, we could build an array of small transmitters whose phase can be adjusted to generate beams on all the target stars. The total EIRP needed still grows as R^5, but this can be generated by several modules. Adding modules increases the EIRP as the product of the area and total power. The beams get smaller as the array grows.

Beaming is always efficient if the locations of receivers are known. We often assume that ETIs beam the stars because we assume that they live near stars, but this assumption is not necessarily correct. Perhaps their advanced technologies enable ETIs to live in deep space, far from stars. Then, just beaming stars would miss them, whereas omnidirectional beacons would not. Even today, omnidirectional beacons are plausible.

If beaming is feasible, the required power is focused in some region less than or equal to the celestial sphere. If one wishes to beam to so many receivers that the entire sphere is covered, one requires the equivalent of an omnidirectional beacon. Such transmission is possible with a phased array. One can imagine covering the sky with many transmitters using independent beams. If the beam size is fixed, the power increases as R^5. Of course, this means that at some distance, the cost of beaming exceeds the cost of an omnidirectional transmitter, which increases only as R^2. However, this is only true for a fixed beam size. For real transmitters, this limit does indeed exist, even for phased arrays. For these reasons, the conclusions of the *Project Cyclops* report concerning the feasibility of beam beacons need to be reexamined.

Targets of Opportunity for a Beacon in Frequency

ETIs with advanced technology may be able to simultaneously transmit at many different frequencies, and at the lower microwave frequencies there is a benefit from doing so. Interstellar multipath scattering causes amplitude fading that is time and frequency dependent [Cor91a], [Cor91b], [Cor91c]. Transmitting at multiple frequencies ensures that at any time, the receiver will still have access to unfaded signals [Coh95]. The uniqueness of the 'water hole' from 1.42 to 1.72 GHz, and its symbolic importance to life as we know it, first made it a preferred region for searches. More than 100 different molecules have now been detected in the interstellar medium [Wot00], many of which are organic, so the special character of the 'water hole' has diminished.

The transmitter may also wish to choose frequencies for transmission that involve constructs of natural emission lines and fundamental constants such as π, e, or the fine structure constant [Bla93]. At optical frequencies, the stellar background for narrowband signals will be diminished if transmission takes place within the dark Fraunhofer lines of the host star. Likewise, the so-called anti-maser line of formaldehyde (H_2CO) in interstellar clouds offers a frequency at which the noise floor is less than the 2.7 K from the cosmic background radiation [Mor73]. The list of 'magic' frequencies has expanded over the years, and now there is no longer a very compelling target of opportunity for transmission.

Special Targets of Opportunity for a Beacon

There are several special strategies available to the transmitter. Illuminating the nearest target stars continuously or sequentially has already been discussed. At microwave frequencies, clouds of interstellar gas and dust provide free amplifiers in space. Masing molecules such as hydroxyl (OH), water (H_2O), and methanol (CH_3OH) in interstellar clouds can provide 70 or 80 dB of nonlinear amplification to background signals at the correct frequency [Gol76]. A sender might decide to transmit through nearby maser clouds, in order to produce an extremely loud signal for any receiver in the line of sight on the other side of the cloud.

Beacons transmitted from a star, and directed radially inward and outward from the galactic center are more likely to be intercepted by ETIs who are just discovering observational astronomy at wavelengths longer than the near-infrared. The signal will be detected when the young civilization begins to study the galactic center at the right frequency. Similarly, beacons might be transmitted in the direction towards or away from significant cosmic events such as supernovas, or novas [Mak80].

The galactic center is a unique location for the operation of a beacon by an ETI with a sufficiently advanced technology. Beacons whose beams are confined to the galactic plane can illuminate 90% of the stars in 20% of the sky. While there is a distinct advantage to this strategy at longer wavelengths, absorption by dust at optical frequencies favors directions out of the galactic plane. Extremely powerful beacons with broader beams, could be directed at clusters of stars and portions of nearby galaxies to increase the number of potential targets within the beam.

Temporal Targets of Opportunity for a Beacon

For beacons that are temporally sparse, it is advantageous to find ways to synchronize transmission times with probable reception times. Supernovas present such a synchronizing opportunity [Mak80], [Lem94]. For astronomical purposes, these events are studied intensely over a wide range of frequencies, during their early stages of development. The transmitter can take advantage of the supernova event, counting on the interest of the potential receiver.

5.7.2 Targets of Opportunity for the Receiver

Because our primitive technology is not symmetric with that of an advanced transmitter, the targets of opportunity are also asymmetric. In this section we will consider frequency, spatial, temporal, and serendipitous opportunities for a receiver.

Frequency Targets of Opportunity for a Receiver

'Magic' frequencies apply to the receiver and transmitter alike. Digital signal processing capacity is increasing so rapidly that we will probably soon be able to search all accessible frequency regions. We will not have to second-guess the transmitters. Using our relatively primitive technology, and still conducting our searches from the ground, we are also restricted in frequency by atmospheric absorption. Our searches are also limited to frequency bands not already polluted by satellite and terrestrial transmitters. If it is possible to establish a SETI observing facility within a science preserve on the far side of the moon [Hei00], and if spectrum management effectively protects this unique site [Par80], then both of these restrictions can be lifted. At best, this will not occur before about 2020.

Spatial Targets of Opportunity for a Receiver

Limiting a sky survey to the galactic plane captures 90% of the Milky Way target stars in only 20% of the sky area [Sul97]. If success is predicated on the high power of the ETI transmitter, then there may be a shortcut to the sky survey. By selecting nearby star clusters or galaxies as targets, it is possible to search a large number of target stars in one beam, at one time. Kardashev Type II or III civilizations should be detectable. A few such searches have been conducted [Sag75] over a limited range of frequencies. It would not be unreasonable to extend the frequency coverage of these searches and set aside a small fraction of any systematic program to such targets of opportunity.

The class of targets of opportunity can be substantially expanded beyond nearby galaxies, placing additional emphasis on those stars with known planetary companions. The exoplanets known so far do not seem hospitable, but are being discovered steadily and some may harbor large habitable moons.

Pasteur said, "In the field of observation, chance favors the prepared mind." In this spirit, spatial targets of opportunity can make use of the fact that additional objects of interest may occur within the observing beam, beyond a foreground target. If a search of nearby stars is biased to those that lie in the galactic plane, the average distance to such targets will be increased, but the chances that a second target lies in the background will be improved [Sho99].

An extreme version of this strategy would exploit the amplification from a stellar gravitational lens, which occurs when the alignment of foreground and background stars is precise [Esh97]. At any given moment, looking in the direction of the galactic bulge, there is about a 1 in 10^8 probability that an alignment will amplify any signal from a given background star by a factor of 10. A simple way to exploit this is to follow the alerts generated by the optical microlensing experiments. These experiments watch millions of stars in the galactic bulge almost every night. They often catch gravitational lensing events well before their peak magnification [Ben96]. Such stars could easily be made the subject of "target of opportunity" investigations in the radio.

Since the stars near the galactic center are 30 times farther away than local targets, stronger transmitters would be required for us to be able to detect them. Limiting a search to the center of the Galaxy assumes that, since the stars there are far older than the Sun, the ETI civilizations are probably far older and potentially more advanced than ours. Thus, they are more likely to have the stronger transmitters that are required.

Extended observations in the direction of the galactic center will also enhance the probability that multipath scattering by electrons in the interstellar medium will produce an amplification, by as much as a factor of 10, for a distant point-source signal [Cor91b], [Ric98]. This should produce structure in the frequency-time domain with scales of kilohertz-minutes that would currently elude detection by schemes optimized for narrowband signals or broadband short pulses. Augmenting the search algorithms to look for this scintillation structure may be a cost-effective way of searching for featureless *pink* noise from very advanced civilizations. It relies only on the intrinsic small angular size of the source, and not on detectable spectral characteristics.

Temporal Opportunities for a Receiver

Just as the ETIs may decide to transmit in the antipodal direction from a new supernova, we can also use these spectacular events to select the time for observations of known targets. An ellipsoid of revolution with the Earth and the supernova as its foci is the locus of points from which a signal transmitted to us at the time that the supernova is detected by the transmitting civilization will arrive at a time t after we first see the supernova. This ellipse begins as a straight line when we first detect the supernova, but expands continuously with t. A simple timing strategy consists of observing any particular target when it lies on the expanding ellipse defined by a recent supernova event [Lem94].

Serendipitous Opportunities

Finally, the list of targets should be augmented with the sources of unexplained or anomalous events from astrophysical observations at all wavelengths. A plausible transmission strategy might be to generate a signal that looks almost exactly like a certain kind of astrophysical phenomenon. Then, this signal could be discovered during routine astronomical exploration without recourse to unique instrumentation. However, as soon as detailed observations of the source were conducted, some anomaly would be detectable that would suggest a technological origin.

One example of this strategy might be an apparent pulsar signal whose spindown rate remains zero within the limits of precision. Five such sources can be found in the pulsar catalog [Tay95], [Tay93]. There may be some interesting new astrophysics that explains these sources, but they are an interesting target of opportunity for optical and radio searches. Observers working at all wavelengths should be encouraged to make public such anomalous sources, and to consider a technological explanation.

5.7.3   Commensal Opportunities for the Receiver

At microwave frequencies, it is possible to amplify the received signal and split it up for multiple uses without loss in S/N. This offers the opportunity to add a

SETI backend to a facility being used for other astronomical observations. As fully steerable telescopes develop multiple beam feeds and focal plane arrays for survey work, the possibilities for commensal searches expand [Sta97], [Deh99]. The multiple beams serve as coincidence detectors for RFI, and permit some RFI excision schemes.

The lack of telescope control for instantaneous follow-up, and off-source observations, will limit the percentage of SETI resources that should be allotted to commensal searches. Nevertheless, we should seriously consider the idea of installing a joint-use search capability on every large telescope, particularly as the price of signal processing continues to drop.

At microwave frequencies, the best opportunity for multibeaming in the next two decades appears to be with the construction of the Square Kilometer Array (SKA), which will have a one square kilometer collecting area and frequency coverage from 300 MHz to 20 GHz. (See Section 6.8 on page 180). Section 6.8.1 on page 181 examines what such an instrument might do for SETI. The final design and configuration will be optimized. The current straw man design specifies 100 simultaneous beams, some of which could be used for SETI. The instrument will probably cost many hundreds of millions of dollars, which appears to be possible if conducted as an international project within the next decade or two. Thus, it is important for the SETI Institute to remain connected to this loose international collaboration for the SKA, coordinated by the International Union of Radio Science (URSI) Working Group.

There will be pressures for construction at multiple sites, and wholesale replication is not out of the question. Project Phoenix demonstrated that, for signals containing coherent components, an interferometric detection could be made, and interfering signals could be excluded by differential Doppler techniques. This RFI excision technique required the use of widely spaced elements, with collecting areas that differed by up to a factor of 10.

Many of the scientific questions to be addressed by the SKA favor a hierarchical structure in the spacing of the elements that would allow exploitation of this RFI excision technique. To be able to function during the coming decades at microwave frequencies, in the face of increased satellite downlink and crosslinked beamed transmissions, active RFI excision techniques will have to be developed that will benefit SETI as well as radio astronomy.

Multiple beams for simultaneous use seem to be the key to success in any search that contemplates sharing resources with the astronomical community. Comparative numbers for infrared/optical searches versus microwave searches look most favorable when large 10 m apertures are assumed for the infrared/optical receiving systems. Today these large instruments are often used to observe faint objects at the limit of detection. Given the oversubscription of these instruments by the traditional astronomical community, it is unlikely that even a few nights per year will be available for SETI.

One promising possibility is the installation of multifiber focal plane masks and robotic repositioners. Within the fields of view being studied by infrared/optical astronomers, there may be spare fibers that could be positioned on SETI target stars. The cost of the back end processing will determine whether this is feasible. For lack of a better term, we refer to this strategy as "Spaghetti SETI". This idea is elaborated in Section 6.11 on page 221 in the discussion of infrared/optical technology. It may also be feasible to reanalyze stored data from these large instruments to look for subthermal emission lines. (See the next section for a discussion of spectral line searches.)

Analogous to the SKA funding method, it might be feasible to create a consortium to build a new telescope with private funds and with funds from foreign scientific organizations that as yet have no access to ≈10 m class instruments. So far, no investigation of this possibility has been attempted.

For the case of short optical pulses and other searches that explore the domain of strong transmitter powers, observations can be made with many smaller instruments, perhaps even by amateur astronomers. Here the opportunity for amateurs to contribute to the professional astronomical community becomes a real possibility. The smaller aperture telescopes typically used by amateurs are sufficient to detect short optical pulses, even though they are not sufficient for studying faint objects.

Radial velocity searches for extrasolar planetary systems with precisions of about 100 m/s often target solar-type stars. Such surveys can tolerate a beam splitter to enable searches for short pulses, as has been demonstrated at the Oak Ridge Observatory [How00].

5.7.4 Transmit/Receive Matrix

This section organizes the various microwave and infrared/optical search strategies that have been discussed. The results are summarized in two tables (transmit/receive matrices), Table 5.7 on page 144 for microwave searches, and Table 5.8 on page 144 for infrared/optical searches.

Microwave Frequencies – Possible Search Projects

The characteristics of the receiving programs are shown in Table 5.5. For simplicity of calculation, the receiver types are only representative of the real programs, with about the right flux sensitivities.

Many simplifying assumptions went into Table 5.5:

- *Looks.* The number of looks were computed using 1,000 second looks and

Program	Antenna Size (meters)	Band-Width (Hz)	Margin (S/N)	T_{sys} (K)	Looks (3 years)	Flux (W/m^2)
Astronomy	70	300	6	20	10^6	1.8×10^{-22}
Directed	300	1	1	20	10^4	5.6×10^{-27}
SKA	1,000	0.1	1.5	20	10^6	2.5×10^{-28}
META/BETA	15	0.05	25	60	all-sky once	8.3×10^{-24}
OSS	5	1	25	60	all-the-time	1.5×10^{-21}
SERENDIP	150	1	35	20	all-sky once	7.8×10^{-25}

Table 5.5: Receivers – Search Capabilities.

an approximately three year program lifetime (actually 10^8 seconds). For astronomy, we assumed there were 10 major radio telescopes worldwide, and that they were in continuous use. For the targeted search, we assumed 1/10 of the available observing time on one telescope would be available full time for SETI. For the directed search on the SKA, we assumed that 10 beams would be available full time for SETI.

- *Margin.* The S/N ratio needed for a detection. We assume an astronomer would notice a 6σ anomaly, although this is optimistic. Directed search is assumed to use re-observation and confirmation from another site to reduce the margin. Sky surveys and omnidirectional survey systems have bigger margins, since they are unable to use these techniques. SERENDIP has even bigger margins since it also cannot use on/off pointing.

- *META/BETA* and *SERENDIP* are assumed to cover the whole sky once. *OSS* is assumed to be all-sky, all-the-time.

- *Signals.* Only narrowband CW signals are considered, and these are not optimum for radio astronomy receivers. For pulse searches, the radio astronomy surveys would have sensitivity comparable to the other SETI programs.

The transmitter characteristics are shown in Table 5.6 on page 143. The last three rows in this table refer to the three different ways a transmitter might choose to illuminate N targets: as N beams (pincushion), as $M(< N)$ beams in sequence, or as one beam sequentially moving from target to target.

Program	EIRP (watts)	Probability It Is On
Radio Silence	0	1
Leakage	10^7	1
Omnidirectional Beacon	10^9	1
Pincushion	10^{10}	1
M of N	3×10^{11}	0.03
Sequential	10^{16}	10^{-6}

Table 5.6: Transmitter Types.

The matrix which follows (Table 5.7) attempts to organize the various search strategies that have been discussed. Possible transmitter strategies from *weak and continuous* to *strong and temporally sparse* are listed as the column headings in the matrix. Receiver strategies from the purely serendipitous to the full time, dedicated searches are listed as rows in Table 5.7. The last two entries refer to future systems discussed in Chapter 6 on page 147, a targeted search using multiple simultaneous beams on the SKA, and a hypothetical OSS. For each combination, the entry in Table 5.7 represents the limiting detection range in light years (ly).

Although the range limits are very different, the matrix is fully populated, implying that microwave searches have reached a modest state of maturity. This is not the same thing as saying that success is assured. There may be no transmitters, or those that do exist may lie beyond the range of searches being considered.

Infrared/Optical Frequencies – Possible Search Projects

The matrix in Table 5.8 is comparable to the one for microwave searches Table 5.5, except that the strength of the source and the sensitivity of the receiver are both measured in photons/m^2 at the receiver (per pulse for pulsed beacons, and per second for CW searches). The entries are relative signal strength, since distance is not a factor at these wavelengths (the technology exists, or will soon exist, to generate only spots of the desired size, assumed to be 10 AU here). This matrix is much more sparse, because optical SETI has only just begun to be explored. The success of optical SETI will depend upon gaining access to larger apertures, or on greater transmitter EIRPs.

The best current line of search is a byproduct of the work by Marcy and Butler,

	TRANSMITTER					
	← − →					
	weak and continuous			strong and temporally sparse		
RECEIVER	Radio Silence	Leakage	Omni-directional Beacon	Pin-Cushion	M of N	Sequential
Radio Astronomy	0	0.007	0.069	0.2	1.2	220
Sequential	0	1.3	13	40	220	40,000
Multiple Beam	0	6	60	190	1,100	180,000
Sky Survey	0	0.03	0.3	1	6	1,000
All-Sky All-The-Time	0	0.03	0.3	0.9	5.1	920
Commensal	0	0.1	1.1	3.4	19	3,400

NOTE: For each entry, the limiting detection range is given in light years.

Table 5.7: Microwave Transmit/Receive Range Matrix.

	Optical Pulse	Optical Continuous Wave
Visible Line Search	n/a	100
Visible Pulse Search	1,000	n/a
Pulse On Keck	10	n/a
Solar Power Plant	1	n/a

NOTE: Entries are relative signal strength required for detection.

Table 5.8: Optical Relative Power Matrix.

searching for planets around nearby stars by carefully measuring the stellar radial velocities. They use large (Keck-sized) telescopes to observe nearby stars, since photon statistics is one of their major limitations. A Keck-sized telescope aimed at a solar-type star 100 ly away gives about 10^{10} photons/second. Assuming a resolving power of 5×10^4, this gives a mean value of 2×10^5 photons/s in a spectral bin. In a 1,000 s observation, this yields 2×10^8 photons. The standard deviation is about 1.4×10^4 photons per observation, so a line that stands out at 6σ would need about 8×10^4 extra photons. One photon/m^2/second would be sufficient.

The most sensitive pulse search is that of Howard et al.[How00], using a 1.5 meter telescope and searching for multiple photons arriving in the same short time period (≈ 1 ns). Assuming about 40 photons are needed for reliable detection, then 10 photons/m^2/pulse would be sufficient.

One possible improvement is a pulse search with a Keck-sized telescope and an improved detector. If the threshold could be brought down to 10 photons/ns, this would be a hundred times more sensitive (25 times from increased collecting area, and 4 times for a better detector).

Another possible improvement is to use a solar power plant at night to get a larger collecting area. From the discussion in Section 6.10 on page 219, it seems that 0.01 photon/m^2/s might be possible, although sky coverage would be limited.

For beacons, assume 10 kJ per pulse or 10 kW CW, which is about representative of our technology today. To get the number of photons/m^2/s, we assume the beam size is chosen to give a 10 AU spot at the receiving system. A 10 KJ pulse over a 10 AU disk gives about 6×10^{-21} J/m^2. A visible photon at 310 nm has about 6.4×10^{-19} J, so this represents about 10^{-2} photons/m^2/pulse. Similarly, a 10 kW continuous visible light laser gives about 10^{-2} photons/m^2/s. Consequently, the number in Table 5.8 is about the improvement that we are assuming the ETIs have over our current technology. Depending upon the technologies used, the improvements could range from 1 to 1,000.

5.8 Game Theory and SETI

We have identified several plausible strategies for making contact across interstellar distances. The following all look technically feasible:

- Narrowband Optical;
- Pulsed Optical;
- Pulsed Microwave;
- Continuous Wave Microwave.

For each of these possibilities we could transmit a signal, or look for a signal transmitted by others. Similarly, ETIs have the same choice. How is that choice made?

Because the SSTWG was diverse, no single view about how to do SETI predominated. The Working Group, generally, pursued approaches that were plausible and that contained elements of technical opportunity. However, the Working Group decisions did not evolve from an overall theory of SETI, perhaps because we do not yet know enough about the SETI game.

In playing a known game, one tries to maximize the usefulness of the game to the player. This is called the utility function. Generally, the utility is a combination of a strategy with high expected payoff and one that minimizes risk. Maximizing the value of the expected payoff is a good strategy when the relative expected payoffs are known. Pick the highest paying option and put all the resources into that. Minimizing the risk, is a good strategy if one knows and can pursue all the useful strategies. Getting equal value from all strategies ensures that one at least gets something.

Maximizing the expected value is very risky, because the "best strategy" may not pay off. Minimizing risk is very expensive; one may spend all the resources trying a strategy that one cannot afford. Since the Working Group was diverse, the tendency was to try all reasonable, relatively equivalent strategies. The Working Group discussion reflected this, and plausibly reflected the group dynamic. However, the Working Group as a whole certainly did not agree on the utility function. Different parts of the SETI community tend to use their local assumptions about what is an optimum strategy mix. Thus, although the SETI Institute was advised to develop major projects in a few areas, it was also suggested that a grants program be subsidized. Perhaps, in this way, we can minimize our risks.

Chapter 6

Technology for SETI

This chapter begins with a brief review describing the principal historical and technological developments relevant to SETI, and the appropriate current and likely future technologies.

Computing – Computing requirements are considered in the context of Moore's Law and the probable continuing decline in processing costs.

Microwave – The following issues are discussed:

- Technologies applicable to microwaves and RFI mitigation, signal tolerance, adaptive nulling, and post-processing are described.
- A straw man design for an OSS is discussed, and its likely performance is rated.
- The reasoning and appropriate techniques for a "Next Generation Sky Survey" are explored together with focal plane arrays on large dishes and multiple beams on a phased array. A constructive process of phased implementation is described.
- A similar approach is made for a "Next Generation Targeted Search". The design of the planned SKA is discussed, together with details of its search techniques and the necessary computing.
- The proposed 1hT is explained as a perfect prototype for the SKA.

Optical – The technology for infrared/optical searches is explored, including detectors, pulse search instrumentation, inexpensive photon collectors and single pixel pulse detectors. Dual-detector coincidence systems are described as a defense against false alarms. Infrared pulse detection is considered with emphasis on array detectors, and contrasted with continuous-signal detection. Cerenkov radiation detectors are suggested as useful search tools, and multiple target instrumentation is considered.

Other Factors – Careful consideration is given to important factors such as insulating against technological risk, foreseeable technological advances in computing, material science, nanotechnology, artificial intelligence, inexpensive access to space, etc.

6.1   Approaches

In the last chapter we concluded that pulsed or continuous sources, in the bands from the microwave to the optical, were our best way of looking for ETIs. What technology can we apply to this task, both now and for the next twenty years?

We approach this by looking at straw man designs for five search types:

- directed searches in the microwave;
- sky surveys in the microwave;
- all-sky, all-time, wideband microwave searches for strong signals of low duty cycle (a new and unexplored area);
- optical pulse searches;
- optical continuous wave searches.

Many of these approaches are limited in their extent, or dominated in their cost, by computing power. For the last thirty years, computing power capabilities and costs have been improving at a rate known as Moore's Law. When we look at what we might build over the next 20 years, we explicitly assume that this improvement will continue. Since this assumption forms such an important part of our strategy, we include a more detailed discussion of Moore's Law in Section 6.2.

We recognize that technology is changing quickly, and therefore include a short discussion of possible radical technical changes. These changes may become important to SETI in the future, but they are too new and uncertain to directly incorporate them in our strategy.

6.2   Exponential Growth of Technology

It is well established that most scientific advances follow technical innovation, e.g., Harwit [Hrw81]. In 1960 de Solla Price [Pri63], reached this conclusion from his application of quantitative measurement to the progress of science. His analysis showed that the normal mode of growth of science is exponential. Historical examples include the rate of discovery of elements, and the number of universities founded in Europe.

Some more recent examples of exponential growth and their doubling times are power consumption (10 years), volume of overseas telephone calls (5 years), particle accelerator beam energy (2 years), and Internet hosts (1 year). To this we can add the now famous Moore's Law for computing devices, or more precisely for the number of transistors on a chip, with an 18 month doubling time

Figure 6.1: Dr. William J. Welch, holder of the Watson and Marilyn Alberts Chair in the Search for Extraterrestrial Intelligence at the University of California at Berkeley, toasts a good feed.

maintained for the last 30 years. These are all much faster than the underlying growth rates, such as the USA population (50 years) and Gross National Product (GNP) (20 years). Figure 6.2 on page 150 illustrates some technology doubling rates versus these underlying growth rates.

Such exponential growth cannot continue indefinitely, and de Solla Price [Pri63], [Pri86] noted three possible outcomes when the growth begins to slow:

1. Progress in the area of development becomes chaotic.

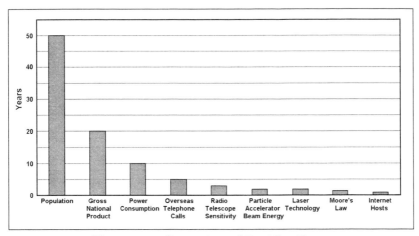

Figure 6.2: Comparison of Doubling Rates.

2. The area of research dies out.

3. There is a reorganization or paradigm shift in technology that results in a new period of exponential growth in which research flourishes.

A rather simplified conclusion is that "any field which has not maintained an exponential growth has now died out". Therefore, all current active areas are still in their exponential phase. Furthermore, as the refining of existing technology levels off, the continual introduction of new technology is required to maintain the exponential phase.

A famous example is the rate of increase of the operating beam energy in particle accelerators, as illustrated by Livingstone and Blewett [Liv62], and updated by Sessler [Ses88]. Starting in 1930, each particle accelerator technology initially provided exponential growth up to a ceiling where the growth rate leveled off. At that point, a new technology was introduced. The envelope of the set of curves is itself an exponential curve, with an increase in energy of 10^{10} in 60 years. This example, originally presented by Fermi, has become known as the *Livingstone Curve* and is shown in Figure 6.3.

The sensitivity of radio telescopes has grown exponentially, with an increase in sensitivity of 10^5 since 1940, doubling every three years. As with the previous example, particular radio telescope technologies reach ceilings and new technologies are introduced. In particular, there was a transition (about 1980) from huge single dishes to arrays of smaller dishes.

Can we continue the exponential growth in sensitivity? Is a new technology available to be used? One new technology that helps is the combination of transistor amplifiers and their large-scale integration into complex systems, which

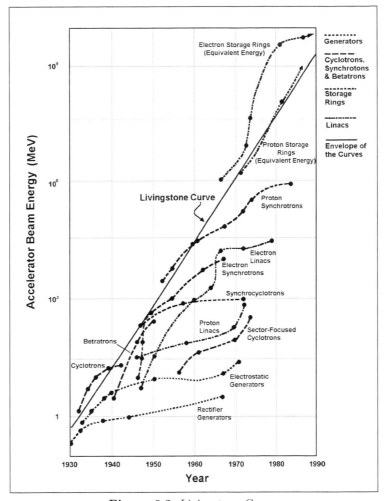

Figure 6.3: Livingstone Curve.

can be duplicated inexpensively. Another essential technology is our recently acquired ability to apply digital processing at high bandwidth. This enables us to realize processes, such as multiple adaptive beam formation and active interference rejection, in ways not previously conceivable.

If the improvement in sensitivity has reached a ceiling, the rate of new discoveries will decrease, and the fields of radio astronomy and radio/microwave SETI will decline. On the other hand, if we can adopt new technologies and reorganize our resources, the exponential increase in sensitivity will continue. Then by 2010, we should have built an instrument corresponding to an antenna with a collecting area of one square kilometer.

6.3   Moore's Law

This section is a quick discussion of Moore's Law. It discusses why we think it will continue to be valid over the next 15 years, and why we think we can continue to take advantage of it.

Moore's Law predicts that the number of transistors available on a single chip (of approximately constant cost) will double every 18 to 24 months. This exponential growth has continued over the last 30 years, resulting in 20 doublings or a factor of about a million in chip capacity. How much longer can we expect this progress to continue?

Moore's original predictions on the number of transistors has been shown to hold for many different components in the information processing technologies, and is now used to predict cost/performance into the future. As Stuart Brand points out [Brn99], Moore's Law looks benign when viewed in semilog coordinates, as it was originally drawn. When drawn in the linear coordinates of our human experience, the law becomes a wall, and may forecast a singularity.

6.3.1   Discussion

Moore's Law is really a consequence of economics, not a law of physics. The chip size referred to is the size of the largest commercially available chip, not the largest that can be physically produced in a laboratory setting where cost is no object. Thus, to continue this growth, the physics must support the needed changes. Then, the manufacturing technology must improve so that the denser devices can be produced for about the same cost.

From the physics point of view, to continue this increase we need smaller devices and bigger chips, all operating reliably. This appears not to be a problem; the required devices work fine in the lab. The larger barriers are economics and development. Much expensive development must occur to make these larger chips economical to produce. The consensus is that because of the enormous size of the worldwide electronics market (about 10^{12} per year), the necessary development will be done.

Since this question goes to the heart of a huge industry, it has received an enormous amount of study. On its Web site, Semiconductor Manufacturing Technology (SEMATECH) stated:

> "The National Technology Roadmap for Semiconductors (NTRS) is a description of the semiconductor technology requirements for ensuring advancements in the performance of integrated circuits. Sponsored by the Semiconductor Industry Association (SIA) and

published by SEMATECH, this report is the result of a collaborative effort of the semiconductor community including industry manufacturers and suppliers, government organizations, consortia, and universities." [Sema]

Currently, the update cycle for NTRS is two years, with scheduled checkpoints during the intervening period.

The 1997 study addresses the period through 2012, and predicts that Moore's Law will continue to hold during this period. To give an idea of the contents of the NTRS, the chapter titles are listed below:

- Introduction
- Resources for Addressing the Roadmap
- The Grand Challenge
- Overall Roadmap Technology Characteristics
- Design and Test
- Process Integration, Devices, and Structures
- Front End Processes
- Lithography
- Interconnect
- Factory Integration
- Assembly and Packaging
- Environment, Safety, and Health
- Defect Reduction
- Metrology
- Modeling and Simulation
- Appendices

Beyond 2012, the situation is less clear. The physics currently used will not work very well at the device densities then expected, because quantum effects will begin to dominate their behavior. The same quantum effects, however, may allow other smaller devices to achieve the same functions with even better performance. Even then, we would be several orders of magnitude above the apparent ultimate physical limits for such devices, for example, at least one electron per device.

Another interesting example of exponential growth is known as Neilsen's Law of Internet Bandwidth. It states that high-end user connection speeds will double every 21 months (slower than Moore's Law). However, this added speed cannot be used for several years, because the physical infrastructure and access servers must be upgraded first.

As we plan for the next generation of systems, we need to retain control of all the resources and connectivity we use. This will allow us to optimize the benefits available from the various technological exponential growth factors.

Figure 6.4: Dr. Gordon Moore in 1998, accompanied by a plot of Moore's Law for the Number of Frequency Channels vs. Time.

6.3.2   SETI-Specific Processing and Moore's Law

Next, we need to examine whether the specific algorithms used by SETI will continue to follow Moore's Law. The question is one of chip speed versus density, and serial versus parallel algorithms. A closer look at the NTRS shows that processing *speeds* are not expected to grow as fast as processing *density*. For example, in the next 10 years, we will able to build a chip 64 times as big, but it will run only 10 times as fast.

A purely *serial* algorithm, is one in which one step must be performed before another. Such an algorithm speeds up as the chip speeds up. Thus, we will only get a tenfold improvement for a serial algorithm, because we only get the speed advantage, but the additional hardware (chip density) does not help.

A *parallel* algorithm, on the other hand, is one that can be split into small parts, which can be solved in parallel without loss of efficiency. Such an algorithm speeds up in proportion to the amount of hardware used (chip density) as well as chip speed. Thus, we will get a full 640 fold improvement (64 times the number of processors, each running 10 times as fast).

Most algorithms consist of both a serial portion and a parallel portion. If you try to use more hardware to run the algorithm faster, the serial portion eventually becomes the limiting factor. This is because additional hardware can speed up the parallel portion of the algorithm, but not the serial portion. Consequently, it is the serial portion of the algorithm that ultimately determines how fast it can run.

Fortunately, the main SETI algorithms, such as beamforming and FFTs, have extremely small serial portions and huge parallel portions. For example, in beamforming we must compute the delays (independent of bandwidth), and then sum the amplitudes over the desired bandwidth. To the first order, computing the delay for a single antenna is the only serial portion of this algorithm.

We can compute the delays and amplitudes for all antennas in parallel. Since the highly parallel parts of the algorithm completely dominate the processing cost, then, as new and faster hardware becomes available, we can use it directly to do beamforming over wider bandwidths. FFTs can take advantage of additional hardware to increase their throughput. Finally, a large bandwidth can be analyzed as many smaller bands in parallel, with only a tiny penalty for overlap at the band edges. This works because we are looking for a signal whose bandwidth is small compared to the search bandwidth.

As shown in Figure 6.5 on page 156, the number of channels in SETI systems has followed Moore's Law since 1960. Since the SETI algorithms are so easily parallelized, we must conclude that they will continue to take full advantage of the additional computing power promised by Moore's Law.

6.3.3   Network Computing and SETI

The growth of computer networks has closely paralleled the growth of computers. Some large computing projects, such as those factoring large numbers and conducting cryptographic analysis, have taken advantage of the easily par-

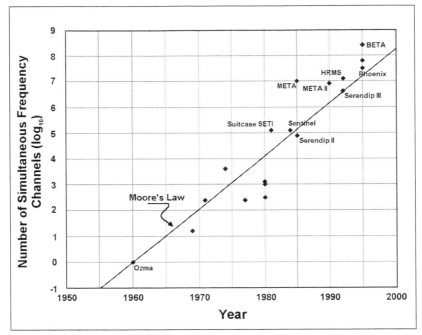

Figure 6.5: Number of Frequency Channels vs. Time.

allelized nature of such problems. To help solve these problems quickly, the projects have recruited the help of hundreds or thousands of otherwise idle machines among the millions of machines tied to the Internet.

To determine if this style of computing can be used for SETI we must answer three questions:

- Do the algorithms allow distributed processing? – The algorithms allow distributed processing, and they are highly parallel.

- Are the computer resources available? – The processing power is available. SETI@home is a project that allows people to dedicate idle cycles on their computers to SETI. By May 1999, about 400,000 people indicated an interest in SETI@home. If each provides a PC capable of handling 100 kHz of real time bandwidth, then it is theoretically possible to analyze 40 GHz in real time. This will presumably increase as computers become more common, more idle time is available, more people become interested in donating time, and the performance of PCs increases.

- Can the network support the necessary bandwidth? – In 1999, the Internet could not handle this amount of traffic. Supporting a 40 GHz bandwidth would require the radio telescope to produce, and post to the Internet, 80 billion samples of data per second. This is about six orders of

magnitude faster than Arecibo's Internet connection can support. Therefore, SETI@home, with vast processing power, but limited I/O capacity,
ships data by a physical tape from Arecibo. It uses its computing power
to search for many signal types and drift rates in a narrow bandwidth,
rather than searching for fewer signal types in a wide bandwidth.

Clearly the processing power is available, and networks are rapidly improving.
Thus, this computing technique is becoming steadily more practical for a wide
variety of SETI tasks.

Figure 6.6: Microsoft Corporation's, Dr. Nathan Myhrvold is a technology master.
(Photo courtesy of N. Myhrvold.)

6.4   Microwave Technology

In the next four sections, we consider three different types of microwave searches and the RFI problem.

- Section 6.5 addresses RFI mitigation.

- Section 6.6 on page 162, considers an all-sky, all-the-time search system denoted the Omnidirectional SETI System (OSS). The OSS would be well suited to look for beacons that are very powerful, but also very intermittent.

- Section 6.7 on page 170, considers sky surveys. The same technologies needed to build the SKA would lend themselves to a useful search for beacons of exceptionally high power anywhere in our Galaxy. All these plans will be to no avail if terrestrial interference cannot be dealt with effectively.

- Section 6.8 on page 180, considers targeted searches. Here, an exciting opportunity exists to collaborate with the larger radio astronomy community in building the SKA, which is an array of vastly greater collecting area than any present instrument.

6.5   RFI Mitigation

The broadband, and possibly all-sky requirements of SETI observations make them particularly vulnerable to interference from man-made and natural RFI. Everything from satellite and ground-based broadcast transmissions to solar radio emissions and lightning will mask portions of frequency, time, and direction space. Strong RFI carries the potential for rendering a SETI instrument inoperative unless the instrument is designed from the start to handle such interference without overload. Weaker but pervasive RFI makes the task of identifying weak ETI signals very challenging. However, both the Doppler drift of CW signals and the dispersion of pulsed signals provide discrimination against signals of terrestrial origin.

Most of search space, with frequency, time, and direction coordinates, is empty. In principle, therefore, we can throw away a small portion of data and use the rest. In practice, however, throwing away data causes a problem of dynamic range. Dynamic range can be defined in terms of antenna beamwidth and sidelobe levels, spectrometer response function, linearity to large signals, time resolution, etc. Improving each of these is expensive. Also, each can be treated dynamically with techniques such as adaptive nulling, but this requires much processing power, which is even more expensive. RFI mitigation is an important and integral component in the economic optimization of instrument design.

Other users of the radio spectrum will be improving the dynamic range of their own systems with the intent of adding more transmitters at more frequencies in more places. The SETI instrument needs to remain state-of-the-art in dynamic range, but it must also exploit the differences between RFI and expected intelligent signals. However, SETI must not become a slave to the RFI environment by only searching far from major sources of interference. Systems based on arrays, with a large number of elements, hold the greatest potential for dealing with RFI.

6.5.1   The RFI Environment

To optimize the SETI instrument in the presence of interference, we must have as much quantitative information about the RFI environment as possible. This requires an uncertain guess about how interference will evolve in the next 30 years, and the accuracy of our predictions will depend on how well we understand our current conditions.

Most, if not all, of what we know about interference comes from samples that are very sparse in frequency and time. Furthermore, much of it is calibrated in a way that is hard to use when designing a new instrument. First, we must assess the existing information about RFI. Then, we probably will need to design and build a measurement system that will provide the data needed to proceed with the design of the SETI instrument.

The selection of a site for the SETI instrument will depend upon the *nature* of the RFI environment. If the primary source of interference comes from satellites, then we only need to avoid the major coverage areas, which are typically population centers. If ground-based transmitters remain a major source of interference, then site selection is very important and more complicated. Good quantitative information on site candidates will be critical. This information will be most valuable if it is obtained very early in the design process.

6.5.2   Strong Signal Tolerance

It is hard to generalize about the strong signal tolerance of a radio telescope, because so much depends on the details of its design. This has not been a big issue for most instruments built in the last few decades. Most of the frequency spectrum was excluded, because signal concentration was done with passive reflectors, the feeds had a limited frequency range, and filters could be added early in the amplifier chain.

A wideband, all-sky or electronically steered SETI instrument will be a very

Figure 6.7: The National Radio Astronomy Observatory's radio frequency interference mitigation expert, Dr. J. Richard Fisher, contemplates strategies to eliminate RFI.

different challenge. The first amplifying elements will need to amplify at least a decade of frequency spectrum. Considerable signal concentration will occur in sensitive electronics, before band-limiting filters are practical to use. There will be a high potential for signal distortion due to amplifier nonlinearities. In each stage of the signal-combining network, it will be necessary to carefully balance the gain. Sufficient gain is needed to overcome losses and noise from the following transmission line and electronics, but the gain must be kept as small as possible to stay away from amplifier nonlinearities. Some sections of an all-sky instrument will inevitably be overloaded in the later stages of a beamforming network, because in some directions, the RFI will be coherently enhanced along with the desired signals. The area of the sky that is blinded by RFI will be reduced for every decibel of extra strong signal tolerance that is achieved. To a more limited extent, this also holds true for an array design where the initial signal concentration is done with a passive reflector.

Most beamforming techniques, envisioned elsewhere in this book, require extensive digital signal processing. Digital signals can be processed quite linearly, if the accompanying white noise is adequately sampled, and if enough bits are carried to accommodate the largest interference signals at each stage of the processing network. The problems are much the same as with analog electronics, except that the failure threshold can be more abrupt.

6.5.3 Adaptive Nulling

If an interfering source is not in the same direction as an ETI source, then in principle, we can tailor the spatial response of our radio telescope to null its reception of the interference, without losing much gain in the ETI signal. Nulling techniques must be developed as key components of the SETI instrument design, but some important considerations must be accounted for in its implementation.

Adaptive nulling requires good linearity of the electronics preceding the null detection and null generation components. Strong signal tolerance is enhanced only for those parts of the beamforming network that follow the nulling electronics.

Adaptive nulling can require a significant amount of processing power, comparable to the processing needed for beamforming. The cost depends on the number of signals to be nulled, their intrinsic bandwidth, and how fast they change direction. Forming a null is very similar to forming a beam. Some of the same electronics can be shared for the two purposes, but some cannot, particularly when the interference is beyond the delay space of the primary beam.

The electronics and antenna literature contains many papers on adaptive nulling, but most of them deal with desired signals of moderate strength. Adaptively reducing signals to S/N levels of -30 dB or less requires a more accurate sample of the interfering signal than is often assumed. This means that a significant amount of antenna gain must be directed toward the interfering sources, which adds to the cost and complexity of the antenna structure. Further R&D work must be done on adaptive nulling for both radio astronomy and SETI antennas, before we can assess its effectiveness and cost.

To the extent that the primary sources of interference at remote, quiet sites are civilian satellites with published orbits, it may be possible to generate the nulling beams without recourse to adaptive techniques. In addition, since the sidelobe levels of an array are inversely proportional to the number of elements, arrays with many elements may remain in the linear regime for most RFI sources. Rapid prototyping and experimentation will be extremely valuable in assessing the real and projected difficulty of dealing with RFI.

6.5.4   Post-Processing

The SETI Institute has much experience with post-processing algorithms for separating extraterrestrial signals from RFI. This is an excellent starting point for a new instrument. It gives us confidence that, even without major advances in adaptive nulling or other techniques, a reasonable fraction of the radio spectrum is still available for searches. Field experience shows that a very long baseline pseudo-interferometer can eliminate almost all interfering signals, given that:

- the interfering signals are not too numerous for the processing power available;
- the interfering signals do not raise the effective noise floor so much that the observing system is desensitized [Tar97].

Nevertheless, much innovation in post-processing is possible and will be required. A multibeaming instrument will add an enormous amount of simultaneous information about signals that occur only in the direction of candidate stars. The search algorithms must be expanded beyond the signals with expected frequency drift rates, but which are still distinguishable from man-made interference. The system's RFI memory and intelligence must become more sophisticated in what it knows about the interference environment. As a search proceeds, it should be possible to make better and better decisions about what is an *unusual* signal in the vast sea of RFI.

Pulse searches can take advantage of the wavelength dispersion imposed by the interstellar medium to separate extraterrestrial signals from signals which have not propagated through space. This technique is already extensively used by pulsar searchers.

6.6   Omnidirectional SETI System

As described in Appendix B, on page 303, there is a promising class of high power beacons with expected fluxes of approximately 10^{-23} W/m^2, but they may only be directed in our direction for a small fraction of time. To search for such signals effectively, we ideally want a system that looks at the entire

sky, at all frequencies, all of the time. The closest we can hope to come to
that is the type of phased array antenna that Dixon is developing for the SETI
League's Argus project [Dix95]. We denote this type of system as an OSS,
to avoid confusion with the Argus project. An OSS is similar to the type of
phased array feeds that Fisher is working on [Fis00] (but turned upside down).
It also shares many features with the 'digital lens' concept pioneered at Waseda
University [Dai96].

The OSS path through the SETI Decision Tree is shown in Figure 6.8 on page
164.

6.6.1   OSS Overview

The OSS will consist of thousands of fixed, low-gain elements, forming a classical
synthetic aperture. With N_e elements, the entire sky will be imaged simultaneously, at the full bandwidth rate, into N_b beams, where $N_b \approx N_e$. Astronomical
sources will move through this sky image at the sidereal rate. Since the total
collecting area of each element will be small, the entire array will also be rather
small (less than 10 m across), and the synthetic beams will be large (approximately one degree).

The effective area of each low-gain element will be just $G_e \lambda^2 / 4\pi$, where G_e is
the gain of a single element. The best sky coverage that we can obtain from the
ground is a half-sphere, corresponding to $G_e = 2$. We would obtain better results
by using elements that have a pattern that illuminates *less* than a half-sphere,
since we can then orient the pattern to minimize thermal radiation from the
ground, as well as interfering sources in the plane of the horizon. A reasonable
value for G_e might be 3 (4.8 dBi), with each element then having an effective
area of $(2.2 \times 10^{-2})\, f^{-2}$ m^2, where f is the frequency in GHz. The number
of elements required is then on the order of 10^4 for $f = 1$ GHz, and 100 times
greater (10^6) at 10 GHz. As will be seen in the Section 6.6.2 on page 165,
the signal processing cost scales as the number of elements, and the cost is not
trivial even at 1 GHz. Thus, at this time, it does not seem practical to consider
OSS designs for frequencies very much above 1 GHz.

The task of forming the sky image is best accomplished using a two-dimensional
FFT. In general, to form N_b beams from N_e elements by direct, brute-force
methods requires on the order of $N_b \times N_e$ complex operations per image. It
requires $N_e \log_2 N_e$ using an FFT. To fully image the sky, we want $N_b \approx N_e$, so
the FFT method is more efficient by a factor of $N_e / \log_2 N_e$, which will be large.
The FFT is a phase-based algorithm, hence it will be necessary to break up the
IF bandwidth of each element into channels that are effectively monochromatic.
Then each channel has a bandwidth Δf satisfying $(\Delta f)(D_{array}/c) \ll 1$. Fortunately for SETI, we need to break the bandwidth up anyway, as part of our
narrowband detection process. Since $N_b \approx N_e$, the order in which we do the

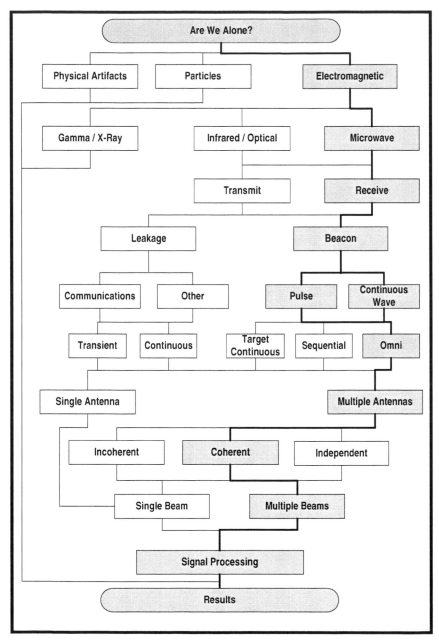

Figure 6.8: SETI Decision Tree – OSS Path.

imaging and frequency-division operations does not affect the overall computation load very much.

6.6.2 OSS Straw Man Design

In this section, we try to work out some of the details by developing a straw man design. Assume an array operating at 1.5 GHz, the center of the 'water hole', with a bandwidth of 1 GHz. Let N_e be 4,096, yielding a total collecting area of 39 m^2 at 1.5 GHz.

The antenna elements need to work over an octave of frequency, which is reasonable. Dual-polarization elements are required, and they need to have a broad, upward-biased pattern directed at the zenith. In support of the SKA, the Netherlands Foundation for Research in Astronomy (NFRA) is studying various possible element designs, of which the most promising at present is a 'rabbit ear' with an "end-fire" pattern. These are linearly polarized and can be crossed to give a dual-polarization element. The Low-Noise Amplifiers (LNA) can readily be mounted on the supporting dielectric planes to produce a sturdy and inexpensive antenna. From the data in Appendix C on page 321, we estimate that the room-temperature LNA has $T_r = 15$ K. By the use of ground screening, it seems reasonable that T_{ant} could be kept between 10 K and 30 K, resulting in T_{sys} around 35 K. The element itself should be very inexpensive, but the dual-polarization LNA is likely to cost about $800 each in quantities of 1,000, including post-amps, case, power supply, etc.

Since the array is only approximately seven meters across, it could be mounted on the roof of the structure built to hold the electronics. It may be least expensive to mount the analog to digital converter (A/D) directly beneath each element, although the idea of transmitting the amplified analog signal into the building via optical fiber is attractive from an electromagnetic interference (EMI) perspective. Using an 8-bit A/D might be prudent, given the possibility of severe interference in the RF band. With today's technology, it would be best to break the analog signal into approximately 100 MHz bands before digitizing them. In a few years, however, it may make more sense to just digitize the entire 1 GHz to minimize the number of components, especially the filters. Flash A/Ds (8-bit) are already available for a Geosynchronous Satellite Positioning System (GSPS).

Using Monolithic Microwave Integrated Circuits (MMIC) for the entire analog chain would be attractive from a price perspective. However, it remains to be seen whether they will have enough dynamic range to support a wideband, omnidirectional, noise-limited system. Another possibility would be to use operational amplifiers, which already operate over 1 GHz and may reach 2 GHz, for the downstream gain. The cost of the analog signal chain and digitizers can probably be kept well under the cost of the LNA. We estimate the total cost per element to be about $1,000.

As previously mentioned, we will need to break the 1 GHz band down into N_{chan} narrower channels, each with BW/N_{chan} bandwidth before we can use the FFT to form sky images. While the image forming operation rate for each channel is reduced by N_{chan}, we need to form images for each of the N_{chan} channels. Hence, the total image-forming load is independent of our choice of N_{chan}. We need to choose a value of N_{chan} that is large enough, so that the channel bandwidth is $\ll c/(array\ size) = 47$ MHz, and therefore so that the channels are effectively monochromatic. On the other hand, if we want to search for radar pulses with ≈ 1 MHz bandwidth, we need to have the image outputs available at this wide a resolution for the pulse detectors.

Let us then adopt $N_{chan} = 1,024$ as the number of channels, so that the IF from each element is divided into channels with 1 MHz bandwidth before doing the image formation. Since we will be using these channels to form finer resolutions for CW detection later, we need to be careful to form a decent quality digital filter bank. It is probably fair to estimate the operation load from a plain FFT and double the result to allow for this. With $N_{fft} = 1,024$ as the number of elements, each FFT takes $6N_{fft}\log_2 N_{fft}$ operations to be done in N_{fft}/BW seconds for a rate, per element and polarization, of $6BW\log_2 N_{fft}$. Doubling this, for two polarizations, we get 1.2×10^2 Gop/s/pol/element for the spectral analysis prior to beamforming. For the array of 4,096 elements and two polarizations, the aggregate rate is 9.8×10^5 Gop/s costing \$2 M at our standard tariff.

Next, we form the images by means of two-dimensional FFTs. For simplicity, we just assume that the 4,096 antennas are arranged in a 64×64 grid. The 2D FFT will take approximately $6 \times N_b \times \log_2(N_b)$ operations per polarization, at the bandwidth rate. For both polarizations, the total computational load for the imaging is then 5.9×10^5 Gop/s, costing \$1.2 M. In an actual design, questions of optimum array configuration and weighting would need to be considered. These questions are particularly important with regard to 'scalloping' effects, as a source moves through the sky image, and to grating lobe responses.

Each beam (i.e., image pixel) will be approximately two degrees across, although their shapes will be functions of both their position in the sky, and frequency. Each beam will be fixed in the Earth system, with a star drifting through the beam in approximately $500\ b$ seconds, where b is $1/\cos(declination)$. We can compensate for the Doppler drift from the Earth's acceleration by applying suitable, time-varying phases (different for each channel and for each beam) to the output data. However, these compensations will only be correct at the center of each beam.

The change in the Doppler shift over a beam will be ≤ 5 kHz at 1.5 GHz. This does *not* mean that there will be a Doppler drift of $(10\text{ kHz})/(500\text{ s})$, because what matters to the Doppler shift is the *dot* product of the observatory velocity and the unit vector towards the star, which changes much more slowly. The change in the Doppler drift rate will be $\leq 3 \times 10^{-3}$ Hz/s across the beam.

To detect a CW beacon, we want to transform down to a very narrow bandwidth, assuming that the beacon is fully compensated for Doppler effects at the transmitting end. As just calculated, we expect a small uncompensated residual drift rate due to the synthesized beam size. This limits our bandwidth for the final FFT to be about $\sqrt{3 \times 10^{-3}} = 0.05$ Hz or larger. As discussed in Section 2.7 on page 60, other effects will limit our bandwidth to approximately 0.01 Hz in any case, so this limitation is not too severe. While we could incoherently accumulate the 50 final FFT outputs, yielding an improvement of approximately three, a simpler system that thresholds directly on the output 0.05 Hz powers is more attractive. It would allow us to track the response of the source as it moves through the beams at the sidereal rate. If it does not move, flag it as RFI.

For the straw man design, we adopt a simple threshold at 24σ. This yields about one false alarm per polarization per beam from each observation. Noise alone, of course, will not produce a second false detection, and we can keep a database of the events that we are tracking, as the sky rotates through the beams. With this threshold the flux sensitivity at 1.5 GHz will be 10^{-23} W/m^2. An alternative, higher threshold could be used to search for transient events. The final FFT to 0.05 Hz will have an aggregate calculation rate of 2.4×10^6 Gop/s (where we have allowed another 'fudge' factor of two for baselining and other tasks) at an estimated cost of \$4.8 M.

It is unclear just how to build a good detector for wideband pulses. The best we can hope for is a matched detector, for which the noise energy in a pulse bin is $k \times T_{\mathrm{sys}}$. With a threshold of 10σ, this ideal detector could find a single pulse from our hypothetical Near Earth Asteroid Radar mission (NEAR) out to a range of 1 kpc. (See Table 5.4 on page 131.) Actual commercial detectors could easily be orders of magnitude worse, yet it is plausible that useful detection ranges could still be attained. As a cost estimate, we just assume that the pulse detector will cost the same as the CW detector, about \$4.8 M.

With an aggregate computation rate of approximately 4×10^6 Gop/s, power consumption and heat loading may be a problem. Today, Digital Signal Processing (DSP) chips consume about 3 W/Gop/s. Assuming, perhaps slightly optimistically, a factor of 10 improvement over the next decade, the DSP processing will consume 1 MW. Assuming a total efficiency of 50%, including conversion losses and the cost of removing the heat, we arrive at an estimate of 2 MW, and an annual cost for electricity of \$1 M at \$0.05/kWh.

Given the large power and heat requirements, we allow \$1M for the facility. In addition to these costs, a substantial amount of manpower will need to be devoted to developing the system, especially its software. The estimated cost for the labor-intensive software development is approximately \$2 M - \$4 M.

Beam Width

For any receiver with wide beams, another effect is important. If the transmitter is not observed on the beam axis, the compensation for terrestrial Doppler drift will not be precisely correct.

The FWHM of the beam for the OSS given on the previous page is approximately 0.035 radians. The acceleration of the observatory produces a drift rate of approximately 0.1 Hz/s/GHz $\approx$ 0.15 Hz/s at 1.5 GHz. Since the beam axis and the direction of the acceleration are not necessarily aligned, the worst-case difference in drift at the half maximum of the beam will be $(0.5)\,(0.035)\,(0.15) = 5 \times 10^{-3}$ Hz/s, and the maximum coherence time will be $\sqrt{1/(2.6 \times 10^{-3})} = 20$ s, integrations of 20 s are feasible.

6.6.3   Performance of the OSS

When built, the OSS would be far superior to existing telescopes for the class of occasional, strong signals. Appendix H derives a complex figure-of-merit that can be used to compare unrelated searches. For the current discussion we consider a simple measure of search merit, namely the product of the number of detectable stars (as a function of the assumed transmitter power in Watts EIRP) and the number of octaves of frequency (the number of bands) observed. To estimate how much better the OSS would be, we must modify the concept of EIRP transmitter power to apply to transient signals. We do this in a very simple way, by keeping the average power constant. Thus, if a source is on 1/N of the time, then it has N times the EIRP. For a pulsed source, the simple measure of search merit is reduced by 1/N for a directed search system, because the source probably will not be switched on when we look. For both systems, we increase the sensitivity by N, to account for the brighter source. For a pulsed microwave system, a reasonable value for N might be 10^6, which gives the results shown in Figure 6.10 on page 170. Clearly, the OSS is many orders of magnitude better than the existing alternatives. Note that the simple measure of search merit of the searches flattens above 10^{12} W average power. This is because the OSS can detect an occasional, strong signal from any star in the Galaxy, and additional transmitter power does not help until extragalactic sources are reached.

6.6.4   Summary of OSS

An Omnidirectional SETI System seems feasible, with the parameters summarized in Table 6.1 on page 171. The OSS would watch one third of the sky at once. With a detection flux of 10^{-23} W/m^2, it would be sensitive enough to detect CW signals out to perhaps 1,000 ly from phased array beacons used with

Figure 6.9: Dr. Arnold van Ardenne, Chief Engineer at the Netherlands Foundation for Radio Astronomy, brings international perspective to the SSTWG's effort.

the single-beam, sequential illumination strategy (discussed in Appendix B, on page 303 and summarized in Table 5.2 on page 130). Pulses from asteroid-detecting radars might be detectable over useful distances. (See Section 5.7 on page 132 and Table 5.4 on page 131.

The cost of an OSS depends almost entirely on the cost of signal processing. At an assumed cost of $2 per Gop/s, as expected in 10 to 12 years, the straw

man OSS design would have capital costs of approximately \$20 M, as detailed
in Table 6.2 on page 171. These costs are very uncertain. Not only are the
estimates crude, but the cost of the digital signal processing is expected to be a
very strong function of time. If the price seems too high, just waiting one or two
years should solve the problem! Also, we would need to build a smaller-scale
prototype of this type of system, with less sensitivity or bandwidth, in order to
identify and resolve any problems.

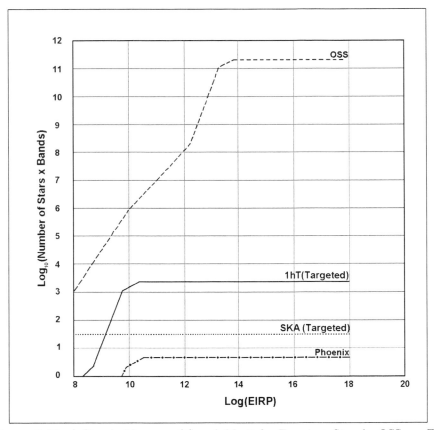

Figure 6.10: A Simple Measure of Search Merit for Transient Signals, OSS vs. Existing Searches.

6.7   Next Generation Sky Survey

The same advances in silicon technology that enable better targeted searches
and an OSS, also enable better sky surveys. This section looks at what might
be possible with a next generation sky survey. It also specifies under what
conditions a sky survey succeeds, while both a targeted search and OSS fail.

Parameter	Value
Frequency	1.5 GHz
Bandwidth	1.0 GHz
Element Spacing	0.5 λ
Number of Elements	4,096
Diameter of Array	6.4 m
Beam Width	2.0 degrees
Beam Drift Time	470 s
Solid Search Angle	4.2 sr
Total Effective Area	39 m^2
T$_{sys}$	35 K
Flux Detection Limit	1.4×10^{-23} W/m^2

Table 6.1: OSS Parameters.

Item	$M
Elements	4.1
Channel Formation (first FFT)	2.0
Beam Formation (2D FFT)	1.2
CW Detector (final FFT)	4.8
Pulse Detector	4.8
Facility	1.0
Miscellaneous (10%)	1.8
Total	19.7

Table 6.2: OSS Hardware Budget.

6.7.1 Motivation

Why conduct a sky survey for ETIs? What can a sky survey offer that targeted searches and all-sky, all-time searches cannot?

Targeted searches, such as Phoenix, aim for maximum sensitivity, but search only a relatively small number of nearby stars (i.e., a maximum of about 10^6

stars within about 1,000 ly of Earth). Thus, only 10^6 of the 10^{11} stars in the Galaxy, or one out of 100,000 stars, are examined. A targeted search also suffers from several potentially serious selection biases, since it assumes that life occurs only on Earth-like planets near solar-type stars.

Omnidirectional, all-time systems such as the OSS have no selection biases. They examine all the stars in the Galaxy, but have low sensitivities since they have apertures of only a few square meters.

Sky surveys have higher sensitivities than OSS-like systems, and a much greater number of stars examined than targeted search systems. On the other hand, they are *less* sensitive than a targeted search, and *not* sensitive to low duty cycle signals, unlike OSS. Existing sky surveys include META [Hor93], BETA [Lei00], and SERENDIP IV [Wer00].

In other words, a targeted search assumes that there are at least 10^5 beacons in the Galaxy, so one will be found in the nearest 10^6 stars, within about 1,000 ly. Sky surveys and OSS assume that there is at least one beacon in the Galaxy that is relatively bright (in pulses for the OSS, in CW for a sky survey). Plugging in rough numbers for systems that might be on-line in the year 2010, OSS needs about 2×10^{16} W EIRP at 1,000 ly, and up to 10^{20} W EIRP at 70,000 ly. A sky survey needs about 2×10^{13} W EIRP at 1,000 ly, up to 10^{17} W EIRP at 70,000 ly.

Therefore, a sky survey is the best strategy if there are between 1 and 10^5 beacons in the Galaxy, if they have EIRPs between 10^{13} and 10^{17}W, and if they are using continuous wave transmission. This seems to be a relatively likely combination of parameters, given the limited information we have.

6.7.2   Can the Beacons we are looking for Exist?

For the sky survey to be useful, the beacon must be on almost constantly, since we will be looking in any given direction only a tiny fraction of the time. Furthermore, the beacon must be at least 1,000 ly away, otherwise a directed search will find it more easily. This implies that the EIRP of any beacon useful to a sky survey must be quite high. It must be about 2×10^{13} W at 1,000 ly, and up to 10^{17} W at 70,000 ly, assuming a 100 m receiving antenna. Can even a more advanced civilization than ours build such beacons?

This is discussed in detail in Appendix B, on page 303. The short answer is that it does not look impractical from a physics point of view, although it is too expensive for us. A beacon with 10^{14} active elements of 1 W each could illuminate each of the 10^{11} stars in the Galaxy with 10^{17} W EIRP. This could be detected by the 100 meter sky survey out to 70,000 ly, i.e., anywhere in the Galaxy. Such a transmitting array would be large (1,000 to 10,000 km on a side,

Figure 6.11: Dr. Jaap D. Bregman, an interferometry expert from the Netherlands Foundation for Radio Astronomy, discusses a point with a table companion.

depending on the density of transmitters), but not impractically large.

It must be pointed out that such a beacon could also reach an OSS system at approximately the same range. If the beacon hits each star with a 1,000 s pulse each 10^6 s (about once every 12 days), then the power gain is 10^3, which makes up for the 10^3 lower sensitivity of an OSS.

6.7.3   Techniques

The fundamental difficulty with sky surveys is that of covering the whole sky with sufficient sensitivity. High sensitivity implies big antennas, but big antennas imply small beam sizes, which in turn require many pointings. For example, a 3 GHz (10 cm) beam on a 100 meter telescope has a beam width of about 10^{-3} radians, subtending about 10^{-7} of the whole sky. The best sensitivity requires looking for about 100 s per pointing, which would take 10^9 s (32 years) for one beam to cover the sky just once. Typically, we would like to 'look' several times, perhaps for RFI mitigation, or for reduction of scalloping, or to reduce our processing burden by not processing all the bandwidth at once. A higher frequency limit also makes matters considerably worse. For a fixed antenna size, the required number of pointings scales as the square of the frequency.

In order to finish in a reasonable amount of time, either we need to use many beams, or we need to reduce our time per pointing and accept the loss of sensitivity. The following sections examine three possible solutions, namely:

- a transit instrument with reduced time per pointing;
- a focal plane array involving multiple beams on a conventional telescope;
- forming multiple beams electronically on the 1hT.

6.7.4   A Transit Sky Survey

BETA uses drift scans at constant declination to cover a large part of the sky. However, BETA suffers from a small antenna and a less than optimal system temperature. How big a telescope can we afford? For drift scans, we can use a transit instrument. The 300 foot telescope at Green Bank was about the size we need, but unfortunately it collapsed and has not yet been rebuilt. However, based on the 1961 construction cost of \$1 M, we estimate we could rebuild it for about \$6 M, not including the site or electronics.

Since we have only one telescope, it would make sense to put a maximum effort into lowering T_{sys}. Based upon Appendix C on page 321, a cryogenic front end cooled to 20 K, coupled with optimized optics (probably Cassegrain), might yield a system temperature as low as 14 K at the lower microwave frequencies. We would have to restrict ourselves to the lower frequencies in any case, because the time to complete the survey increases linearly with frequency. Furthermore, the beam transit time (which limits the final bin width), causes the sensitivity to decrease linearly with frequency. Also, T_{sys} increases as the frequency increases.

For illustrative purposes, we adopt a one octave frequency search span, extending upward from the 1.2 to 2.4 GHz 'water hole'. With such a simple system, compared to those previously considered, we might as well assume that we can search the entire octave in one pass.

The FWHM of the telescope will be about 0.08 degrees at the upper end of the search range. Thus, for a telescope at the equator, a source will drift through the beam in about 18 s. We will thus be limited to a final FFT bin width of about 0.1 Hz, assuming Hanning windowing. A threshold of 25σ will produce approximately one over-threshold event per observation due only to noise. To avoid excessive scalloping on the sky from the beam response, we assume that the declination tracks are spaced at half the FWHM of the beam.

To speed up the observations, we assume that we can provide a cluster of beams, similar to those now in use at Parkes, and have three beams pointing at adjacent declinations. Other beams can be spread in hour angle if we adopt an RFI suppression scheme similar to that used by BETA. With these choices, and allowing for three looks at each position to improve performance with respect to ISS, we can complete a sky survey of the declination range -30 to +60 degrees in about six years with a flux sensitivity of 10^{-25} W/m^2. The parameters are summarized in Table 6.3. Since this is a 'three-look' survey, it will only be sensitive to beacons that are essentially 'always on'. Referring to Table 5.2 on page 130, we see that, as expected, we would not be likely to detect the kinds of beacons that we could currently build ourselves, although we might have a chance of detecting more 'advanced' beacons.

6.7.5   Focal Plane Array on Large Dishes

If we want more integration time for better sensitivity, we need more beams. Assume 100 s and two looks per pointing, an effective aperture size of 100 m, and a frequency of 10 GHz. Then, about 100 beams are required to survey the whole sky in six years.

Equipping a 100 m dish with 100 feeds should be possible and even practical. The Parkes radio telescope [Sta97] is using 13 feeds at 1.4 GHz, so 100 feeds at 10 GHz should be sufficient. This approach, although requiring a big antenna, makes the receiver design relatively easy, because all feeds and receivers can share a common cryogenic system. Also, Doppler compensation is easy since it can be done for all frequencies simultaneously by tweaking the sampling rate.

The big drawback to this technique is that it requires the dedicated use of a 100 m antenna for many years. The latest such antenna cost about $100 M. Presumably, this work would not require such a fancy antenna, but even if the cost was $20 M, it would be more expensive than other possibilities.

6.7.6   Multiple Beams from a Phased Array

Many beams could be generated by signal processing from the output of an array of small dishes, such as the 1hT, described on page 244 in Section 8.2. If the array contains 500 dishes, and each beam is 2 GHz (to minimize the number of pointings needed), then it requires about 4,000 Gops to form each beam by direct summation. In early 1999, that was quite expensive! At \$250 per digital signal processing Gop, it amounted to \$10 M per beam for one polarization. The cost was doubled if both polarizations were needed.

However, the processing costs are dropping rapidly. Assuming that Moore's Law continues to hold, then in about 2008, the processing will cost only \$2/Gop. Even now, each beam is only \$8,000, or \$1.6 M for all 100 beams with both polarizations. This is a much more reasonable price.

Furthermore, if the dishes are placed in a regular array, it may be possible to use FFT beamforming instead of direct summation. Since FFTs are narrowband, we would have to break the spectrum into narrow bands before creating the beams by FFTs. This would not affect the operation count much, since it just places the band creation before the beamforming, instead of after it. With 500 antennas and 100 beams with both polarizations, the savings would be $6 \log_2 500 \approx 54$ computations per input as opposed to the 100 required for direct summation. While this is not a huge savings, it would still be helpful.

The cost of analyzing the beams is small, compared to the cost of generating them by either summation or FFT. For example, performing FFT on a 2 GHz data stream in one second blocks requires the input to first be divided into two gigasample blocks. Each FFT takes about $6N \log_2(N)$ operations, or about 400 Gops for the stream. Even if baselining, band splitting, peak detection, and other tasks increase the processing load by a factor of two, it is still well under the beamforming requirements. At \$2/Gop, processing 800 Gops/s costs \$1,600 per beam, and for 100 beams and two polarizations, the cost is \$320,000.

RFI rejection is relatively straightforward in such a multibeam system. See Section 6.5 on page 158.

6.7.7   Starting Sooner – A Phased Approach to Processing

Table 6.4 on page 179 illustrates a phased implementation of systems. First a sky survey could be built as a relatively low bandwidth system. Then, it could be expanded as the electronics portion becomes less expensive due to Moore's Law.

Consider the analysis of the beams, which is the same in either the single dish or the arrayed version. An N point FFT takes about $6N \log_2(N)$ operations.

Assume that we have 4 million point FFTs, and that an equal amount of effort is devoted to dividing into bands and looking for peaks in the results. Then we have only about $6 \times 22 \times 3 \times 10^9 = 396$ billion operations per beam per GHz.

If we have a single dish, this is the total processing cost. If we have an array (e.g., 144 dishes of eight meters each), then we also need to consider beamforming costs. For a 1 GHz bandwidth we need 2 billion samples per second, times 2 operations per sample, times 144 sources. This amounts to 576 Gops per beam. The four beams of the first configuration would then take about 2,300 Gops, so the total cost would be increased by a factor of 2.5.

Since we can use a number of evenly spaced beams, FFT techniques may help. The difference is not dramatic with only 144 antennas: $6 \log_2 256$ is 48 operations per input point, down from 144, for a savings of about a factor of three. This would bring the beamforming costs down to slightly less than the analysis costs. However, the individual operations of the FFT approach are more complex than the operations of the direct method, so careful study would be necessary to calculate an exact cost.

Note that the same analysis engines could be used for both the OSS and the sky survey. If OSS is the primary new instrument, then as soon as the next generation is built, the old spectrometers could be moved to a sky survey instrument. Thus, both searches could be performed with one set of analysis hardware, staggered by about four years.

6.7.8   Performance Comparison

The transit instrument, discussed in Section 6.7.4 on page 174, would be limited by transit time to about 0.1 Hz final bandwidths. However, with a very good feed we could get a low system temperature and a sensitivity of 10^{-25} W/m^2.

Both the multibeam phased array or 1hT survey, discussed in Section 6.7.6 on page 176, and the multibeam conventional telescope discussed in Section 6.7.5 on page 175, should have approximately equivalent performance. If we assume a 35 K system temperature, 100 m diameter aperture, an aperture efficiency of 50%, 0.01 Hz resolution, and that an S/N of 20 is required for detection [Lei00], then the sensitivity is about 2.5×10^{-26} W/m^2. From a distance of 1,000 ly, this requires an EIRP of 2×10^{13} W.

The multibeam phased array (1hT survey) and the focal plane array could be pushed upwards in frequency. Either more beams, or a longer time to complete the survey would be required, but there is no fundamental problem. The transit instrument, however, would lose sensitivity as its beam size becomes smaller. Consequently, about 2 to 3 GHz seems to be a practical upper limit for such an instrument.

Figure 6.12: Dr. Anthony A. Stark of the Smithsonian Astrophysical Observatory, demonstrates equanimity while awaiting dessert.

6.7.9   Conclusions

The same technology that helps targeted searches and enables OSS can also considerably improve sky surveys, for a reasonable cost.

A sky survey is a good idea if there are between 1 and 10^5 beacons in the Galaxy, with EIRPs between 10^{13} W and 10^{17} W, and if they are using CW transmission. Three possible approaches are summarized in Table 6.3.

A simple measure of search merit, namely the number of stars times octaves

Proposal	Beams	Mechanical Cost	Beam Forming Cost	Signal Proc. Cost	Sensitivity W/m^2	Upper Freq. (GHz)
Transit	3	> \$6 M	n/a	\$16 K	10^{-25}	2.4
Many Beams	100	> \$20 M	n/a	\$320 k	2.5×10^{-26}	10
1hT	100	> \$5 M	\$1.6 M	\$320 k	2.5×10^{-26}	10

Table 6.3: Comparison of Sky Surveys.

searched for a given EIRP, is shown in Figure 6.13 on page 180. The figure shows the 1hT and BETAmax, along with SERENDIP IV, the most sensitive current sky survey. Also included is the 1hT targeted survey, since it covers a significant fraction of the sky. The focal plane multibeam survey is not shown, because it has a figure-of-merit similar to that of the 1hT survey but is considerably more expensive.

Table 6.4 on page 179 shows a summary of sky survey processing costs. All the proposed sky surveys are considerably better than SERENDIP IV, the current state-of-the-art system. They cover about two orders of magnitude more stars for the same hypothetical EIRP, and an order of magnitude more bandwidth. Interestingly, the 1hT targeted survey has a very similar figure-of-merit to BETAmax. The greater sensitivity of the 1hT survey makes up for its smaller sky coverage.

The most economical approach, by far, is to wait for signal processing costs to come down, and then use the 1hT approach, which gives the best performance. If we want to begin sooner, we can use the phased approach.

Frequency Range	Beams	Operations	Year	\$/Gops	Cost
1-2 GHz	4	$1,600 \times 10^9$	2000	256	$\$400 \times 10^3$
2-4 GHz	16	$8 \times (1,600 \times 10^9)$	2004	32	$\$400 \times 10^3$
4-8 GHz	64	$64 \times (1,600 \times 10^9)$	2009	4	$\$400 \times 10^3$
8-16 GHz	256	$512 \times (1,600 \times 10^9)$	2013	0.5	$\$400 \times 10^3$

Table 6.4: Sky Survey Processing Costs.

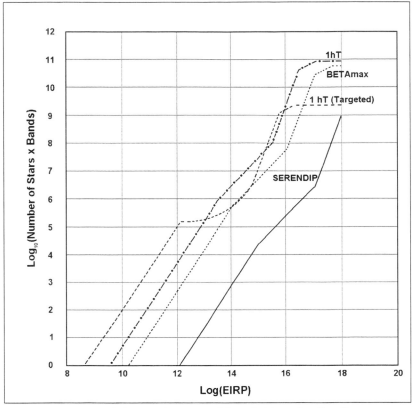

Figure 6.13: A Simple Measure of Search Merit for Sky Surveys.

6.8   Next Generation Targeted Search

Current SETI systems are not sensitive enough to be able to detect leakage from the Earth at a distance of even 4 ly. Thus, if they are used in the all-star multibeam mode, they are unlikely to be able to detect the big beacons discussed in Appendix B, on page 303.

Can we hope to build an array, at a cost per unit collecting area that is orders of magnitude less than that of the VLA, by exploiting technologies developed since it was built? Many radio astronomers think we can. The Large Telescope Working Group of URSI has proposed building the SKA, an array with 1 km^2 of collecting area, with a budget well under $1,000 M.

R&D for the SKA are being carried out by a consortium of interested parties, including the SETI Institute. A basic design must be selected, and technical feasibility demonstrations must verify the basic elements of the design. Then, international funding will be sought for construction, probably beginning after 2005. Until at least the preliminary design is completed, the cost of the SKA is very uncertain.

The preliminary specification for the SKA calls for a value of $A_e/T_{sys} \geq 2 \times 10^4$ m^2/K, and a frequency range of 0.2 to 20 GHz in the main array. This would be supplemented by a low-frequency extension to 0.03 GHz, probably with a lower value of A_e/T_{sys}. The bandwidth would be approximately 1 GHz, and the SKA would have the ability to form 100 simultaneous beams, spread over at least 1 square degree at 1.4 GHz.

All of these features would make the SKA very suitable for SETI. The wide frequency range permits a better, more comprehensive look at each target. If 10 of the 100 beams were available for SETI, we could afford to make long observations to maximize our sensitivity, examine a large number of stars, and still finish our first survey in a reasonable time. The one square degree field offers us the flexibility to use these beams all the time, while sharing the telescope with other users.

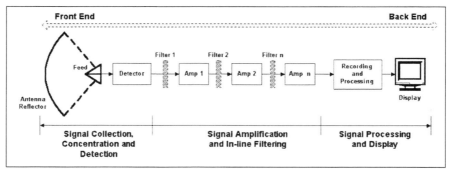

Figure 6.14: Conceptual Drawing of a Typical SETI Receiver.

6.8.1   First Cut SETI Survey with the SKA

It is informative to try to plan a first cut SETI survey with the SKA. Just making a list of the suitable stars in our neighborhood is not easy. We will assume that F, G and K main-sequence stars are the best targets. If we have the appropriate data about their planetary systems, if any, then we can further refine our selections based upon our knowledge of the planetary orbits and the age of the target star. Fortunately, a large part of the necessary ground work has just been completed by the Hipparcos project. Using these data, we can probably assemble a first cut target list of about 10^5 stars.

By the time the SKA is built, we can anticipate that second-generation astrometric projects will have increased the number of available F, G and K main sequence stars even further. Vigorous observations now being pursued to identify planetary companions will also give a much more complete set of data about the planets orbiting many of these stars. We might even be able to select at least a few stars that have planetary systems similar to our own. For our first survey, we therefore adopt a goal of 10^5 F, G and K stars, with distances ranging out to about 500 ly (assuming we can observe about half the sky).

Within the 0.2 to 20 GHz range of the SKA, observations much below 1 GHz will suffer from increased noise and RFI, while above 4 GHz, the available patch of sky may become too small for efficiently using multiple beams. Thus, a two-octave search from 1 to 4 GHz seems to be a reasonable plan for our first survey. Naturally, dual polarizations would be used. It would be inefficient to waste any of the available instantaneous bandwidth, so we will assume that the SETI detector system can handle 1 GHz (in each polarization) at a time. For a pure beacon search, we use the narrowest safe final channel width, 0.01 Hz (See Section 2.7 on page 60). Since we have multiple beams, we can also afford to make observations of 1,000 s, where each observation will be the incoherent power accumulation of 10 power spectra with 0.01 Hz resolution. In addition, because of the increase in the number of statistical 'tries' we will make per observation, we will need to use a threshold of about 11σ. (Note that the statistics are not exponential.) The resulting flux limit is 2×10^{-29} W/m^2, more than two orders of magnitude better than the best current systems.

Comparing this flux limit to the values in Table 5.2 on page 130, we see that this survey could detect any but the first, tiny beacon out to respectable distances. At 200 ly we could detect approximately 1 GW EIRP, and less than 1 MW at the distance of the nearest star. With an ideal pulse detector, the SKA could detect pulses from a LEO radar (Table 5.4, page 131) out to 10 ly, and the pulses from the postulated NEAR mission anywhere in the Galaxy. But this sensitivity is almost surely to no avail, since radar leakage will rarely be aimed at us and the SKA does *not* observe the whole sky at once as the OSS does.

Another good idea is to look at each star, at each frequency, more than once. First, the hoped-for transmitter may have a duty cycle which is high, but less than 100%. Second, for the more distant targets, interstellar scintillation will usually reduce the observed flux, but sometimes will increase it. Several 'looks' gives us a better chance of *benefiting* from ISS [Cor91c].

With these assumptions, with 80% throughput, and with 10 beams used for SETI, this first survey will take about 3.6 years to complete.

Table 6.5 summarizes the survey parameters. Figure 6.15 on page 184, shows how much better the result would be than the current observing program of Project Phoenix, using the simple measure of search merit, namely the number of stars times the number of octaves of frequency bandwidth observed. This first cut survey improves on Phoenix by more than two orders of magnitude.

Parameter	Value
Ae	5.0×10^5 m^2
T_{sys}	25 K
Bin Width	0.01 Hz
Integration Time	1,000 s
N Incoherent	10 spectra
Threshold	11 σ
	1.2×10^{-23} W
Detectable Flux	2.4×10^{-29} W/m^2
Number of Stars	10^5
Lowest Frequency	1 GHz
Highest Frequency	4 GHz
Detector Search Bandwidth	1 GHz
Number of Repeat Looks	3
Number of SETI Beams	10
Total Time for Survey	3.6 yr

Table 6.5: Parameters for First-Cut SKA Survey.

The fact that the SKA targeted search has a poorer figure-of-merit than the 1hT targeted search for transmitters more powerful than 10^{15} W EIRP, is the result of the smaller beam of the SKA. All the background stars in the beam to the edge of the Galaxy are already detected. It should be feasible to operate the SKA as 100 independent stations with broader beams for this targeted search mode. This would be the equivalent of 100 parallel 1hT arrays.

In subsequent surveys we can expect many more than 10^6 good targets to be identified, with ranges beyond 1 kpc. The GAIA project, for example, expects to produce a catalog of at least 5×10^7 stars with accurate distances and spectral types before 2010. In addition to making possible an even better search, this larger target list would permit the more efficient sharing of "beams for SETI" with "beams for other purposes", especially at the higher frequencies.

6.8.2 Design of the SKA

The approximately 10^6 m^2 area of the SKA would be distributed over 30 to 100 stations, each with a diameter of 200 to 300 m, in a region 100 to 300 km in diameter. Outliers could be as far away as 1,000 km. Many concepts for the economical design of the elements for these stations have been proposed. Some examples are listed below.

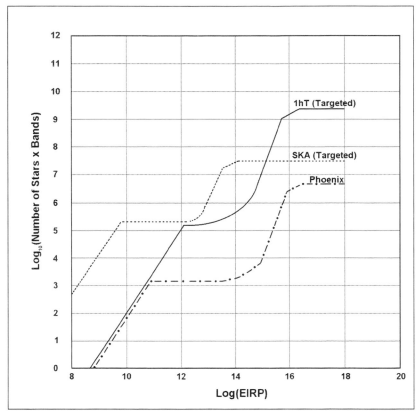

Figure 6.15: A Simple Measure of Search Merit for Targeted Searches.

- At one extreme, a pure phased array consisting of many low-gain elements (similar to the OSS described in Section 6.6, page 162) is a promising concept. It permits independent multibeaming over a large part of the sky and potentially offers the opportunity for RFI suppression by means of adaptive beamforming. The NFRA in The Netherlands is working on this concept.

- The SETI Institute is investigating the use of small, inexpensive, parabolic dishes of the type used for commercial Television Receiving Only (TVRO) applications.

- The National Center for Radio Astrophysics (NCRA) in India is investigating the extension of the rope truss, wire mesh, parabolas used in the Giant Metrewave Radio Telescope (GMRT) to higher frequencies (2 GHz), with an element approximately 30 m in size.

- The Commonwealth Scientific Industrial Research Organization (CSIRO) in Australia is exploring options including an array of Luneburg lenses which will give multiple instantaneous beams over a large area of sky.

- A novel way to implement a 200 m station is to use a large, almost flat, ground supported parabolic reflector with a very large, Focus/Diameter (F/D) ratio. The surface panels would be moved over a small vertical range for steering. At the prime focus, high off the ground, a phased array feed system could be supported by an aerostat. This approach is being developed in Canada by a team under the auspices of the Herzberg Institute.

- At the other extreme of element size, a single station could consist of one Arecibo-like spherical reflector, possibly with an active surface to eliminate aberrations. This approach is being studied by a team led by members of the Beijing Astronomical Observatory.

The details of the ten or more concepts that have been suggested, and the substantial amount of design work that has already been done, are beyond the scope of this book. The feasibility, and especially the costs, of the various designs are still uncertain. Nevertheless, we have attempted a crude, preliminary analysis of a range of array options with respect to their application to SETI. Table 6.6 summarizes the eight options that were analyzed. It is followed by eight lists of assumptions, one list for each option.

Type of Element	Diameter of Element (m)	Number of Elements	Frequency Range (GHz)
Tile	2.3	1.8×10^5	0.2 – 1.4
TVRO	3	3.4×10^5	1.0 – 20
Fiberglass	7	2.4×10^4	0.4 – 15
VLA	25	873	0.2 – 20
GMRT	25	2716	0.2 – 1.7
Parabola	100	55	0.1 – 20
Canadian	200	22	0.2 – 20
Arecibo	300	8	0.2 – 10

Table 6.6: Options for the SKA.

Assumptions for the Tile Element

- A tile consists of approximately 250 elemental receptors, plus beamforming circuits.

- A_e will be larger at low frequencies and smaller at high frequencies.

- Upper frequency may be limited to 1.4 GHz or 2 GHz, at reduced A_e.

- Room temperature LNAs use the same assumptions as for the OSS, which may be optimistic.
- The element and receiver costs are NFRA goals, which may be optimistic.

Assumptions for the TVRO Element

- Use a single log-periodic feed as described in Appendix D on page 329.
- Limit the minimum frequency so that $D > 10\lambda$.
- Small commercial TVRO dishes should reach 20 GHz.
- T_{sys} is taken from Appendix D on page 329.
- The element cost is based on the 1997 retail unit cost, plus twice for the mount.
- The LNA cost is $600, from Appendix C on page 321, slightly reduced for larger quantity orders.

Assumptions for the Fiberglass Element

- Use a fiberglass dish, commercially made for CATV 'head-ends'.
- Assume three horn feeds, because Ap_{eff}=60% may not be ideal for A_e/T_{sys}.
- Limit the minimum frequency so that $D > 10\lambda$.
- Limit the maximum frequency to the surface accuracy of large commercial dishes.
- The cost of the element is high enough to justify 80 K cryogenics.
- T_{sys} is taken from from Appendix C on page 321, plus 10 K for uncooled, Ortho-Mode Transducers (OMT).
- The element cost is based on the 1997 unit cost, plus twice for the mount.
- The cost of three LNAs and an 80 K dewar are from Appendix C on page 321.

Assumptions for the VLA Element

- Use a standard construction Cassegrain telescope.
- Assume total Ap_{eff} is 70%, based on shaped reflectors with three horn feeds.
- T_{cryo} was 20 K since the element cost is very high.
- T_{sys} taken from Appendix C on page 321, with cooled OMTs.
- The element cost is taken from NRAO model.
- The receiver cost is taken from Appendix C on page 321.
- N beams are determined by the number of elements per station.

Assumptions for the GMRT Element

- Use a GMRT rope truss with wire mesh improved for 1.7 GHz operation.
- Ap_{eff} assumes prime focus with three horns.
- T_{cryo} is 80 K because 20 K is too expensive for this cost element.
- T_{sys} was from Swarup's talk at the 1997 Sydney meeting [Swa00].
- The bandwidth is reduced because of the limited frequency range.
- N beams are determined by the number of elements per station.

Assumptions for the Parabola Element

- Use a standard construction Cassegrain telescope.
- Assume total Ap_{eff} is 70
- T_{cryo} was 20 K since the element cost is very high.
- T_{sys} is taken from Appendix C on page 321, with cooled OMTs.
- The element cost is taken from the NRAO model.
- The receiver cost is taken from Appendix C on page 321.
- This has a single very small beam, but it might be possible to use a multibeam feed to produce multiple beams.

Assumptions for the Canadian Element

- Ap_{eff} is used for an 'under-illuminated' primary to reduce spillover.
- Assume Cassegrain for high frequency, and prime focus array feed for low frequency.
- T_{cryo} is 15 K for Cassegrain, ambient for prime focus.
- T_{sys} boosted by 10% by spillover onto ground.
- The element cost is \$300 per geometric m². The goal is 100 to 400 elements.
- A 25 m feed dish per station for Cassegrain operation is included.
- The receiver cost is multiplied by 10 for a phased array at low frequencies.
- This has a single very small beam, but it might be possible to use a multibeam feed to produce multiple beams.

Assumptions for the Arecibo Element

- A_{eff} is that expected for Arecibo after its upgrade.
- The maximum frequency is based on the Arecibo Observatory panels.
- The element cost is based on the National Atmospheric and Ionospheric Center (NAIC) study of southern AO and it may be optimistic.
- The assumed facility is less expensive due to the small number of sites.

In this analysis, we have spanned the range of proposed element diameters, without trying to optimize a given concept. We have also made certain global assumptions, which are:

Figure 6.16: Dr. Douglas Bock, fresh from Australia, is a new recruit at the Radio Astronomy Lab at the University of California at Berkeley. (Photo courtesy of D. Bock.)

- Eventually there will be about 10^6 stars that are known to be good SETI targets. This is ten times more than what was assumed for the first cut, straw man, SKA SETI survey in the previous section.

- Each system will achieve $A_e/T_{sys} = 2 \times 10^4 m^2/K$.

- SETI will have the use of 10% of the instrument, either as 10% of the beams or 10% of the time for single beam options.

- There will be 100 stations, except when the element size does not permit this number.

- The cost of calculations will be \$2 per Gop/s, which is appropriate about a decade in the future.

Tables 6.7 and 6.8, on pages 190 - 191, show the assumptions and results of the analysis for each of the eight options.

Option	Tile	TVRO	Fiberglass	VLA
Element Diameter (m)	2.3	3	7	25
Total Number of Elements	1.8×10^5	3.4×10^5	2.4×10^4	873
Elements per Station	1.8×10^3	3.4×10^3	243	9
Aperture Efficiency	0.80	0.50	0.60	0.70
Element $A_{eff}(m^2)$	4.00	3.5	23	344
Frequency Range (GHz)	0.2 to 1.4	1.0 to 20.0	0.4 to 15.0	0.2 to 20.0
Wavelength Range (cm)	150 to 21	30 to 1.5	75 to 2	150 to 1.5
Primary Beam FWHM (deg)	120 to 120	7.2 to 0.36	7.2 to 0.20	4.3 to 0.04
Stars per Primary Beam	3×10^5 to 3×10^5	977 to 2	977 to 1	352 to 0
Instantaneous Bandwidth (GHz)	0.5	1	1	1
T_{cryo}	RT	RT	80	20
T_{sys}	35	60	28	15
Total Effective Area (m^2)	7.0×10^5	1.2×10^6	5.6×10^5	3.0×10^5
$A_e/T_{sys}(M^2/K)$	2.0×10^4	2.0×10^4	2.0×10^4	2.0×10^4
N Beams (Total)	100	100	100	9
N Beams (for SETI)	10	10	10	1
Station Beamforming Method	2D FFT	2D FFT	2D FFT	Direct
Station Beam Operations per Element (ops/s)	6×10^{10}	1×10^{11}	1×10^{11}	1×10^{11}
Array Beamforming Methods	FFT&Direct	FFT&Direct	FFT&Direct	FFT&Direct
Array Beamforming Operations (all beams) (ops/s)	3×10^{15}	6×10^{15}	6×10^{15}	5×10^{14}
Element Cost (each) ($ 1997)	500	800	8,200	1.1×10^6
Receiver Cost per Element ($)	750	550	5,000	25,000
Other Cost per Element ($)	130	500	1,000	10,000
Station Beamforming Cost per Element ($)	800	282	190	210
Total Element Cost (M$)	88	272	199	936
Total Receiver Cost (M$)	131	187	121	22
Total Other Cost (M$)	23	170	24	9
Total Station Beamforming Cost (M$)	140	96	5	0
Array Beamforming Cost (M$)	6	12	12	1
Facility Cost (M$)	50	50	50	50
Total Cost of Array (M$)	432	774	399	1,017
FOM1: $A_e/(T_{sys}Cost)$ [VLA = 1]	33	18	36	14
FOM2: SETI [Cyclops = 1]	4,090	7,023	16,162	717

Table 6.7: Comparison of Various Array Concepts – Part 1.

Option	GMRT	Parabola	Canadian	Arecibo
Element Diameter (m)	25	100	200	300
Total Number of Elements	2,716	55	22	8
Elements per Station	27	1	1	1
Aperture Efficiency	0.60	0.70	0.50	n/a
Element $A_{eff}(m^2)$	295	5,498	15,708	30,000
Frequency Range (GHz)	0.2 to 1.7	0.2 to 20.0	0.2 to 20.0	0.2 to 10.0
Wavelength Range (cm)	150 to 18	150 to 1.5	150 to 1.5	150 to 3
Primary Beam FWHM (deg)	4.3 to 0.51	1.07 to 0.01	0.54 to 0.01	0.30 to 0.01
Stars per Primary Beam	352 to 5	22 to 0	5 to 0	2 to 0
Instantaneous Bandwidth (GHz)	0.5	1	1	1
T_{cryo}	80	20	20	20
T_{sys}	40	15	17	15
Total Effective Area (m^2)	8.0×10^5	3.0×10^5	3.4×10^5	2.4×10^5
$A_e/T_{sys}(M^2/K)$	2.0×10^4	2.0×10^4	2.0×10^4	1.6×10^4
N Beams (Total)	27	1	1	1
N Beams (for SETI)	3	0.1	0.1	0.1
Station Beamforming Method	Direct	n/a	n/a	n/a
Station Beam Operations per Element (ops/s)	2×10^{11}	0	0	0
Array Beamforming Methods	FFT&Direct	FFT&Direct	FFT&Direct	FFT&Direct
Array Beamforming Operations (all beams) (ops/s)	8×10^{14}	2×10^{13}	3×10^{12}	4×10^{11}
Element Cost (each) ($ 1997)	7.5×10^4	4.5×10^7	9.4×10^6	1.0×10^8
Receiver Cost per Element ($)	6,000	25,000	6×10^6	25,000
Other Cost per Element ($)	2,000	10,000	10,000	50,000
Station Beamforming Cost per Element ($)	326	0	0	0
Total Element Cost (M$)	204	2,437	204	800
Total Receiver Cost (M$)	16	1	22	0
Total Other Cost (M$)	5	1	0	0
Total Station Beamforming Cost (M$)	1	0	0	0
Array Beamforming Cost (M$)	2	0	0	0
Facility Cost (M$)	50	50	50	30
Total Cost of Array (M$)	276	2,489	276	831
FOM1: $A_e/(T_{sys}Cost)$ [VLA = 1]	52	6	52	14
FOM2: SETI [Cyclops = 1]	1,908	34	302	68

Table 6.8: Comparison of Various Array Concepts – Part 2.

The following list gives more details about the rows in Tables 6.7 and 6.8.

Element Diameter (m)	Input variable.
Total Number of Elements	$(2 \times 10^4\ m^2/K)(T_{sys})/(Element A_{eff})$
Elements per Station	$(Number\ of\ Elements)/100$
Aperture Efficiency (Ap_{eff})	Input variable.
Element $A_{eff}(\mathbf{m^2})$	$Element\ A_{eff} = (Ap_{eff})(\pi/4)(Element\ Diameter)^2$
Frequency Range (GHz)	Input variable. The specification for the SKA (main array) is 0.2 to 20 GHz.
Wavelength Range (cm)	$wavelength\ =\ c/frequency$
Primary Beam FWHM (deg)	$(factor)(Wavelength)\ /\ (Element\ Diameter)$, where $factor\ =\ 1.25$ for prime focus, and $factor\ =\ 1.06$ for Cassegrain focus.
Stars per Primary Beam	$(\pi/4)(beam)^2(number\ of\ targets\ in\ sky)/(4\pi)$
Instantaneous Bandwidth (GHz)	Input variable. Instantaneous bandwidth.
$\mathbf{T}_{cryo}$	Input variable. Cryogenic cooling temperature.
$\mathbf{T}_{sys}$	Input variable. System Temperature.
Total Effective Area (m²)	$A_e(Number\ of\ Elements)(Element\ A_{eff})$
$\mathbf{A}_e/\mathbf{T}_{sys}(\mathbf{m^2/K})$	The specification for the SKA is $2 \times 10^4 m^2/K$.
N Beams (Total)	Input variable. This is the total number of beams. The specification for the SKA is 100. When the FFT forms more than 100 beams, the excess are discarded.
N Beams (for SETI)	$(N\ Beams\ Total)(SETI\ Fraction)$, where $SETI\ Fraction$ represents either the fraction of the beams used for SETI in the multibeam case, or the fraction of the available observing time used for SETI in the single beam case.
Station Beamforming Method	FFT or Direct complex weighted summation.
Station Beam Operations per Element (ops/s)	Operations per second per element, both polarizations. $= 2 \times 6 \times Log_2(Elements\ per\ station)(BW)$ for FFT $= 2 \times 6 \times (N\ Beams)(BW)$ for Direct
Array Beamforming Methods	Cross-correlation or Direct.
Array Beamforming Operations (all beams)(ops/s)	Operations per second, for all beams, for both polarizations. $= (NBeams) \times 2 \times 6 \times (Number\ of\ Stations)^2/2 \times BW$ for Cross-correlation $= (NBeams) \times 2 \times 6 \times (Number\ of\ Stations) \times BW$ for Direct

Element Cost (each)($ 1997)	Input variable. Cost of element antenna.
Receiver Cost per Element ($)	Input variable. Cost of LNA and postamp.
Other Cost per Element ($)	Input variable. Cost of A/Ds and other electronics for the elements.
Station Beamforming Cost per Element ($)	Cost per element to form a station beam. = (Station Beam Ops per Element)($2 per Gop/s)
Total Element Cost (M$)	(Number of Elements)(Element Cost)
Total Receiver Cost (M$)	(Number of Elements)(Receiver Cost)
Total Other Cost ($)	(Number of Elements)(Other Cost)
Total Station Beam-forming Cost (M$)	= (Number of Elements)(Station BeamformingCostper Element)
Array Beamforming Cost (M)$	(Array Beam Form Ops)($2 per Gop/s)
Facility Cost	Input variable. Cost of physical plant.
Total Cost of Array	(Total Element Cost) + (Total Receiver Cost) + (Total Other Cost) + (Total Station Beam Form Cost) + (Array Beam Form Cost) + (Facility Cost)
FOM1	The simplest figure-of-merit for the array, which is $(A_e/T_{sys})/Cost$ normalized to the VLA value.
FOM2	The figure-of-merit for SETI observing is defined below.

The last two rows of Tables 6.7 on page 190 and 6.8 on page 191 show figures-of-merit values. These values are from two attempts to assess the relative merits of the different designs. The first, simple figure-of-merit (FOM1) is just (A_e/T_{sys}) for the array, divided by the total cost of the array, and normalized to the VLA value. This parameter ignores the bandwidth, the frequency range, and the number of available beams, and these are all very important to SETI. The second figure-of-merit (FOM2) tries to take some of these parameters into account. It is defined as follows:

$$(6.1)$$

$$FOM2 = \frac{(A_e/T_{sys})(N\ SETI\ Beams)(Inst.\ BW)Log(F_{high}/F_{low})}{(SETI's\ Share\ of\ Cost)}$$

The equation is based on the following assumptions:

A_e/T_{sys}	Sensitivity is good.
N SETI Beams	More beams are like more telescopes.
Inst. BW	More beam-width is good. We can trade off the size chunk of spectrum we can handle in one observation against the number of beams. The assumption is that we process the entire available bandwidth, but we think that is reasonable.
$Log(F_{high}/F_{low})$	The more spectrum that we can search, the better. If we can search a significant portion of the spectrum of a target, then we have had a chance to find at least some of its brighter leakage sources. Thus we get diminishing returns by searching in frequency without limit. We have tried to capture this view by introducing a term that covers how many decades of spectrum are available.
SETI's Share of Cost	Cost is bad. However, we do not plan to pay for the whole array. We assumed that we would pay for 10% of the cost and be able to use 10% of the beams.

The FOM2 values in the tables are normalized to the parameters of the original Cyclops system proposal. Although the Cyclops system would have been far more sensitive than the SKA, it would not be very cost-effective compared to any of the modern designs.

The model parameters used in the analysis of Table 6.6 on page 185 are still quite uncertain. The costs, in particular, might vary by factors of over two. Still, we can draw some tentative conclusions:

- Big, parabolic dishes are still too expensive.
- A pure phased array is too expensive, at $2 per Gop/s, to use at microwave frequencies over about 1.4 GHz. Thus, it would need to be coupled with another array that could reach the upper frequencies economically.
- The GMRT concept has an attractive cost, but limited frequency range. It might be adequate for purely SETI work.
- The Canadian aerostat concept is also promising, but the costs are even more uncertain than for most of the other designs. Moreover, the extremely small beam is unattractive for SETI work.
- Again, while promising, the costs for the Chinese KARST concept are very uncertain, and its small beam is not very useful for SETI work. (KARST is the Chinese version of the SKA, named for its location in a karst formation.)

Small commercial dishes seem the most promising for SETI. This option is discussed in more detail in Appendix E on page 335, where the optimum seems to be a dish of about 7 m diameter with an 80 K cryogenic LNA. The 3 meter dishes have the advantage of larger primary beams, but appear at present to be more expensive for a given A_e/T_{sys}. At the high frequency end of the band, it will probably not be practical to use all of the "SETI beams". Thus, SETI surveys at frequencies above a few GHz may need to wait until target catalogs are available with substantially more stars than the 10^6 assumed here.

Figure 6.17: The Netherlands Foundation for Radio Astronomy's Dr. Harvey R. Butcher in a classic pose. (Photo courtesy of H. Butcher.)

In addition to the small, inexpensive, commercial dishes available today, this option requires inexpensive, reliable mounts; an inexpensive signal interconnection system; and inexpensive, reliable 80 K cryogenics. It may be possible to exploit the inexpensive linear actuators used on small TVRO dishes for the mounts, but it seems more likely that a fully customized mount will need to be developed. The cost of developing a fully customized mount is the main risk factor in this design option. There are many possibilities, but we do have the square-cube law working on our side. The ideal interconnection system would use analog optical fiber. While already available, we need an order-of-magnitude reduction in their cost before they become practical.

The prospect of so many cryogenic systems is daunting, but perhaps not more daunting than the prospect of 300,000 moving antennas without cryogenics! Fortunately, unlike the situation for 20 K cryogenics, there is a widespread commercial demand for inexpensive, reliable 80 K cryogenics for wireless communication, sensors, computing, and other applications. Units claiming a service life in excess of five years are already available, which corresponds to 5 to 10 replacements per day for the array. With proper design for easy swapping of the cryogenics system, such a replacement rate would not require a large labor force. There are also designs in the research stage that have no moving parts and the potential for an extremely long service life.

Design, software, and integration costs of the SKA are likely to be large, but seem unlikely to exceed 10% to 20% of the capital costs shown in Table 6.6 on page 185.

6.8.3   Computation for the SKA

Some thought was given to the idea of distributing the computing functions within a large integrated array, which performs both signal capture and processing. The most obvious major difficulty is that in order to achieve the maximum benefit from the incoming signal, it is necessary to sum all the signal on the array at a single point. Signals could be digitized or even transformed (FFT) at individual dishes and returned to the central point on optical fiber. Except for the complicated filtering and connectivity problems, this could alleviate the need for very high sample rates.

Two problems, however, become apparent with this scheme. First, the antenna temperature is critically dependent upon the cooling at the amplifiers. Depending upon how many antenna elements are used, cooling to cryogenic temperatures might become impractical. Thus, instead of having antenna temperatures of 20 K, we might have to make do with temperatures twice or three times this high, which would require a proportionate increase in the number of elements for phasing. It is not clear where the trade-offs are in this case, and additional study is needed to define the optimum configuration.

Secondly, if filtering and digitizing occur at the antenna's array elements, then the distributed microchips performing the task might emit frequencies in the range from 1 to 10 GHz. This would add RFI within the bands of interest to SETI. It is possible, however, that the processing required for mechanical control of the antenna subelements could be distributed, because the necessary processing speeds are not likely to be in the range from 1 to 10 GHz. Thus, it seems that the whole idea of distributed processing requires a short study of its own.

Beamforming for the SKA

To get an idea of where the large computing demands may lie for the next generation of SETI hardware, consider a special case for the antenna array. Assume that there are 100,000 elements, each element with an area of 10 m^2. Each element would be about 3 m in diameter, and have a circumference of about 10 m. (Clearly, the numbers are being rounded off for simplicity of understanding.)

Consider the gain of such an antenna at 3 GHz or 10 cm. The gain of a circular antenna, such as a satellite dish, is the square of the number of wavelengths in the circumference, about 100 in this case. Thus, each individual antenna element has a gain of 10,000 and sees approximately 0.0001 of the sky.

The gain of an array is proportional to the number of elements phased together. This is because the signal amplitude grows as the number of antennas focused on a desired source, while the individual noises add incoherently, which yields a square root growth behavior. Thus the S/N ratio, measured in power, grows as the number of elements, or as the area of the antenna.

This is not true, incidentally, for *optical* SETI, where the major noise is not separate for each antenna. Rather, it is a stellar point source that is background noise to an intelligent transmission. In the optical case, the number of photons, the power, and the S/N all grow as the square root. In the radio case, however, the overall antenna gain is a billion, or 100,000 times that of the individual antennas.

This immense gain requires two things. First, all the antennas must point to the same 0.0001 part of the sky. Second, their signals must be added together to focus on one particular source within the primary beam of the antennas elements. In this case, focusing means adding up presumably digitized data streams to sufficient accuracy. For example, if we want to focus with frequencies as high as 3 GHz with some safety margin, then we sample the data from each antenna element at 10^{10} samples per second, forming one sample stream. To form one beam, we must add 100,000 sample streams with a particular set of weights, which implies about 10^{15} operations per second.

Further, suppose that we want to form all the beams for all the target stars within 1,000 ly (about a million solar-type stars). Then we need an average of about 100 beams per element of sky for our small antennas. This would require about 10^{17} computations per second for the beamforming operation.

The computation per beam for a gigachannel FFT with 1 Hz channels requires about $10^9 \times 30 \times 4$ or 10^{11} operations per second. This is a factor of 10,000 below the computation required per beam for beamforming, and thus can be neglected. (Signal processing is usually a couple of orders of magnitude less than this smaller number.) Thus, if we can afford the beamforming computation, we can afford the other computational load.

Today, one million operations per second costs $1.00 in commercial processors and $0.10 in special purpose DSPs. The MIPS per dollar increases as $e^{0.5t}$. Hence, within 20 years, we should expect a decrease of 10,000 in cost of hardware, and hence in a commercial processor, 10^{10} operations per second per dollar. Thus, even a conservative 10^{18} operations per second of beamforming will cost about $100 M using commercial processors, and somewhat less if special purpose DSPs are used. This is considerably less than the total mechanical costs for the antennas and mounts. Thus, computing costs appear to be feasible for the SKA.

The powerful SKA receiving array could be used in reverse to act as a beacon. At the focus of each antenna a digitally controlled signal source could be centrally synchronized. The result can be estimated by simply time reversing the receiving process. Each of 100 beams would get a digitized stream that would focus power on one of a million stars, and the sky would be scanned a hundred stars at a time. If each digitized signal source radiated 10 W, then each beam would get 0.1 W per element or 10 kW in all. The gain of this signal would be a billion, yielding an effective radiated power equal to that of Arecibo.

Today, in our most sensitive searches, we could detect Arecibo at distances greater than that of the nearest million stars. If we assume that a civilization used two arrays, or even one array, half for transmitting and half for receiving, this would be a search indeed. Thus, while optimism about leakage is lost in a morass of growing signal complexity and decreasing average power, optimism about beacons has some technical basis.

6.8.4   One Hectare Telescope and SKA

It is clear that the SKA is an ambitious and expensive project that will take some considerable time to realize. How can the SETI Institute help? What should it be doing in the meantime? In this section, we discuss one possibility, which is to construct a prototype of one of the SKA stations. Building a prototype could accomplish two goals. First, it could validate the SKA concept by demonstrating that a large collecting area could be built far more inexpensively than has been possible in the past. Second, it would provide a dedicated SETI facility, from which the expected improvements in detectors and target lists could be exploited.

Consider an array that is equivalent to a 100 m antenna (55% aperture efficiency), which is about the area of one of the stations of the SKA as described above. Since the area is about one hectare, we call this system the 1hT. At present, the most likely technology appears to be the small parabola option. It is close to technical readiness, so the 1hT could probably be designed and prototyped in less than three years, and undergo its first tests by about 2005.

This relatively small system would not fully benefit from the cost savings available to the large-scale SKA. Still, it seems likely that it will cost less than $10 M, based on the estimates of Table 6.6 on page 185. Detector costs are likely to be insignificant compared to the cost of the array itself. The 1hT needs to have specifications close enough to those of the SKA to prove the concepts, but they can be relaxed a little in order to control costs. For the purposes of this discussion, we adopt a frequency range of 1 to 10 GHz, an instantaneous bandwidth of 500 MHz, and 3 beams for SETI (with perhaps others dedicated to astronomical studies). This would enable us to analyze a targeted survey of 10^5 stars similar to that proposed on page 181 in Section 6.8.1 for the full SKA.

The parameters for the 1hT are summarized in Table 6.10. Its figure-of-merit, derived in Appendix F on page 339, is plotted in Figure 6.15 on page 184, where it can be seen to be a substantial improvement over Project Phoenix.

Project Phoenix uses similar collecting areas, so the gain in sensitivity is primarily due to using narrower detection bandwidths at the cost of losing any drifting CW signals. If we added a drifting detector, the 1hT would have a sensitivity about equal to that of Phoenix. The assumed target list is 100 times that used for Phoenix, which raises the merit at the 'plateau' region (where we run out of targets) by the same factor. A side benefit of the much larger number of observations is that the background survey (of non-target stars that happen to lie in the beam) is also much improved over Phoenix. Indeed, we would cover about 1% of the entire sky in the course of such a targeted survey.

Parameter	Value
A_e	4.3×10^3 m^2
T_{sys}	27 K
Bin Width	0.01 Hz
Integration Time	400 s
N Incoherent	4 spectra
Threshold	9 σ
	1.7×10^{-23} W
Losses	1 dB
Detectable Flux	4.9×10^{-27} W/m^2
Number of Stars	10^5
Lowest Frequency	1 GHz
Highest Frequency	3 GHz
Detector Search Bandwidth	0.5 GHz
Number of Repeat Looks	3
Number of SETI Beams	3
Total Time for Survey	6.3 yr

Table 6.10: Parameters for 1hT Targeted Survey.

6.9   Technology for Infrared/Optical Search Systems

This section discusses current and near-term future technologies for performing infrared/optical searches. The discussion is sometimes more extensive than for microwave searches, because the infrared/optical technology has changed much more than microwave technology. Exponential growth of technology has also characterized this wavelength regime. Today, our conclusions regarding the use of infrared/optical techniques for SETI are more favorable than they were in the past. Figure 6.18 on page 201 shows the relevant pathway through the SETI Decision Tree.

6.9.1   Introduction

Sections 3.2 on page 70 and 3.3 on page 74 presented the rationale for optical SETI and the variety of pulsed and CW signals an ETI civilization might emit. In this section, we describe the detectors and instrumentation required to implement searches for these signals.

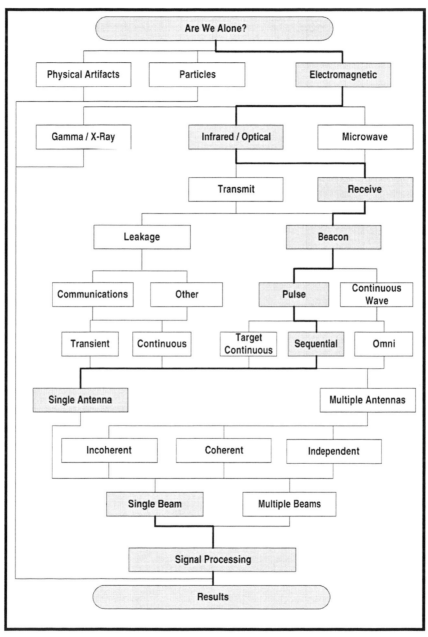

Figure 6.18: SETI Decision Tree – Infrared/Optical Search Issues.

Infrared and optical SETI experiments can be simple and inexpensive. To search for pulses there are several good detectors that can be used on existing telescopes. To search for continuous signals existing high resolution spectrometers can be used. At some wavelengths, even existing data can be 'mined' as part of the search.

Figure 6.19: Harvard University's Dr. Paul Horowitz' SETI interests encompass the entire spectrum.

6.9.2 Detectors

The information available to a photon detector consists of:

- the times of arrival of the photons in the incoming light beam;
- the X and Y positions of the photon arrivals;
- the wavelengths or energies of each of the arriving photons;
- the polarization state of each photon.

Ordinary single-channel sensors, such as photomultiplier tubes, photodiodes, avalanche photodiodes, etc., have serious limitations in recovering all of this information. The detector bandwidth sets a limit on how accurately the arrival times of the quanta can be recovered, and the X and Y positions are ignored. Except for the Superconducting Tunnel Junction (STJ), the photon energy is also ignored.

One way to overcome some of these limitations is to use a position-sensing array detector. They report the X and Y positions, as well as the arrival time of each photon. However, the simplest instrumentation for pulse searches uses single element detectors. These inexpensive experiments, using simpler instrumentation, should be carried out before implementing more complex systems that use array detectors.

6.9.3 Pulse Search Instrumentation

A high intensity pulsed laser, e.g., 4 MJ in 1 ns, at 1 μm wavelength, teamed with a 10 meter telescope used as a transmitting aperture would form a powerful and efficient interstellar beacon. If used by a civilization to send light pulses to a set of chosen stars, it could be clearly detected at those targets. Its light would outshine the host planet's unfiltered starlight by a factor of several thousand during the laser pulse. At a range of 10,000 ly, such a beacon would deliver an ample signal, of about 25 photons per pulse, to another 10 meter telescope used as a receiver.

Such a signal would stand out significantly from the background starlight. Starlight would pepper a detector with a one-photon-at-a-time background, but it is an extremely rare event when many photons arrive within a nanosecond. Photons will arrive from a star with a Poisson distribution of arrival times. The probability, p, of N or more photons arriving within an interval t, is:

$$p = (t <r>)^N/N! \quad \text{(for } r < t > \ll 1)$$ (6.2)

where $<r>$ is the mean photon rate.

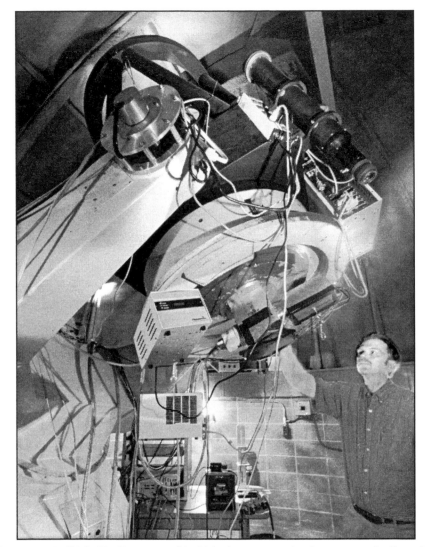

Figure 6.20: Dick Treffers, uses this 30 inch automated telescope at the Leuschner
Observatory at the University of California at Berkeley for an optical
SETI project.

When R is the *false alarm* rate of N or more coincident photons from background starlight, then:

$$R = <r>^N \times t^{N-1}/N! \tag{6.3}$$

To minimize the false alarm rate, the integration time of the instrument, t,
should be made as small as possible. Of course, the pulse is assumed to be
shorter than t. This can be done by using detectors, amplifiers, and electronics,

which all have very fast rise and fall times.

With current technology, pulse searches are considerably more sensitive than continuous signal searches. Continuous signals may have very narrow bandwidths, so the number of channels desired for a high sensitivity search is of the order of 10^{12}, much higher than presently available (see Section 6.9.9 on page 213). The number of drift rates to search can also be enormous at such high spectral resolution. Nanosecond pulses have intrinsically higher bandwidths. They may not require a spectrometer for detection, because high-power pulses could easily outshine a star during the pulse.

6.9.4 Collecting Photons

It is possible to consider using *less than optimal* mirrors, because there is no requirement to image the stellar target. It is only necessary to collect all of the photons that fall onto the telescope for analysis by the pulse detection systems. Optical 'light buckets', used to collect photons, are being produced for several different purposes, e.g., those built for Cerenkov detectors used for cosmic ray searches. (See Section 6.10.1 on page 220). There may be opportunities to use these same 'light buckets' for SETI work.

6.9.5 Single Pixel Pulse Detectors

In this section, we consider a variety of single pixel pulse detectors that could be used to detect nanosecond optical pulses. Table 6.11 summarizes the key characteristics of these detectors.

Although PMTs have low quantum efficiencies (10 to 25%), they have high gain, are very linear, and some types have fast rise and fall times (approximately 0.5 ns). PMTs must be cooled to achieve low dark count rates. Cooling is not needed for targeted SETI experiments however, because the target stars are very bright, and the dark count rate is overwhelmed by the star. These factors make PMTs very easy to use for an optical SETI pulse detection experiment. (See Section 6.9.6 on page 208.)

Avalanche Photodiodes (APD) have substantially better quantum efficiencies than PMTs, especially at red and near-infrared wavelengths. (See Table 6.11.) APDs have low gains in the linear mode, so high gain amplifiers are needed. Unfortunately, high bandwidth amplifiers produce substantial noise, making it extremely difficult to detect a single photon event from an APD operating in the linear mode. (This could only be accomplished with a low-noise, cooled amplifier and a long time constant.)

Some APDs can be operated in the Geiger mode, where the gain is high enough

Device	Quantum Efficiency	Gain	Rise Time	Notes
Photomultiplier Tube (PMT)	15%	10^6	500 ps	Low Quantum Efficiency (QE)
Avalanche Photodiode (linear mode)	80%	10^3	10 μs	Low gain
Avalanche Photodiode (Geiger mode)	65%	10^7	300 ps	Not linear
Hybrid Photocathode & Avalanche Photodiode Target	15%	4×10^4	5 ns	Excellent pulse height distribution, low QE
Solid-state Photomultiplier	80%	10^5	1 ns	Helium cooled, 0.4 to 28 μm
Superconducting Tunneling Junction	80%	0.02	1 μs	0.02 eV resolution, 100 mK cooling

Table 6.11: Single Element Photon Counting Detectors.

to overcome amplifier noise, but this mode saturates from a single photon. The detector is not linear, and one photon produces the same charge as many photons. However, an instrument with a beam splitter followed by two or more Geiger mode APDs could discriminate between one and many photons. (See the next section: Section 6.9.6, on page 208.)

Hybrid photocathode – avalanche diode detectors (e.g., Hamamatsu R7110U)– accelerate a photoelectron from a photocathode to an avalanche diode target, producing about 40,000 electrons per photoelectron. The resulting pulse height distribution is excellent, so it is easy to discriminate a single photoelectron event from multiple photoelectrons. Both PMTs and hybrid detectors use photocathodes, so they suffer from the same low quantum efficiency. Currently available hybrid detectors are not as fast as the faster PMTs. (See Table 6.11.) However, the rise time of hybrid detectors could probably be improved if manufacturers used smaller avalanche diodes, thereby reducing detector capacitance.

STJ detectors may be useful in future SETI experiments and are under active development. The most sensitive devices require cryogenic cooling to temperatures below 1 K, and hence require more complicated and costly cooling apparatus than higher temperature cryogenic systems. STJs consist of two layers of superconductor separated by an extremely thin insulating layer, a few atoms thick. An incident photon breaks Cooper pairs in the superconductor, and the charge is detected by measuring the current of electrons which tunnel through the insulator. STJs are sensitive from 0.1 to 2 μm and are expected to have quantum efficiencies approaching 100%. They can have a spectral resolving

Figure 6.21: Undergraduate Shelly Wright built the photomultiplier detector affixed to the back of the 1 meter telescope at Lick Observatory for use in its optical SETI experiment. This project involves the Lick Observatory, University of California at Santa Cruz, and the SETI Institute.

power in the visible of about 100, and detect photons at rates up to 100,000 photons per second. Further advancements in performance are expected, and STJ development programs are under way at places including the European Space Agency in Noordwijk, The Netherlands, Lawrence Livermore National Laboratory, and Stanford.

Technically, STJs are not bolometers, which work by exhibiting an electrical resistance change in response to heating by photons. However, transition edge superconductor detectors are being developed by several groups, and they show promise for high-energy resolution detection of individual photons.

Solid-state photomultipliers (SSPM) have excellent quantum efficiencies and gains in the infrared/optical, but need to be cooled to liquid helium temperatures. SSPMs have been arrayed, but are only commercially available as single element detectors. The SSPM was developed by Rockwell International Science Center (Petroff et al., [Pet87]). It is a SiAs detector with single photon counting capability over the wavelength range from 0.4 to 28 μm. The detector has an intrinsic time response on the order of 1 ns, and low detector noise. This photon-counting capability gives the SSPM great advantages over other infrared detectors in terms of both sensitivity and time resolution, making it the optimal detector to use to search for pulsed infrared laser signals. (See Section 6.9.7 on page 210.)

6.9.6 Dual Detector Coincidence Systems

The first generation of pulse searches should use single pixel detectors because they are simple and inexpensive. These experiments would lead to a second generation of pulse searches based upon array detectors.

A very simple targeted experiment to search for pulses in the visible could employ a single high-speed PMT at the focal plane of a telescope, followed by a fast amplifier and discriminator. The Russian Multichannel Analyzer of Nanosecond Intensity Alterations (MANIA) optical SETI program [Shv93], used a single PMT, and timed the arrival of each photon with a 50 ns resolution.

A single PMT detector system could search for large pulses (two or more photoelectrons within a few nanoseconds). There are significant problems, however, from occasional high amplitude background pulses. These are intrinsic to PMT detectors and create false alarms.

A typical PMT pulse height distribution for single photon events is shown in Figure 6.22 on page 209. The large peak in the distribution is from single photoelectron events. Because the PMT is a linear detector, if several photons hit the photocathode at once, e.g., within 0.5 ns, the pulse height is proportional to the number of photoelectrons. These large pulses, from multiple photoelectrons, are easy to detect using conventional discriminator circuits. (The threshold should be set above the pulse height for single photons.)

The unwanted high amplitude background pulses from a PMT detector (false alarms) can be seen in the tail of the PMT pulse height distribution (Figure 6.22). These occasional large pulses are due to scintillation in the PMT glass, cosmic rays, ion feedback, and the radioactive decay of K^{40} in the PMT glass. They can be minimized by using a box and grid structure PMT to reduce ion feedback, by using small tube PMTs to minimize glass area, by using positive high voltage to reduce scintillation, and by shielding the detector with lead. However, experiments show that the rate of unwanted large pulses cannot be easily reduced below about one per minute [Wer97].

A simple way to discriminate against these background events is to use a beam splitter that feeds the light from the telescope into a pair of detectors, followed by a pair of high-speed amplifiers, a pair of discriminators, and a coincidence detector (figure 6.23 on page 210). Such an instrument has been built and tested by Werthimer [Wer97], using a pair of 0.6 ns rise time photomultiplier tubes, 2 GHz bandwidth amplifiers and high-speed ECL comparators. The technique works well.

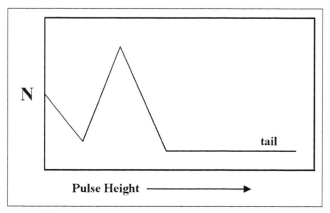

Figure 6.22: PMT Pulse Height Distribution.

The dual detector system described effectively eliminates background events from the detectors, but it cannot eliminate pulsed interfering events from the sky, e.g., very high-energy cosmic ray showers, lightning or artificial pulsed light sources. Preliminary tests by Werthimer [Wer97] indicates that such sky background interference is not seen by a system searching for nanosecond pulses. However, if further experiments show that the sky background is a problem, such events could be rejected by using a third detector looking at another nearby part of the sky. Events seen with this third detector, in coincidence with the 'on source' detectors, would be treated as false alarms.

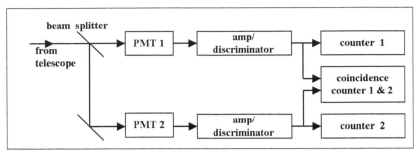

Figure 6.23: Simple Nanosecond Visible Pulse Detector for SETI.

Substantial, yet inexpensive, improvements can be made by placing A/D converters and 'time tagging' electronics on each detector channel. Present inexpensive technology can time-tag events to an accuracy of approximately 100 ps. More complex systems achieve resolutions of 10 ps. This would allow one to study:

- the pulse height distributions from each detector, to search for an excess of high amplitude pulses;
- the arrival time of each photon, to search for repetitive single photoelectron events;
- the lengths of times between photon arrivals, to search for non-Poisson distributions.

Paul Horowitz [Hor93] built a sophisticated electronics package for a dual detector system. It can measure four levels of pulse amplitude, arrival time, pulse width, and the overlap time of each coincident signal detected. This system is fed from a pair of hybrid APDs that view the target star via a beam splitter. Within the initial six months of operation on a 1.5 m telescope, about 4,000 observations were made of about 1,500 stars. This search was commensal with a radial velocity survey of solar-type stars.

6.9.7   Infrared Pulse Detection Systems

An SSPM is an ideal detector to use to search for pulsed carbon monoxide (CO_2) laser signals in the 10 μm band. A simple first-order SETI experiment would consist of a cold Fabry-Perot filter, with a resolution of 10^5 (3×10^8 Hz bandwidth) in front of the SSPM, and an output pulse width of 10 ns. The narrow bandpass, Fabry-Perot filter is necessary because the high thermal background photon emission from the telescope (10^8 photons/s/m^2 μarc second^2) would otherwise saturate the SSPM. The sensitivity of the SSPM would then be background limited. The disadvantage of this scheme is that it is limited to only one channel of bandwidth.

Eikenberry, Fazio, and Ransom [Eik96a] built a ground based, high-speed, single element, near-infrared photometer for pulsar research using a 200×200 μm back illuminated SSPM. Their results show a clean separation between single photon events and the low level electronic noise. Single photon events produced typical pulse amplitudes of approximately 110 mV, while the RMS noise was typically much less than 15 mV. By placing a discriminator with an appropriate level (typically around 40 mV), they were able to count the individual photons, thus measuring the photon flux with complete freedom from electronic or read noise. Note that the SSPM must be operated at temperatures near 6 K using a liquid helium dewar.

Another option is to use an SSPM at near-infrared wavelengths (1 to 2.2 μm). In that region, the thermal background is negligible, and high-power pulsed laser sources exist at 1.06 μm. The quantum efficiencies of commercially available SSPMs are low (a few percent), but custom devices could be made with very high efficiency in the near-infrared.

A final option is to use a two-dimensional array of SSPMs behind an echelle spectrograph. (See Section 6.9.9 on page 213.) This would permit a large number of high-bandwidth channels to be searched in a pulsed mode, and would be the ideal infrared SETI detector system. While two-dimensional SSPM arrays have been constructed, they are not yet commercially available.

6.9.8   Array Detectors for Pulse Detection

Silicon photodiode array sensors, or CCD sensors provide up to $2,000 \times 2,000$ pixels, and the number grows larger every year. Quantum efficiencies are high across the visible and near-infrared wavelength bands. However, the spatial resolution comes at the cost of response time, because the arrival time of a photon (or photon pattern) cannot be recovered to an accuracy better than the frame rate of the image. The frame rate is typically 1/30 s for commercial, uncooled, video cameras. For static targets this is not a limitation, but this may be a serious limitation for potentially modulated, 'information bearing', optical SETI signals.

Avalanche photodiode arrays are still very limited by the number of pixels available in a single device. To a lesser degree, they are limited by the output connector and downstream electronics.

STJ array detectors are similarly limited by the number of available pixels. As this is written, the largest functioning array is the European Space Research and Technology Centre (ESTEC) 3×3, whose center pixel is dead. Nonetheless, the development of STJ arrays is of great interest, because the concept is inherently able to recover photon energy, and the time resolution is good in a fully parallel implementation.

Microchannel Plates (MCPs) make it possible to build array detectors with large 1D and 2D spatial resolution, while maintaining fast response speed. These position sensitive detectors report X and Y values promptly, using an anode that encodes the position of the centroid of the amplified photoelectron cascade. Using the best available readout system, the planar delay line, 250×250 pixel fields can be recovered, with a timing precision on the order of 20 ps [Lam91]. However, the position encoder and decoder limit the detector throughput to maximum count rates on the order of 10^6 - 10^7 counts per second. This can restrict their usefulness to stars fainter than $m_v = 5$ on a 1 m telescope. MCP detectors have quantum efficiencies from about 10% to 25% in the visible.

Transition Edge Sensors may take advantage of MEMS technology, and can be made into large arrays.

Positional information from array detectors can be used for optical SETI in one of three ways:

- **Targeted Search** – Individual stars are examined according to a schedule. One or both position coordinates could be made to correspond to wavelength by interposing a prism (dispersing the starlight in one dimension) or an echelle spectrometer (dispersing the starlight in two dimensions).

- **Untargeted Search** – Examine the dark sky, globular clusters, or galaxies. The positional information might be best used in a direct image mode. One could look for hot spots, from which multiple pulsed patterns appear, or for object points from which large amplitude multiphoton events originate.

- **Coincidence Observations** – This involves looking for coincident photo events between two detectors, and is limited by the accidental (random) rate of coincidences. Having spatial resolution in one or two dimensions can vastly reduce this random coincidence rate, because a significant 'event' becomes the product of several probabilities:
  - The two events must coincide in time, to within T_{resol}.
  - The two events must coincide in X, to within X_{resol}.
  - The two events must coincide in Y, to within Y_{resol}.

6.9.9   Searches for Continuous Signals in the Visible

In this section, we discuss the possibility of using high-resolution spectroscopy in the visible or near-infrared to detect continuous laser lines transmitted by ETIs.

High Resolution Echelle Spectrometers

High-resolution spectroscopy with large wavelength coverage in the visible and near-infrared is possible using echelle spectrometers. Recently they became available at many observatories. Echelle spectrometers have the following typical characteristics:

- The wavelength coverage is approximately one octave.
- The spectral resolution is $R = \nu/\Delta\nu \approx 5 \times 10^4$.
- The S/N per resolution element is ≈ 300. This is achievable in 10 minute exposures for the nearest 1,000 solar-type stars, using a 3-m class telescope.
- The time resolution is arbitrarily high, and limited by the detector. The readout time is several seconds for cooled 'scientific' CCDs, but there is a trade-off between shutter speed and the resulting S/N ratio in the spectrum.

Good reviews of the characteristics of echelle spectrometers are given by Pilachowski, et al. [Pil95] and Vogt, et al. [Vog94]. Throughout the world, about a dozen echelle spectrometers are commonly used on 3 m class telescopes. On the Keck 10 meter telescope, two echelle spectrometers are, or soon will be, available. They are the High Resolution (HIRES) spectrometer for the visible, and the Near-Infrared Spectrometer (NIRSPEC) for the near-infrared.

The advantage offered by echelle spectrometers is the large wavelength coverage and high resolution, which permit searching for laser lines at a wide range of wavelengths. The typical number of spectral resolution elements (number of channels) monitored simultaneously is $R \approx 5 \times 10^4$. The echelle spectrum is acquired within approximately 50 spectral orders, each containing approximately 1,000 resolution elements.

Laser–Line Detectability

As previously discussed, short pulses from a laser having modest average power may successfully compete against the stellar luminosity. In this section, however, we assume no pulsing advantage. (Perhaps the ETIs are not using a pulsed laser.) Instead, we consider the advantage offered by high spectral resolution to dilute the background noise from starlight.

An optical laser operating at frequency ν, can compete with the stellar luminosity, within a frequency bin $\Delta\nu$, in three ways:

- Ratio of solid angles where $\Omega_{Laser}/\Omega_{Star} \approx \lambda^2/D^2$. Here we assume a

   diffraction-limited laser beam at wavelength λ, from an emitting aperture
   of dimension D.

- Ratio of emitted spectral bandpasses $\Delta\nu/\nu$, where $\Delta\nu$ is the resolution of
  our spectrometer. We assume that the laser line is considerably narrower
  than the resolution of our spectrometer, so that all of the laser light falls
  onto one resolution element.

- The laser flux competes only with the photon noise of the stellar spectrum.

These three effects suggest that only a small fraction of the stellar luminosity
streams within the cone of the laser beam, and that it falls within the same spectral resolution element as the laser light. Our telescope subtends a small fraction
of that cone, yielding a stellar spectrum with an emission feature superimposed
due to the laser line.

The fraction, F, of the total stellar luminosity contained within the laser cone
and within $\Delta\nu$ is:

$$F = \lambda^2 \Delta\nu / D^2 \nu \tag{6.4}$$

We assume diffraction limited emission, and a stellar spectrum width on the
order of ν. Then, the cone contains a stellar power of:

$$P = FL_* \tag{6.5}$$

If we adopt, as a benchmark, the values of a 10 m transmitting telescope, operating at optical wavelengths, and $\nu/\Delta\nu \approx 5 \times 10^4$, then the fraction (F) is about
5×10^{-20}. If we adopt the solar luminosity as the benchmark stellar luminosity,
then the cone contains a stellar power P of $5 \times 10^{-20} \times 3.8 \times 10^{26}$ W $= 1.9 \times 10^7$
W.

However, it is only the photon noise of the detected stellar flux that competes
with the laser because the laser line will exhibit a distinct spectral width. It will
be narrower in frequency than the stellar features, which all exhibit thermal and
turbulent Doppler broadening. This broadening corresponds to approximately
10 km/s, and is resolved spectrally. However, the laser line will be broader
than cosmic ray hits directly on the detector (CCD or infrared array), because
cosmic ray hits do not reflect the point spread function (PSF) shape of the
spectrometer. Furthermore, the cosmic ray hits are not repeatable, whereas the
laser line must be repeatable if it is to be confirmed.

For the laser line to be detected, it needs to have a brightness that is no greater
than that of the star spectrum by some fractional amount within the spectral
resolution element. For a given S/N ratio of the stellar spectrum, and setting
a threshold of 4σ for the laser blip, that fractional amount is $4/(\text{S/N})$. This
discrimination threshold will not provide a definitive detection, but will serve
as an alert to search for repeats in subsequent spectra. Thus, the laser beam in

the cone can be detected if its total power into that beam is:

$$P = 4FL_*/(S/N) \tag{6.6}$$

Applying the benchmark numbers used above, along with a spectral $S/N = 300$, we find that the laser power necessary for a 4σ detection is:

$$P = 1.9 \times 10^7 \text{W} \times 4/300 = 3 \times 10^4 \text{W} \tag{6.7}$$

Thus, high-resolution spectroscopy can detect CW lasers having power of approximately 10^5 W for the nearest 1,000 solar-type stars if directed by a diffraction limited telescope. In deep stellar absorption lines such as the Ca II H and K or Na D lines, the contrast is improved by a further factor of five, but this is not included in these calculations.

A laser line detection program requires the repeated acquisition of spectra from the same star, and comparison with the spectra of other stars having the same spectral type. Time resolution is also possible by using fast shutter speeds, or by translating the detector during acquisition of the spectrum.

One open issue is the degree to which laser lines can be discriminated from the stellar spectrum, and hence where the noise threshold should be set. Another issue is the possibility of obtaining time-resolved echelle spectra. Existing spectra can be used for the initial reconnaissance of laser lines. The existing spectra can come from the various Doppler searches for planets, or from the spectral surveys of nearby stars.

Searches for Continuous Signals in the Infrared

A search for infrared signals has an advantage over visible searches for sources beyond about 1,000 ly because interstellar extinction is lower in the infrared, and the contrast to the primary star is greater. A wide variety of infrared searches for continuous signals could be done at this time, such as a simple line search using existing spectrometers. There are a number of fundamental lines in the infrared, such as the ground state, pure rotational line of molecular hydrogen at 28 μm.

The best spectral resolution to use for the observations would be equal to the line width of the transmitted signal. This would be limited by the Doppler width of the transmitter which would certainly be much less than the observed bandwidth of any existing spectrometer. The search should use the highest feasible spectral resolution to reduce the background from the primary star and the noise background from other sources, such as thermal emission from the telescope.

For the near-infrared region between 0.8 and 2.2 μm, much of the background is from OH night sky emission, which occurs in very sharp lines. Thus, observing with a resolution of greater than $R = 10,000$ between the lines gives a great S/N advantage [Mai93]. For example, the background between the OH lines approaches that of a cold telescope in space. Ultra-low dark current infrared detectors can be used, such as those being developed for the Next Generation Space Telescope (NGST).

Betz and Bester et al. [Bet93] used a heterodyne detector to search for signals in the 10 μm band without success. They observed 50 stars for 20 minutes each on the P8 and P16 lines of $^{12}CO_2$, using 450 channels each of 2.4 MHz bandwidth.

A number of searches could be done with existing spectrometers if we assume that the signal would be beamed to us in key spectral lines so much stronger than the stellar continuum that, even with a modest resolution of $R = 10,000$, the line would be equal to the continuum signal. Observations could be made on a number of nearby stars, the choice depending on the lines to be observed and the telescope and instrument to be used.

These projects could include ground-based observations using, for example, the Infrared Telescope Facility (IRTF) and the Cryogenic Echelle Spectrometer (CSHELL). They could observe at the many atomic and molecular lines in the region from 1 to 5 μm. Airborne observations could be carried out using the Stratospheric Observatory for Infrared Astronomy (SOFIA) [Bck97] and its proposed spectrometers. They could observe at several key lines including the molecular hydrogen ground state line at 28 μm.

In the 1 to 5 μm region, the best instruments today are the CSHELL spectrometer on the IRTF ($R = 40,000$), CGS4 on the U.K. Infrared Telescope (UKIRT) ($R = 10,000$), and Phoenix on the National Optical Astronomy Observatory (NOAO) telescopes ($R = 100,000$).

NIRSPEC on the Keck II telescope is a another good instrument to use for these searches [McL95]. It can be used to observe many lines at once, to a very high S/N ratio, and in a very short time. For example, a main sequence G star at a distance of 100 pc can be observed with a resolution of $R = 30,000$ in the 2 to 2.4 μm region, with an S/N of at least 100 in the stellar continuum, and with 100 s of integration. The short observation would allow emission lines to be detected in almost the full region from 2 to 2.4 μm.

For sensitive searches in the ground-based 10 μm window for many lines, there is at least one good instrument. It is an echelle spectrograph built by Lacy and Achtermann [Lac94], called IRSHELL and is used mainly at the IRTF.

SOFIA and the two spectrometers, Airborne Infrared Echelle Spectrometer (AIRES) ($R = 10,000$) [Ham98], and the Echelon Cross Echelle Spectrograph (EXES) ($R = 100,000$) [Rch98], are the best for almost any search for the many infrared lines beyond 5 μm that are blocked by the Earth's atmosphere. However, the task will not be easy, since a sensitive measurement of the continuum near the 28 μm rotation line of molecular hydrogen can only be done on the few nearest main sequence G-type stars. For further information, consult Appendices L on page 381, M on page 427 and N on page 435.

6.10   Future Optical Searches

The first requirement for any optical search is to collect enough photons. Clearly, a traditional telescope will work. Since good angular resolution is not required for searches for pulsed signals however, a less expensive collecting area may suffice, provided that the sky background is sufficiently dark. Groups searching for Cerenkov radiation from high-energy gamma rays face a similar problem, and have developed considerable instrumentation of this type. Following are short descriptions of three alternatives for higher sensitivity searches.

Big Optical Telescope

A big optical telescope, such as Keck, has about 100 m^2 collecting area. With the good spatial resolution of such a telescope the background rate is dominated by the target star. Since there is only one detector, it can be highly optimized, and may well have a quantum efficiency of 50% or more. A solar-type star 1,000 ly away gives about one count every 20 ns, so 8 counts in a nanosecond gives a reasonable false alarm rate. This requires about 16 incoming photons, or 0.16 photon/m^2. A sky survey is not an option with this system because of the small field of view.

Cerenkov Detectors

Groups using Cerenkov detectors for cosmic ray searches use large, inexpensive mirrors. They can do this because the flash they are seeking is a short pulse spread over a wide area of the sky. Typically, they use a few hundred pixels of 0.15 to 0.5 degree each and each pixel is viewed by an individual photomultiplier tube. They image a small portion of the sky with an array of these pixels, and look for coincidences among the pixels to signal the events for which they are searching. SETI searches for optical pulses could not use such a telescope directly because their large field of view means that the background photon rate is too high.

More advanced Cerenkov detectors, such as the Very Energetic Radiation Imaging Telescope Array System (VERITAS) [Bon00] or the High Energy Stereoscopic System (HESS) [Hof00], use multiple telescopes, placed a few tens of meters apart, to get stereoscopic views of air showers. Such a detector could also serve as an excellent SETI pulse detector by looking for single pixel coincidences among the multiple telescopes. The collecting areas are large. The

Figure 6.24: The University of California at Berkeley's Daniel Werthimer, engineer and entrepreneur of SETI@home fame, also does optical SETI.

VERITAS group is building an array of seven 10 m telescopes, and HESS proposes to build sixteen 10 m telescopes. For the next generation SETI projects, HESS looks better than VERITAS for three reasons: better pulse width mirrors, more collecting area, and more uniform pixel coverage.

These advanced telescopes make it possible to conduct a sky survey with appreciable sensitivity. To estimate the sensitivity, consider what we could do with the HESS mirrors, photodetectors, and dedicated signal processing. The HESS advanced configuration consists of 16 telescopes, each with a 10 m diameter. The mirrors are not parabolic (because of cost constraints), and they have about a 3 ns time dispersion, depending on where the photon strikes the mirror. Each telescope has a 768 pixel camera with 0.15 degree pixels and a 4 degree field of view, and each pixel is a photomultiplier. The total system efficiency, counting mirrors, photomultipliers, and other losses, gives an effective area of about 10 m^2 for each telescope.

The sensitivity is dominated by the background photon rate. From the VER-ITAS proposal [Verit], we see that an existing 0.25 degree pixel on a 10 m telescope gives 0.43 photoelectrons/ns, or one photoelectron every 2.33 ns. A 0.15 degree pixel and similar sensitivity should give 0.16 photoelectrons/ns, or one photoelectron every 6 ns. Next, assume a time gate of 3 ns. (We cannot use a shorter time gate, since the non-parabolic mirrors introduce a 3 ns spread depending on where the photon arrives on the mirror.) Then, summing the corresponding pixels from all the cameras, we would expect 0.5 photoelectrons per interval per camera, or about 8 photoelectrons per time gate.

If 40 photoelectrons arrived in the same interval, it would constitute a very unusual signal, happening only once every few months by chance. This is two photoelectrons per telescope, each of which has an effective area of 10 m^2, so 0.2 photons/m^2 will do. Detectors with twice the quantum efficiency would reduce this threshold to about 0.14 photons/m^2.

These telescopes have a very wide field of view, thanks to their large pixels. This makes a sky survey possible with just a few thousand pointings.

Solar Power Plant

A third possibility is to use the much greater collecting area of a solar power plant at night. Such a project is currently under construction by groups searching for the Cerenkov pulses caused by gamma rays striking the atmosphere. Because of the unusual (for astronomy) optical setup, the pixels are quite large (about 0.7 degrees) and few in number (perhaps 200 to 300), and all look at the same spot on the sky. Because of the large area covered by each pixel, the background counting rates are correspondingly high (about 400 MHz per pixel). Nonetheless, because of their enormous collecting area (about 10,000 m^2) they can be quite sensitive.

Imagine dividing the incoming photons into 0.5 ns bins (if the optics make this possible and the electronics can be improved). Each bin has a 1/5 chance of getting a background count, so with 200 pixels we would expect 40 counts in each bin, with a standard deviation of about $\sqrt{40} \approx 6$ counts. A 10σ event is required to keep the background rate down to something practical, so about 60 extra counts are required. Factoring in the losses in mirrors and efficiency of photomultipliers, about 500 photons are needed, or about 0.05 photons/m^2. This is about a factor of three more sensitive than a conventional telescope of 100 m^2 collecting area. It requires only 2×10^3 J/pulse, but the pulse must be short (0.5 ns).

6.10.1   Optical SETI and Cerenkov Detectors

The similarity of the optical hardware, photon detection, and pulse processing for atmospheric Cerenkov telescopes and optical SETI brings up some obvious questions. If short optical pulses are being transmitted by ETIs, would Cerenkov detectors, built for cosmic ray searches, have seen them? Would an optical SETI program see Cerenkov pulses as false alarms? Fortunately, the answer is no. In both cases, the trigger criteria for each kind of search will efficiently reject the other event. The reasoning is as follows:

- When a high-energy cosmic ray, or a gamma ray, enters the atmosphere, it creates a shower of particles. These particles, in turn, emit Cerenkov radiation because they are traveling faster than the speed of light *in air*. This creates a pool of light, perhaps 200 by 100 m in diameter, on the ground. From the ground, if you are in one of these patches, you see a flash on the night sky, occupying an oval about two degrees across. All photons arrive within about 5 ns, and there are about 20 to 30 photons/m^2. Each event excites several adjacent pixels, and the triggering circuits rely on this. So a SETI event, which has more than the expected number of events in precisely one pixel, would be explicitly rejected.

- Since the detectors are photomultiplier tubes, and exhibit occasional large dark current glitches, this rejection is required for the same reasons that optical SETI needs multiple photodetectors. A Cerenkov experiment will not see even very bright SETI events within the field of view of the telescope. Because of the large glitch problem, it is not possible to use a single Cerenkov telescope for optical SETI.

- The next case is for multiple telescopes which are close enough together so they are in the same light pool. In this case the SETI source does produce a signal distinguishable from noise. The photomultiplier in the same location on each telescope would fire, but no adjacent ones would. Again, this is rejected by the triggering circuits that look for diffuse patches in the sky. Likewise, a SETI trigger would demand that the adjacent pixels have no more than the statistically expected number of photons. This should efficiently reject the Cerenkov pulses.

- Some solar power plants currently are used at night as Cerenkov gamma ray detectors. The light from each heliostat is focused on a photomultiplier. A SETI signal would result in all photomultipliers in the array being triggered, independent of their position. This would be rejected by the gamma ray detection algorithms. These algorithms assume that the light pool for gamma ray events should be considerably smaller in size than the overall dimensions of the heliostat array, whereas the SETI detection algorithms look for a pulse that is uniform across the array. Thus, the SETI detection algorithms would accept events that are rejected by the gamma ray detection algorithms.

6.11 Multiple Target Optical SETI Instrumentation

Unlike the single dish radio telescopes, optical detectors, such as photographic plates and CCDs have traditionally measured large numbers of simultaneous, independent pixels. These detectors are too slow to find the very short pulses envisioned as possible optical beacons. However, two-dimensional optical technologies include devices that could be easily fitted with individual pulse detectors, thus providing an efficient targeted search instrument.

One such device is the Multi-Object Spectrograph (MOS) built by Jean Brodie and her team at the Lick Observatory. This consists of optical fibers whose ends are attached to elements that can be arbitrarily positioned in the focal plane of the telescope, normally at the projected positions of stars or galaxies. A robot arm positions the pucks between observations. Since the MOS is used for redshift surveys, the far, or 'business' ends of the fibers are connected to spectrographs. However, the fibers could be fitted with pulse detectors for $50 to $100 each. There are now 67 fibers on the Lick MOS, and as many as 100 could be accommodated.

Following is an example of what might be done both quickly and inexpensively. Consider that the field of view of the 120 inch reflector, upon which the MOS is now used, provides a 1 degree field of view over the 15 inch extent of the MOS. By adding 33 additional fibers connected to pulse detectors, a concurrent targeted search could be initiated that would not impact the current redshift surveys. A 'spaghetti SETI' scrutiny of single, solar-type stars out to approximately 1,000 ly would provide the 33 stars per square degree that could be observed. The number of available targets in this 'one shot per square degree' mode is over a million (both hemispheres), and the cost for detectors is only a few thousand dollars. The cost of adding 33 new fibers to the Lick MOS is on the order of $20,000, including labor. The pulse detectors increase this cost by approximately $3,000. By making these modest investments, and by carrying out the search concurrently with existing observing programs, a large scale optical survey could be quickly initiated.

6.12   Technological Risk

Whenever we spend money on technology, either hardware or software, we are taking a technological risk. Technological risk comes in many forms:

- We can fail to accomplish the mission. For example, space probes often have catastrophic failure modes.
- We can do the job, but not enable expansion to future requirements. This is typically the fate of non-mainstream technologies. BETA format video recorders are a good example of this.
- We can spend too much money for too few results. This is the most common technology failure.
- We can reject a project because it seems too difficult or expensive. This is usually a failure of technical imagination.

We want to insulate ourselves from these risks as far as possible. Some are best treated politically. For example, we can reduce the risk of showing inadequate progress and insufficient technological audacity by funding competing efforts, and by funding small projects using speculative but promising technology. We can minimize the remaining risks by ensuring that our system is as flexible and expandable as we can make it. The following list contains four good recommendations.

- Adopt an architecture, rather than an implementation. A good architecture can be upgraded many times over the life of the system, whereas almost any implementation will be obsolete in a few years. A good architecture, such as a flexible network using general purpose computing, already incorporates a technological future.
- Depend upon Moore's Law (See Section 6.3 on page 152) and related laws for other processing components. These 'laws' are likely to hold because they have the enormous backing of the world semiconductor, computer, and communications industries. SETI should assume, until proven otherwise, that Central Processing Unit (CPU), memory, disks, and other technologies will continue on the same exponential growth rates that have been seen in the past. The precise rate varies from one category to the next. For example, Dynamic Random Access Memory (DRAM) has shown a 30% per year price/performance improvement, and hard disks have shown a 60% per year price/performance improvement.
- Use the purest, simplest functionality that can accomplish the task at hand. Depend upon Floating Point Operations per Second (FLOPS), since the function is so widely available, and depend upon very fast networks. It is much more risky to bet on niche functionality, or particular implementations, such as a particular proprietary instruction set, or a particular network switching technology.

- Depend upon software for things that might change. The one certainty is that software will have a much longer life than hardware. Software beamforming and related algorithms are a much better choice than specialized hardware.

For example, suppose SETI has a certain technical requirement. First, we should determine if it can be fulfilled by software running on a general purpose computer. If so, that is the best choice, since it will not become obsolete, and it can "ride the learning curve" of a very generic product. If that cannot be done, then the next best choice might be software running on a DSP chip. These change each generation, but usually support the same software with minor modifications. If that still does not provide enough performance, then the next best choice is a Field Programmable Gate Array (FPGA), which at least is a commodity part that can be reprogrammed, although not very simply. If that still is not adequate, then we should consider waiting a few years until the programmable solutions improve. The final resort is to build special purpose hardware. If this approach is taken, however, we must ensure that the architecture can be upgraded as the technology improves.

Another way to say this is that *computer hardware should not be viewed as a capital item*, because *it is an annual expense*. The software and the non-computing hardware (e.g., antennas) are capital items. This is very counterintuitive, but it matches the reality of the hardware industry. With this strategy, there is very little risk if technology moves either too slowly or too quickly, because the functionality can be adjusted accordingly.

The SSTWG recommends that the computing equipment should deliberately be underpowered during the early stages of a project. This is because it improves so much faster than the other elements. If SETI builds an SKA or other telescope array, it is important *not* to buy all the computing hardware on the first day, because some of it will become obsolete very quickly. Instead on the first day, buy enough hardware to do a partial job; for example, to record short bursts of data and process it off-line. In a few years, we will be able to afford to do it all in real time. Then, a few years later, we will be able to do the same task with even greater capacity.

6.13 Technology Wild Cards

Any technology assessment, even if accurate when written, can become obsolete as new ways of performing the same task are discovered. For example, the sections of the 1972 *Project Cyclops* report dealing with optical Fourier Transforms employed technology that was obsolete by 1992.

Similarly, we must expect that much of the technology mentioned in this book will become obsolete. Where possible, we have tried to predict the advance of technology using Moore's Law to extrapolate from current technology. However, it is precisely those new technologies whose rise we cannot forecast that will have the biggest impact on any 20 year plan.

What technology advances can we see on the horizon that might have a significant impact on SETI? In the following sections, we discuss a few of the technology advances that might affect our SETI strategy.

Figure 6.25: Interval Research Corporation's Dr. David E. Liddle, brings Silicon Valley dynamism to SETI research.

Quantum Computing

There are computing projects on the horizon that may dwarf existing approaches. Very small devices, limited only by the speed of light, might be used for complex computations. They would have the advantage that their speed would increase as the inverse of the distance between their components. Such devices will perform complex computations by excitation of their wave function, in which the data will be stored. Experts say prototypes of such devices are just around the corner, though we should probably not depend on them when assessing the constraints for the SKA.

Quantum computing attempts to use the evolution of a quantum mechanical wave function to explore an exponential number of possible solutions in linear time. This might work particularly well for pulse searching and FFTs, since weighted sums (the core of these algorithms), seem particularly amenable to quantum mechanical implementations.

Biological or Self Reproducing Computers

It is also possible that very small machines, based originally on biological models, may be able to grow computers very inexpensively, or even reproduce themselves. If one were to seed a computer to grow exponentially, computing costs would plummet. Self-replicating machines might make it possible to construct huge antennas for little more than the cost of the material. Again, for the purposes of this SSTWG, these thoughts are speculative at best.

Optical Processing Techniques

The extremely rapid development of integrated optics and optical components may result in architectural and technological approaches not yet considered. For example, using optical processing techniques instead of just electronic techniques could provide a wideband (>10 GHz) and cost-effective solution starting as early as 2005.

Material Science

Room temperature superconductors might enable us to design and build better wideband feeds, and systems with lower noise temperatures. It would certainly allow us to build faster chips if we could use superconducting interconnections, and might let us continue the Moore's Law improvement rate longer than is currently probable. Lower noise amplifiers would have less of an impact, since there

Figure 6.26: Dr. W. Daniel Hillis of Walt Disney Imagineering, peers into the technological future.

are many other contributors to system temperature. The need for cryogenics would certainly be reduced.

Artificial Intelligence

It seems likely that sometime soon (on the cosmic scale) scientists will either discover how biological intelligence works, and replicate it on a nonbiological substrate, or some separate line of research will lead to intelligent machines.

This will have a huge number of effects, both social and technical. A civilization, based upon machine intelligence might be completely different from our civilization. This might make many of our speculations about ETIs completely wrong. As such, this is probably the wildest of the wild cards.

A being based on machine intelligence could have many properties that our current biological intelligences do not have. Here are just a few examples:

- personal immortality;
- the ability to create copies of beings;
- the ability to exist on different computing substrates (since any computer can imitate any other);
- the ability to live and work at many different speeds;
- the ability to live in many different environments. (For example, deep space might be preferable to the oxidizing atmosphere of planets.)

Certainly, these differences will make it hard to predict changes in a society of such beings. Machine intelligence would certainly make interstellar travel easier, since:

- Lifetime limits are relaxed.
- Suspended animation is easy to achieve.
- Life support is easier.
- The mass requirements are lower.
- A 'backup' of voyagers for risky mission phases is possible.
- There is the possibility of shipping information instead of matter.

The existence of machine intelligence would require us to rethink some of our SETI plans. For example, the Habitable Zone for a machine civilization might be much larger than that for a biological civilization. Therefore, if we use a tightly beamed transmitter, we might want to beam a larger zone around star.

It is not clear that restricting our strategy to planets with life would be correct, assuming that our telescope technology becomes good enough to find planets. If these advanced machine ETIs carry their own energy supplies, then, being free of planets and stars, they could appear anywhere.

Inexpensive Access to Space

There are many potential advantages to placing SETI equipment in space. Outside the atmosphere we might consider the 10 to 50 GHz band in addition to the 1 to 10 GHz atmospheric window. There is less dispersion at these higher frequencies, and antennas of a given size have higher gain. Large antennas could benefit from the absence of wind and gravity. Ionosphere corrections are not needed, so phasing would be easier. Interferometer spacing could be more than an Earth diameter. Any sources could be tracked almost continuously. Low temperatures are easier to maintain. RFI could be dramatically reduced by operating on the far side of the Moon, which is always shielded from the Earth, or simply by operating far from the Earth. Finally, gravitational focusing requires access to space, since the focal line of the Sun begins at 550 AU for photons.

However, the high cost of building and operating equipment in space has prevented the SETI effort from taking advantage of these benefits. Less expensive access to space might allow solutions that take advantage of these better conditions.

Exobiology, Planet Searches, Etc.

If planets with life are common, then we might want to concentrate our search on those systems. We must remember, however, that just because life is found, it does not imply either the evolution of intelligent species or its survival. For example, although life on Earth may have evolved in the ocean, clearly we do not live there now.

Technologies for Neutrino Handling

Neutrinos have wonderful interstellar propagation properties. They can travel throughout the Galaxy at the speed of light without suffering absorption or dispersion because they are uncharged, have zero or negligible mass, and have little interaction with matter. Unfortunately, these same properties make them extremely difficult to focus or detect. Further, if neutrinos do have mass, then gravitational interactions might be sufficient to make them difficult to target. If the targeting and detection problems could be solved, then neutrinos might become the information carrier of choice for SETI.

Dark Matter, Cosmology, Neutrino Puzzle, Etc.

There are a number of outstanding puzzles in cosmology and particle physics. Where are all the non-luminous baryons or dark matter? What are the properties of the dark matter? What causes the appearance of a cosmological constant? It is difficult to know whether the solutions to these puzzles will have any effect on the SETI program.

Chapter 7

Conclusions and Recommendations

This chapter consolidates the conclusions and recommendations reached in the preceding chapters.

The major recommendations are:

- Build and operate the 1hT using a phased array of many small dishes. This is becoming an increasingly cost-effective way to increase the collecting area of a radio telescope as the impact of Moore's Law reduces the cost of electronic components compared to the cost of large dishes. In addition to being a premier facility for SETI, it can be used for radio astronomy and serve as a prototype for the SKA.
- Build and operate the OSS, an all-sky, all-the-time, microwave search system for transient beacons. It will be capable of exponential scaling, relying on Moore's Law.
- Support optical searches for rapid pulses, and data mining of existing records for CW optical signals.
- Research the next generation of optical SETI systems with higher sensitivity and greater sky coverage.
- Seek alliances to conduct commensal searches on optical telescopes.
- Support proposals to explore new portions of phase space.
- Create a Strategic Plan.

Two major conclusions are:

- Earlier arguments for not exploring the infrared/optical bands have been altered by the rapid evolution of high-speed electronics, which allow very short optical pulses to be detected.
- While our own technology is still leaking distinctive signals, passive listening remains the most cost-effective strategy for discovering ETIs. Therefore, any active transmission strategy must be long lived, and its development lies beyond the horizon of this book.

7.1   Conclusions

The conclusions of the Working Group fall into two basic categories: those referring to the nature of the search, and those referring to the nature of the signals being sought. The first enumerated list below contains conclusions about searching. The second list addresses what we presume we know about any ETI signals.

7.1.1   Searching – General Conclusions

1. The search for ETI life is a legitimate scientific undertaking and should be included as part of a comprehensive and balanced space program. There is significant public interest, and SETI logically fits within the major new NASA initiatives of Origins, and Astrobiology.

2. We maximize the potential for positive results by using a search strategy that assumes either that the ETI is trying to attract our attention, or that it is transmitting information to other ETIs. This book concurs with the earlier *Project Cyclops* report, listing the properties of the ideal information carrier as follows:

   - The energy per bit should be minimized, all other things being equal.
   - The velocity should be as high as possible.
   - The particles should be easy to generate, launch, and capture.
   - The particles should not be appreciably absorbed or deflected by the interstellar medium.

3. It is vastly less expensive to search for and to send signals, than it is to attempt contact by spaceships or probes. This conclusion is not based upon the present state of our technological prowess, but rather on our present knowledge of physical laws.

4. There are orders of magnitude of uncertainty in the estimates of the average distance between communicative civilizations in our Galaxy. This strongly argues for an expandable search system. The search can be started with the minimum system that would be effective for nearby stars. The system is then expanded and the search carried farther into space until either success is achieved, or a new search strategy is initiated.

5. Existing searches optimized for microwave sources are still a sound idea. Modern technology could greatly improve these searches, making them less expensive, faster, and with wider bandwidth and more sensitivity. Since all these goals are shared with the radio astronomy community, this seems like the right time for a combined effort in this area. The technology is ready, the SKA needs development effort, and the SETI community needs

dedicated collecting area. This convergence of "ends and means" makes the 1hT project a consensus favorite of the committee.

6. Other approaches to interstellar communication should be investigated. In particular, pulsed microwave beacons and pulsed optical beacons appear to be at least comparable to CW microwave beacons, and might well be chosen by ETIs wishing to communicate. Since these represent largely unexplored areas of parameter space, they might reveal new natural phenomena. Therefore, an effort in these areas is warranted.

7. The amplitude of any signal may vary substantially with time. If the average length of time ETIs emit detectable signals into space is small, then the distance between civilizations is great. Microwave signals traversing such great distances will have their amplitudes modulated by interstellar scintillation, possibly rising above a detection threshold only occasionally [Cor91a], [Cor91b]. Sequential beacons with low duty cycles are easier to miss. For this reason, it is desirable to be able to search all targets, at all frequencies, all of the time.

8. Covering all targets all of the time with high sensitivity has no foreseeable solution in the infrared/optical portion of the spectrum. It may, however, be possible at microwave frequencies.

9. Having several systems for SETI research, located around the world, would have great value. It would provide complete sky coverage, continuous reception of detected signals, and the capability of conducting long base line studies. International cooperation for such endeavors should be solicited, and will be encouraged by complete dissemination of information. SETI represents an effort of all humans, not just one country.

10. Communicative races may beam signals toward nearby likely stars, at relatively low powers, for very long periods of time. During the search phase, we can neither expect to receive signals intentionally beamed directly to us, nor can we afford to overlook this possibility.

11. The best part of the microwave region seems to be the low frequency end from about 1 to 3 GHz. Our SETI searches should be extended upwards because current antennas are being upgraded to work at higher frequencies, and because interstellar signal propagation is degraded less at higher frequencies. Cost-effective engineering implies that the upper operating frequency of a receiving system should be as high as is economically feasible.

12. The increase in digital signal processing capacity in the past few decades has profoundly altered the capability-to-cost ratios that may be achieved. In particular, continued growth in processing capacity for the same cost doubles every 18 months, according to Moore's Law. This leads to the conclusion that software should be considered to be a capital expense,

just as the mechanical (concrete and steel) components of telescopes are traditionally treated. The actual signal processing hardware should be considered as an operating expense, rather than a capital expense.

13. Any properly designed processing system will be flooded with too much data to process at its inception. It will reach parity with the data rates during its middle age, and have excess data processing capacity towards the end of its life cycle.

14. The chance of a false positive detection should be reduced by requiring each candidate signal to pass three steps:

   - First, at even a single telescope, the source must behave as expected for one located at a great distance. It should disappear when the telescope is pointed off target.
   - Second, the signal must be confirmed at a second site. It should appear at the same spot in the sky, as is proper for sources at interstellar distances. In the case of a narrowband signal, the differential Doppler signature between the two telescopes can be used to rule out any sources on the Earth's surface or in near-Earth orbit.
   - Third, once a signal is discovered and confirmed, the last filter will be provided by the scientific community. A careful description of what was actually observed must be made, and the discovery data must be made available for scrutiny. Before any signal can be accepted as coming from an ETI, the scientific community must be given a chance to provide other explanations for the observed signal.

15. The chance of a false negative should be reduced by regularly performing end-to-end tests of the entire system, in cooperation with the professional and amateur astronomy communities.

16. Listening is the appropriate strategy for emerging technologies such as ours. Deliberate transmission of signals will be undertaken by ETIs that expect to be able to continue transmitting for sufficiently long periods of time to make detection probable.

17. The search will almost certainly take years, perhaps decades and possibly centuries. To undertake so enduring a program requires not only that the search be highly automated, but also requires a long-term funding commitment.

18. For the SETI Institute, the challenge for the coming decades will be to creatively couple private sector philanthropy with the international collaborations among governments, to advance both SETI and astronomical exploration. Patronage is a long-standing tradition for astronomical research, continuing to the present with the remarkable contributions of the Keck Foundation. Within the next twenty years, contributions from very wealthy individuals, who are eager to support SETI, could be comparable

to the contributions that some individual governments are able to make in support of astronomical research.

19. In all its endeavors, the SETI Institute should arrange for ongoing, unbiased, critical reviews by external experts.

7.1.2 Signals – What we might expect

1. Microwave beacons with circularly polarized narrowband signals and spectral widths of less than 1 Hz represent an easily recognizable and, arguably a good technological choice for microwave beacons. Intentional beacons at optical frequencies may well use broadband pulses, permitting very high data transfer rates.

2. An advanced technology might occupy real estate on several planets within its planetary system. It might also engage in vast astroengineering projects to reconstruct its planetary system to provide more livable real estate. This latter possibility can be investigated by conducting infrared searches for Dyson spheres. Doing so requires the development of spectral index discriminants to distinguish Dyson spheres from infrared excesses due to natural dust in the system. Kardashev [Kar73] suggested that a Planckian spectral index of -2 will characterize large-scale engineering projects, whereas the index for dust will be more negative.

3. Of all the wavelengths at our disposal, microwaves still remain an excellent choice. The required peak transmitter power is at a minimum, and the necessary stabilities and collecting areas are fundamentally easier to realize, and less expensive, than those at shorter wavelengths.

4. For nearby stars, signal propagation through the interstellar medium is excellent in the infrared, and is acceptable in the optical. Extinction and scattering may limit the detectability of optical pulses to distances of a few thousand light years. In the infrared, however, they may be detected over much greater distances, and perhaps even from the galactic center. Therefore, search efforts in the infrared/optical parts of the spectrum should be increased.

5. In the infrared/optical regions of the spectrum, very short and intense pulses remove the frequency domain search problem. With much greater difficulty, CW signals might be sought with subthermal resolutions (resolving powers of 5×10^4) covering a wavelength range of about an octave by using echelle spectrographs.

6. Today, it is technologically feasible to build phased antenna arrays that operate in the 1 to 10 GHz region, have very large total collecting areas, and have the capability of forming many simultaneous independent beams.

7. Microwave communication is possible over intergalactic distances using antenna arrays equivalent to a single antenna a few kilometers in diameter at both the transmitting and receiving end. High-speed infrared/optical communication is possible over large interstellar distances. Thus, information transmission is feasible between civilizations.

8. The efficient detection of beacons involves searching in the frequency domain with high resolution (less than 1 Hz). For ground-based microwave searches, the natural limits are set by increased noise from galactic synchrotron emission below 1 GHz, and contributions from atmospheric O_2 and H_2O above 10 GHz. Current data processing methods permit this span to be searched simultaneously with a resolution of 0.01 Hz. This will become much easier in the future.

9. We now know that beacons efficiently targeting many stars could be built, and we expect that advanced technologies may not produce leakage radiation of an easily identifiable form. This causes us to shift the emphasis away from strategies to detect leakage, and toward those to look for beacons. See the Decision Tree path in Figure 3.1 on page 69.

10. High amplitude low duty cycle pulses are not expected from astrophysical sources. They should be sought, however, as continuous all-sky searches become feasible.

11. As we have learned more about the atomic and molecular constituents of our Galaxy, and as our instrumentation has opened up higher radio frequency regions, the 'water hole' and other 'magic' frequencies have lost their apparent uniqueness.

12. An advanced technology may generate less narrowband leakage than does our current terrestrial technology. Broadcast communications may be replaced by point-to-point systems on fiber or cable, or beamed from orbit. Multiple uses of the limited available spectrum may result in various spread-spectrum modulation techniques with the suppression of coherent components and concentration to higher frequencies.

13. For leakage at microwave frequencies, the coherent radiated power will not grow indefinitely because spatial, temporal, and frequency reuse of the spectrum causes the emission to resemble Gaussian noise. This presents a difficult signal processing challenge for the receiver. Currently, only the Karhunen Loeve (KL) transform [Mac94] shows potential for recognizing the difference between the incidental radiation of technology and white noise. The KL transform is too computationally intensive for the present generation of systems. The capability for using the KL transform should be added to future systems when the computational requirements become affordable.

7.2 Recommendations

In the previous chapters, we discussed a wide range of signals that an ETI might send, and an equally wide range of technologies that we could use to search for such signals. Given a limited budget, which of these possible choices should be pursued by the SETI Institute over the next 20 years? The Working Group decisions are enumerated below.

- Build and operate a phased array radio telescope with 10^4 m^2 of collecting area, made of low-cost three to seven meter dishes. This One Hectare Radio Telescope (1hT) will be an excellent SETI instrument in its own right, and serve as a prototype for one possible implementation of the SKA.

- Build and operate an all-sky, all-time microwave search system. This system, called the OSS, will look for relatively strong transient radio sources. This is both an unexplored corner of radio astronomy, and exactly the system needed to detect some plausible, low duty cycle, ETI beacons. Furthermore, such a system is dominated by processing, making it a perfect candidate for a phased exponential scaling construction mode to take fullest advantage of Moore's Law.

- Support optical searches for broadband nanosecond pulses, and data mining for subthermal, optical, CW signals. Concurrently, research should be done for the next generation systems that will have higher sensitivities and greater sky coverage. In order to accomplish this growth, consider ways to form alliances with scientists who observe the sky at optical wavelengths, and implement commensal search strategies.

- Look for ways to expand SETI searches into the optical/infrared. When dedicated observing facilities become warranted, or large-scale collaborative programs are implemented, the searches should be subsumed within the SETI Institute.

- Support innovative new proposals for observation programs and strategies that explore new portions of phase space in a cost-effective manner.

- While our own technology is still leaking distinctive signals, passive listening remains the most cost-effective strategy for discovering ETIs. Therefore, any active transmission strategy must be long lived, and its development lies beyond the horizon of this book.

- The SETI Institute should develop a strategic plan to address all questions pertinent to the implementation of the science and technology plan described in this book. It should include studies of how to proceed with an expanded SETI enterprise after the first signal is detected, and what to do if, after many decades, no signal has been detected. It should address issues of organization, management and development. It also should incorporate strategies for continuing close interactions with the scientific and engineering communities, for education, public programs, and studies of interactions between SETI and society.

- The SETI Institute, along with other interested parties, should develop and refine a plan of how to proceed with an expanded SETI enterprise after the first signal is detected. Government and private funds should be sought to undertake this task. The plan should address how to respond to such a signal, as well as how to expand or refine the detection program.

- The SETI Institute should continue and expand programs to study the societal implications of detecting signals from an ETI. It should also encourage an active interest among scholars in these studies.

- The SETI Institute should extend the life of the SSTWG, or part of it at least. The group would continue to provide assistance to the SETI Institute as the new SETI programs get under way.

- The Development Office of the SETI Institute should strive to raise funds from private sources. There are clear opportunities for capital fund-raising campaigns for the large telescopes.

- The SETI Institute should engage in discussions with NASA and the NSF to encourage them to join in a public/private partnership for direct participation in SETI projects, for technology development, or for other joint ventures, such as the use of the 1hT for communication with interplanetary spacecraft.

- There is an exciting new NASA program called Astrobiology. It seems to be widely agreed in the scientific community that a modestly funded SETI effort should be a component of this program. Since SETI is inherently an international endeavor, collaborations should also be sought with interested organizations in other nations.

- A concerted attempt should be made to encourage the US Congress to support the visionary nature of the SETI endeavor which is of such profound importance to science, and indeed to people from all walks of life. It is worth remembering that Congress did appropriate $78 million for SETI over the years 1975 through 1992. A few million dollars a year is a small price to pay for the chance to find another civilization among the stars.

Figure 7.1: A wide angle aerial view of the VLA Array in New Mexico. (Photo courtesy of VLA, NRAO.)

Chapter 8

Budget and Implementation

This chapter discusses budgetary and implementation issues for the recommendations presented in Chapter 7 on page 229. Economical and political considerations have shaped a plan of action which the SETI Institute might carry out over the next two decades. It presents cursory budgets for the implementations being considered. It states specific proposals, broad implementation strategies for optical and microwave searches, and touches on the reasons for not transmitting in the foreseeable future. It reviews the 1hT and the OSS.

8.1   Selection of Feasible Strategies

From all the concepts discussed by the SSTWG, three areas of focus have been recommended for implementation. These are the 1hT, Optical SETI and the OSS.

8.1.1   The Near Future: 2000 Through 2005

Studies, prompted by the SSTWG meetings, have altered the probable course of Project Phoenix observations over the next five years. In 2002, the first module of a new SETI signal processing system, called the New Search System (NSS) is expected to be installed at the Arecibo Observatory. This will begin the transition to a new system, based upon commercially available components. Eventually, all of the older processors will be replaced by the new fully modular processors. The completion of this transition will mark an appropriate end to Project Phoenix. (The name commemorates the rebirth of the targeted search after Congressional termination of the NASA High Resolution Microwave Survey.) Phoenix will continue at Arecibo until approximately 2003, when the current SETI observing program, totaling 2,600 hours, is completed.[1]

In parallel, the SETI Institute intends to construct the 1hT, the equivalent of a 100 m telescope, using commercially available antennas, and frequency independent, log-periodic feeds. It will be equipped with wideband receivers, cooled to 80 K if possible, and the next generation of digital signal detectors derived from the NSS. The 1hT can serve both as a SETI/radio astronomy facility, and as a prototype of one technology for building the SKA. Surprisingly, to obtain large amounts of observing time, it appears to be more cost-effective to build the 1hT, rather than to rent existing large single dishes.

The surprise is not that a 100 m equivalent telescope, built with techniques intended to meet the "few hundred dollars per square meter" design goals for the SKA, should be affordable. Rather, the surprise is that consumer driven markets are providing an opportunity to attempt to do this next year, not 10 years from now. Because the SETI Institute intends to construct this array in a very few years, placing its components at an existing radio astronomical facility could minimize the environmental impact issues. If fund raising efforts are successful and timely, the partially completed 1hT should be available for use in 2004. The 1hT should be completed by 2005, at which time it will have one hectare of collecting area, and the ability to process three independent beams, each with 100 MHz of dual polarization bandwidth. Since the array is fully extensible, it will continue to be upgraded with additional processing bandwidth and antennas, as funding permits.

During the next five years, relatively small-scale optical searches will be supported and ways to improve their sensitivity, expand into the infrared, and increase sky coverage will be investigated. Most of this research will be conducted by groups outside the SETI Institute. The 2005 time frame now appears appropriate for a formal review of what has been learned, but the timing of that review will be dictated by how rapidly the field advances.

Limitations and new opportunities for infrared/optical searching will need to be analyzed in detail. If it is decided to advance this research area beyond the initial explorations, a definite plan including resource requirements must be formulated. The SETI Institute will have to decide whether to transition these searches into an in-house effort, or to attempt to organize a cohesive, systematic program utilizing researchers outside the SETI Institute, or to seek another sponsoring organization for these searches. Initial discussions of the probable cost for a dedicated 10 m, light-gathering telescope to extend the sensitivity

---

[1]Since the SSTWG meetings were completed, major telescope upgrades at the Jodrell Bank Observatory have slowed the progress of Phoenix observations. It now seems likely that Project Phoenix will continue observing at Arecibo and Jodrell Bank until about 2004, when it is time to begin the transition to the 1hT.

of searches for optical pulses have suggested that $10 million might suffice. Alternatively, collaborations with communities building Cerenkov detectors for cosmic ray searches might prove to be a more cost-effective way to achieve an enhanced collecting area.

Progress can be made on the OSS during the next five years, even though the computational requirements for the system cannot be met for years to come. The first problems to be tackled are those in common with the 1hT, such as beamforming. A tiny array of four elements and minimal bandwidth of 4 MHz can provide a very useful test bed. This is another area where funding for external researchers, as well as in-house staff, will be fruitful. In particular, the NFRA efforts to build the low frequency elements for the SKA, and the Ohio State University Radio Observatory (OSURO) efforts to build the Argus radio telescope can advance the OSS. Since the OSS is so computationally intensive, it is particularly ripe for donations from, and participation by, the technology companies with which the SETI Institute shares the Silicon Valley. By 2005, the first incremental factor of 16 (four times as many antennas, and four times the bandwidth) for the OSS should have been implemented.

8.1.2 After that: 2005 through 2020

This will be the period of peak productivity for the 1hT. In 2005, the 1hT should be functioning sufficiently to process 100 MHz of dual polarization bandwidth from each of three independent beams, over a frequency range of 1 to 3 GHz. The bandwidth can then be increased by changing filters and electronics in the central processing lab, without any changes to the system in the field. Using 0.5 GHz bandwidth and an integration time of 400 s, and looking back at each target three times (to deal with scintillation effects), a targeted search of 100,000 stars could be completed in 6 years.

As the processing capability grows, we can anticipate the ability to observe 12 beams simultaneously. It would then be possible to extend the upper frequency of the search to 10 GHz, while only extending the time to complete the search to eight years. Further improvements in the instantaneous processing bandwidth will reduce this time.

Successful implementation of the 1hT should go a long way toward the realization of the SKA. The US SKA Consortium[2], intends to demonstrate that an array of a large number of small parabolic reflectors is the most cost-effective means of achieving the scientific goals established for the SKA by the international radio astronomy community. Since the design of the 1hT is driven by SETI goals, this symbiosis will ensure that the final implementation of the international SKA is optimal for SETI observations, as well as for traditional

---

[2]The SETI Institute was instrumental in establishing the SKA Consortium, subsequent to the inception of the SSTWG meetings.

radio astronomical observations. The 1hT will permit the current target list of 1,000 stars to be extended to 100,000 stars. However, the sensitivity and time required to reach a larger number of targets at greater distances will require many beams from the completed SKA.

In its strategic plan, the SETI Institute should include those activities that are necessary to convince the world that the SKA is an essential component of the early 21st century astronomical observational armory. Another plausible strategy should be to advocate public/private partnerships that benefit from these enhanced capabilities. For example, consider the advantages of using the high gain SKA to track inexpensive NASA spacecraft carrying mechanically simple, low gain antennas, as opposed to tracking signals from large, high gain microwave antennas deployed in space, as is typically the case today.

An evolutionary approach to building the OSS suggests that the major capital investment occurs near the horizon of this book. The annual cost, excluding salaries, for this activity does not exceed one million dollars until after 2014, according to Table 8.3 on page 261 which is based on the current Moore's Law projection for CPUs. How the OSS is implemented will depend on what technologies actually emerge in this 21st century, and what doubling times they achieve. However, the strategy of incremental growth as outlined in this chapter remains valid, as long as the cost is dominated by signal processing.

The infrared/optical observations to be conducted from 2005 through 2020 are as yet undefined. These strategies require maturation before a road map can be laid out.

From 2005 to 2020, the 1hT and then the SKA, and perhaps the optical programs, will expand the search from the current 1,000 stars to 100,000 and then to a million stellar targets. By 2020, we should have explored the local stellar population of the Milky Way Galaxy out to about 1,000 ly. To examine more stars at greater distances will require continuous sky monitoring by the OSS, sensitive to very powerful transmitters from advanced civilizations that may illuminate Earth periodically but infrequently. Infrared searches could also expand the distance explored and increase the number of targets, if they prove cost-effective.

As forward-thinking as these Working Group meetings were, the participants were unable to anticipate any system that could be implemented by 2020 which would be capable of detecting the equivalent of early 21st century terrestrial technologies throughout the Milky Way Galaxy. This road map takes us only 1% of the way. While it is certainly possible that we might have a positive result during the next two decades, a null result from any research will not be significant. If we have a null result two decades from now, then our counterparts will be faced with the task of expanding the effort based on the previous lack of success. We will have learned that the task is not easy, and we will still be pondering the best way to succeed.

8.1.3 When should we start Transmitting?

Although not strictly within the scope of their charter, the Working Group participants engaged in a lively discussion about the advisability and possible benefits of adopting a *balanced* strategy for achieving contact. A balanced strategy would devote some fraction of SETI resources to transmission. After all, if no civilization is intentionally sending, there may be no contact.

The discussions began by predicting the future course of terrestrial leakage. Earth will not go radio quiet, but it will go radio 'pink' in the future. The power of our almost noise-like leakage will not increase without limit. Time and frequency multiplexing schemes will permit more efficient use of the spectrum. Thus, the most detectable deviations from white noise may be artifacts from the infrastructure of previous spectrum allocation schemes and protected radio astronomy bands. Searches for leakage will not be impossible, but the terrestrial analogy suggests that more sensitive systems and complex signal detection algorithms will be required. As the narrowband artifacts from leakage fade, efforts to replace them with transmissions which are achievable by individuals or small groups will not substantially improve the detectability of Earth at a distance. To be effective, beacons must be powerful, have a high duty cycle, and have a longevity that is measured in astronomical rather than human lifetimes.

The discussions culminated in the following, by no means novel, conclusions:

- If or when we achieve contact with another civilization, it will certainly be more technologically advanced than we are. Contact with a less technologically advanced civilization is not now a possibility. In fact, any civilizations we contact are statistically likely to be far more advanced. When the evolution of planets and their attendant technologies require billions of years, it is unlikely that two technological civilizations will be synchronized to better than a million years.

- If it happens at all, there always has to be a *first* contact between two technological civilizations. Statistically, it is extremely unlikely that our first contact with an ETI civilization will also be its first contact with an ETI civilization. Thus the advanced technology we detect will have experienced this type of encounter many times before. It already may have established a galactic protocol for information interchange, to which *ab initio* transmissions by Earth will have no chance of adhering. Thus, we justify our asymmetrical *listen only* strategy by recognizing our asymmetrical position amongst galactic civilizations. We are among the very youngest!

- Transmitting is a more expensive strategy than receiving. Within the next two decades, the parameter space explored for signals can be extended by the compounded growth rate of many technologies. Transmissions could benefit from these same exponential improvements in technology, but with the limited resources likely to be available during this same period, we could not add significantly to the high power of our leakage radiation. As that leakage abates or becomes more noise-like, this argument loses its force. Transmission will not be rewarded for decades, perhaps centuries, because of the great distances and round-trip travel times for signals. Our resources are constrained, and it is thus prudent to pursue a passive program of exploration that might provide a positive result within years.

- Transmission is a diplomatic act, an activity that should be undertaken on behalf of all humans. We lack the cultural maturity to accomplish such a cohesive action. Some Working Group participants felt strongly that this active strategy should not be embarked upon unilaterally, without consultation and consent. While most of the participants believed that transmitting now would be merely harmless and wasteful, a few members felt that transmissions should not be carried out without international consultation and approval by appropriate international administrative bodies.

For these reasons, it was generally agreed that deliberate transmissions for the purpose of interstellar contact are currently a waste of resources, and unlikely to produce any benefit. Transmission, therefore, belongs in a future substantially beyond the 2020 boundary of this book.

8.2   One Hectare Telescope

Of the three recommended programs (1hT, Infrared/Optical, and OSS), the 1hT is by far the most mature and well developed. This section presents the design goals for that facility.[3] Figure 8.1 on page 245 shows the SETI Decision Tree with the appropriate elements highlighted for the 1hT.

---

[3]Subsequent to the drafting of this book, the SETI Institute engaged in a formal development exercise for designing and constructing the 1hT, now called the Allen Telescope Array (ATA). The resulting budget more realistically accounts for all the R&D and personnel costs and totals \$25 M, to be contrasted with the \$15 M estimate in Table 8.2.

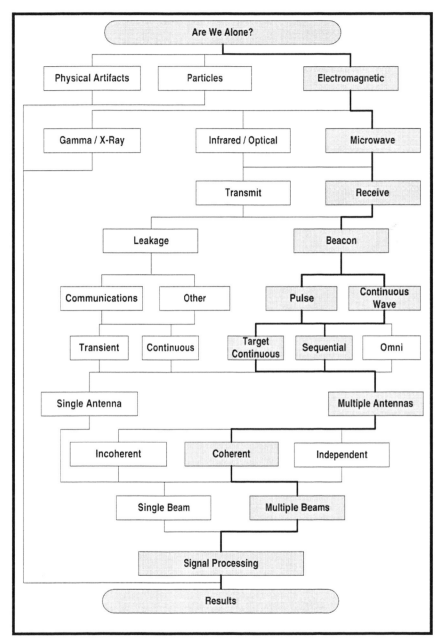

Figure 8.1: SETI Decision Tree – One Hectare Telescope.

8.2.1   Goals

The 1hT has two primary goals:

- The first goal is to provide a dedicated SETI observatory with sensitivity comparable to the world's largest existing radio telescopes.

- The second goal is to provide a significant radio astronomy facility that will encourage synergy between microwave SETI and radio astronomy.

The 1hT also has two important secondary goals:

- The first goal is to increase the likelihood that the SKA will be constructed by demonstrating that a large collecting area at centimeter wavelengths can be built for an unprecedentedly low cost using small parabolic antennas.

- The second goal is to provide a test and development facility for work on RFI excision techniques.

Figure 8.2:   Artist's conception of the 1hT array. (Photo courtesy of L. Ly, SETI Institute.)

8.2.2 Preliminary Specifications

Sensitivity

Sensitivity is commonly measured as the ratio of the effective area, A_{eff}, to the system noise temperature, T_{sys}. Table 8.1 shows the A_{eff}/T_{sys} values for some of the most sensitive centimeter-wavelength radio telescopes, both existing and proposed. The table shows the nominal performance of the SKA. For each telescope, a range of A_{eff}/T_{sys} values for the 1 to 20 GHz range is given, plus an average value.

System Site	Area m^2	Eff. Ap.	A_{eff} m^2	T_{sys} (K) high	T_{sys} (K) low	A_{eff}/T_{sys} low	A_{eff}/T_{sys} high	A_{eff}/T_{sys} avg
NRAO-140 foot	1.43×10^3	55%	787	60	18	13	44	28
Parkes-64 m	3.22×10^3	55%	1,769	80	25	22	71	46
Nançay	6.50×10^3	50%	3,252	80	30	41	108	75
Westerbork	6.79×10^3	55%	3,736	80	25	47	149	98
Effelsberg-100m	7.85×10^3	55%	4,320	50	26	86	166	126
DSN-70 m	3.85×10^3	68%	2,617	21	18	125	145	135
VLA	1.33×10^4	60%	7,952	150	32	53	249	151
EVLA	1.82×10^4	60%	10,897	60	29	182	376	279
GBT	8.64×10^3	70%	6,048	27	20	224	302	263
Arecibo	7.31×10^4	42%	30,360	50	25	607	1,214	911
SKA *	1.00×10^6	50%	500,000	25	25	20,000	20,000	20,000
1% SKA *	1.40×10^4	50%	7,000	35	35	200	200	200
1hT *	1.00×10^4	50%	5,000	35	35	143	143	143
* These are proposed systems, and the specifications may not be final.								
NOTE: Eff. Ap. is the effective aperture.								

Table 8.1: A_{eff}/T_{sys} for Some Radio Astronomy Systems.

While the numbers for the SKA, 1% SKA, and 1hT are all estimates for proposed systems, they give an idea of what is expected. For comparison purposes, the numbers for an array equal to 1% of the proposed SKA are listed. As expected, they are comparable to that of the 1hT.

To have a sensitivity comparable to existing large telescopes, the 1hT should have a value of A_{eff}/T_{sys} of between 100 and 200 m^2/K. Currently, the SKA is conceived as organized into 30 to 100 stations, implying A_{eff}/T_{sys} values of 200 to 600 m^2/K per station. To provide the best proof-of-principle for the SKA, the 1hT should then have a sensitivity of the same order.

The cost of the 1hT is expected to be proportional to the value of its A_{eff}/T_{sys}. A straight scaling from the estimated cost of the small parabola design from Table 6.7 on page 190 yields \$4 M for $A_{eff}/T_{sys} = 200$ m^2/K. This should be about doubled, due to the smaller quantity of elements, and more proximate time. This results in a capital cost estimate of about \$8 M, in addition to the considerable cost of the initial R&D.

How much total funding can be obtained for this project remains to be seen. We conclude that the 1hT should provide a value of A_{eff}/T_{sys} of at least 100 m^2/T_{sys}, and up to approximately 200, if enough support can be found. Fortunately, this type of array is readily expandable, so it would be possible to build the smaller system and expand it later.[4]

Frequency Range

The optimum microwave window for ground-based SETI is approximately 1 to 10 GHz. This range is set by several effects:

- The atmosphere starts to degrade performance beyond 10 GHz.
- The diffuse galactic emission starts to hurt below 0.5 GHz.
- Propagation effects degrade signal coherence significantly at 1 GHz, and grow worse at lower frequencies (approximately as λ^2).

The current specification for the main SKA array is 0.2 to 20 GHz. At present, it seems that the pure phased array technique is best at the low end of this range, while small parabolas are best for the upper end. Another issue is that the surface accuracy of commercially available parabolas, in the three to eight meter range, is only sufficient to provide full ($\geq 90\%$) gain up to 12 GHz, the highest frequency for which these dishes are normally used. Finally, as can be seen in Appendix C on page 321, the contribution to system noise from the LNA rises almost linearly with frequency. Beyond about 10 GHz, the LNA contribution begins to dominate the antenna contribution, and consequently,

---

[4]On the basis of this now detailed budget analysis, the SETI Institute has been successful in obtaining funding for the technology development phase of the project from the Paul G. Allen Foundation. The 1hT name has been changed to the Allen Telescope Array (ATA). The R&D work is proceeding very rapidly in partnership with the Radio Astronomy Laboratory of the University of California at Berkeley. Philanthropic donations from Nathan Myhrvold provided funds for the electronically shielded Myhrvold Laboratory in which the digital signal processing equipment, at the heart of the ATA, will be housed. Successful completion of all technology milestones and a decision to commence construction is expected by mid-2003. One hectare of collecting area, at the Hat Creek Observatory in Northern California, is scheduled for completion in 2005. The Hat Creek Observatory is part of the Radio Astronomy Laboratory of the University of California, Berkeley.

the overall cost-effectiveness of the system is reduced. Hence, we recommend a design range of approximately 1 to 10 GHz.

Beams

As described in Section 6.8 on page 180, only a few beams (~ 10) are needed for use in a Targeted SETI program, because of the small number of known SETI targets in any field of view. The SKA specification is for at least 100 beams, which is rather different. For 100 beams, a two-dimensional FFT technique might be best, but for 10 beams, a direct summation method would be simpler. Many of the possible astronomical uses, such as pulsar surveys, VLBI, and molecular line work, may also require only a few beams.

For some types of astronomical work, the ability to form images is desirable. There are several ways in which this can be accomplished, but they all would require an increase in the complexity and cost of the data processing for the array. The proof-of-principle for two-dimensional FFT operations can better be established for a pure phased array, such as the OSS of Section 6.6, page 162, or the systems under development as SKA prototypes at NFRA.

We recommend that the 1hT first be set up to form a relatively small number of beams, but at least three. Later, when the cost of signal processing drops, it can be upgraded to a larger number of beams, perhaps matching the average number of SETI target stars in the primary beam.

Instantaneous Bandwidth

For SETI purposes, a bandwidth of at least 0.5 GHz will be required for efficient use of the array. The current SKA specification is 0.7 GHz at 1 GHz rising to 2.5 GHz at 10 GHz. At present, the DSP for such large bandwidths would add significantly to the cost, so the larger bandwidths might profitably be delayed until late in the project. It would be best to bring back the entire 1 to 10 GHz range from each 1hT antenna to the central digital signal processing facility, so that the bandwidth can be expanded as the DSP technology improves. Current analog optical fiber can already support this architecture; whether it can be done at a reasonable price remains to be determined.

RFI Abatement

The current Phoenix system has had great success in rejecting interference with a very long baseline pseudo-interferometer. If we wish to preserve this technique with the 1hT, then we will need a second system, with at least 10% of the sensitivity and located at least a few hundred kilometers away. Since the multiple

Figure 8.3: Dr. John E. Carlstrom, an innovative telescope designer from the University of Chicago, stirs up interesting table-talk.

beams used for SETI will be scattered over the primary beamwidth of the small parabolas, this remote site cannot be just a big ($\approx$ 30 m) parabola. The best option probably is a smaller version of the 1hT.

Another option is to split the 1hT area into two identical systems. Each system would be run with a lower threshold, and consequently a higher probability of false alarms from noise. This would be remedied by requiring coincidence detection. For detected powers with exponential statistics, this provides the same sensitivity as for an undivided array.

An alternative approach is to adopt an RFI discrimination system, similar to that now being used by the BETA system. By observing several targets at once, we can use the anti-coincidence principle, in which a signal is considered interesting if and only if it is detected in just one of the beams. If or when full two-dimensional beamforming is implemented those techniques might be used instead.

Systems based on arrays of a large number of elements hold the greatest potential for dealing with RFI. As long as an interfering source is not in the same direction as a SETI signal we can tailor the spatial response of our radio telescope to null its reception of the interference, without losing much gain for the SETI signal. Such nulling techniques need to be developed as a key component of the SETI instrument design, and the 1hT will be an excellent test bed for these innovative studies of RFI elimination.

Preliminary Design, Budget, and Plan

The overall cost of the 1hT will be minimized for some choice of element diameter between three and eight meters, but the minimum is rather broad. As noted in Appendix E on page 335, the best performance, with respect to T_{sys}, and the lowest overall cost, would be obtained by a design using several, octave bandwidth feed horns with accompanying LNAs. Unless we use a complex, turret-style, front-end system, which seems impractical for an inexpensive telescope, there does not appear to be a good wideband solution that covers the 1 to 10 GHz range. Thus, for our preliminary design, we adopt a log-periodic feed (Appendix D on page 329) and a single 1 to 10 GHz LNA. This choice reduces the electronics cost and drives the design to smaller diameters.

Another factor also drives the design to smaller diameters. Preliminary engineering studies of the antenna mounts suggest that the mounts will be somewhat more expensive, relative to the surfaces, than is assumed in Appendix E starting on page 335. In addition, the best pointing accuracy that can be obtained from relatively small, low cost, commercial-grade bearings is about two minutes of arc. For a pointing accuracy of 0.1 beamwidth at 10 GHz, this translates into an antenna size limit of about five meters.

Finally, since the 1hT will be built much sooner than the SKA, we take the cost of computing to be $50 per Gop/s, which is close to current values. Adopting a target collecting area of exactly one hectare results in an array with the parameters and capital costs shown in Table 8.2. Development and personnel costs are not included in this table, nor is inflation.

Item	Value	$ Cost	Notes
Element Parameters			
Diameter	5 m		Solid, RMS 0.6 mm
Beamwidth at F_{min}	3.4 degrees		
Beamwidth at F_{max}	0.34 degree		
Pointing Accuracy	0.03 degree RMS		Wind $\leq$ 15 mph
Area	20 m^2		
Ap_{eff}	50%		
A_{eff}	10 m^2		
Frequency Min	1 GHz		
Frequency Max	10 GHz		
LNA Temperature	80 K		
LNAs/Element	1		
Feed	log-periodic		
T_{sys}	35 K		
Element Cost			
Surface		1,700	
Mount		5,000	
Dual Polarity LNA		1,100	
Cryogenics		2,100	
Feed		100	
Other		2,100	
Total per Element		$12,100	
Array Parameters			
Collecting Area	10^4 m^2		1 ha = 2.5 acres
A_{eff}/T_{sys}	143 m^2/K		
Number of Elements	509		
Number of Beams	3		
Instantaneous Beamwidth	500 MHz		per polarization
Beamforming DSP Rate	9.2×10^{12} ops/s		
Array Capital Costs			
Elements		6.2 M	
Beamforming		0.5 M	at $50/Gops
Array Software		3.0 M	
Other		1.0 M	
Array Labor Costs		4.4 M	
Total Costs		$15 M	

Table 8.2: Straw Man 1hT Proposal.

Figure 8.4: The SETI Institute's Dr. Richard Smegal brings astronomical perspectives and telescope design ideas from Canada, Australia and the USA. (Photo courtesy of R. Smegal.)

The straw man 1hT system, summarized in Table 8.2 on page 252, could be built by 2005. Starting in 1998, this project would consist of three phases:

Phase 1, 1998 through 1999, 1 to 3 full time equivalents (FTEs) plus about $100,000 per year for parts, etc.:

- feasibility studies and conceptual design;
- select site and begin approval processes as needed;
- first element prototypes and preliminary testing;
- prototype LNA and feed antenna control software;
- preliminary testing of one element – Preliminary Design Review (PDR).

Phase 2, 2000 through 2001, 4 to 12 FTEs plus about $300,000 per year for parts, etc.:

- test and revise antenna fiber optics;
- tool design;
- build prototype array of 4 to 16 elements using/testing prototype tools;
- preliminary narrowband beam former;
- develop array control software on prototype array.

Phase 3, 2002 through 2005, 12 to 20 FTEs plus about $2 M per year:

- test prototype array;
- finalize designs;
- revise tools;
- design 500 MHz backends – Critical Design Review (CDR);
- write specifications and bid contracts;
- build array;
- build backend;
- integrate, debug and test array.

These figures do not include inflation. Also, a set of SETI detectors will be needed. Their cost is not included, but will probably be negligible compared to the cost of the array. Adding in the labor and development costs will probably bring the project total to around $15 M. If, as suggested below, the 1hT is built in collaboration with a radio astronomy group, the SETI Institute's share may be less than $10 M.

It should be stressed that these budget numbers are very uncertain. A reasonably reliable budget will not be possible until the preliminary design review at the end of 2000. To reduce the total cost, A_{eff}/T_{sys} could be reduced from the value of 143 m^2/K, assumed in the table, to the minimal value of 100 m^2/K. This option is not very attractive because it would not reduce the development and manpower costs appreciably. It would, however, lead to a system cost that is dominated by development, rather than production. In a sense, the 1hT would be more economical if scaled up to $A_{eff}/T_{sys} \approx 200$ m^2/K, since the production costs would then be 2/3 of the total cost.

Operations and maintenance will require at least four to five FTEs plus about $100,000 per year. The design of the array should encourage ongoing backend development, for example to increase the number of beams. This effort would require several FTEs, and perhaps another $100,000 per year.

Cooperation

Since one of the primary goals of the 1hT is to facilitate the SKA, close ties should be maintained among all of the development teams. The problems of beamforming and RFI suppression are common to all the design efforts, so cooperation in this area should be especially beneficial.

Good opportunities for cooperation also exist with other disciplines. The 1hT presents a fine chance to work with the radio astronomy community, with benefits both technical and political. SETI cannot afford to be off the air very much, but sharing resources with astronomers should be possible, given the flexible, multibeamed nature of the 1hT. The Deep Space Network (DSN) presents another opportunity for technical collaboration, although their mission critical needs would not permit them to actually use the 1hT. In addition to the mainline observing projects sponsored by the SETI Institute, it should be possible to support a variety of other SETI programs, selected by peer review.

It would be highly desirable to build the 1hT in collaboration with an established radio observatory. Not only would this foster cooperation between astronomy and SETI, but it also would provide badly needed expertise in telescope design, construction, and operation. It also offers a chance to share at least a part of the cost of the system.

8.3 Omnidirectional SETI System (OSS)

Figure 8.5 shows the SETI Decision Tree with the appropriate elements highlighted for the OSS.

Goals

The primary goal of the OSS is to continuously observe the entire sky, and search for strong transient signals that would be missed by targeted strategies or all-sky surveys that employ tessellation techniques.

Preliminary Requirements

In Section 6.6 on page 162, we considered the requirements of an OSS facility that would simultaneously and sensitively cover the whole sky at either radio or optical wavelengths. We found that the computing requirements for beamforming alone exceeded 10^{22} operations per second for any conceivable system. Primarily, this was due to the large number of elements needed for a large collecting area in the radio, and the large number of directions in the sky in the optical. For example, at microwave frequencies, a square kilometer of collecting area can plausibly contain a billion elements, the signals from all of which must be combined to form a hemispherical beam. This is beyond our expectations for affordable computing in the next decade or two.

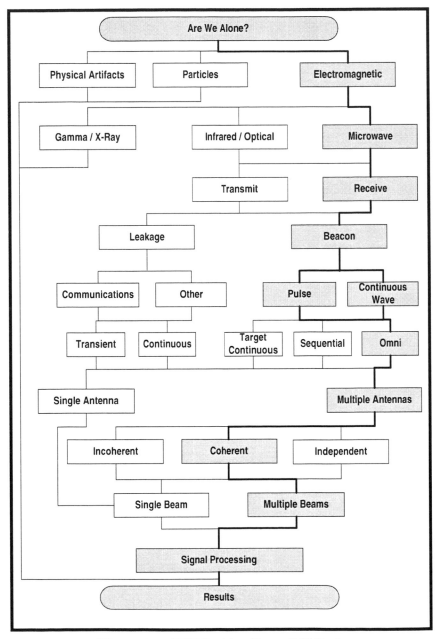

Figure 8.5: SETI Decision Tree – OSS Path.

However, with exponential growth as the currently most plausible model for computer related activities, we would be wise to plan a SETI system with capabilities for all-sky coverage which can grow indefinitely as computation becomes less expensive. At the beginning, such a system would be an R&D effort, primarily because it will require the solution of many unsolved problems. This is particularly true in specialized SETI work, where the optimum solution for narrowband technological signals or rapid optical pulses may be different from those required for broadband radio astronomy. Further, detailed knowledge of the best materials from which to construct a truly omnidirectional antenna is still being gathered by groups outside the SETI Institute.

The situation is the opposite of that for the 1hT, where early on significant money will be spent on developing the collecting area and receivers. The objective of this research effort will be to spend funds on proof-of-concept computing and people until fundamental problems are understood, and only later to spend money on hardware. An exception to this general rule is the SETI-specific signal processing hardware. Such hardware has a long development history at the SETI Institute, and clearly can be used effectively no matter what the final antenna system is. This signal processing hardware will be discussed later.

Real data will be needed in important areas like RFI excision and self-caused interference of distributed networks. Many problems associated with timing and dynamic range, however, can be solved by computer simulation and mathematical analysis. Methods that solve problems for the OSS could solve similar problems for the 1hT, an instrument unique to the SETI Institute. It is possible that even a small instrument could capture ETI signals at an unexpectedly early phase of its development because of the uniqueness of its observations. This is an important reason for building a real instrument as soon as possible.

The unexplored signal domain is still vast. Factors of millions separate the most sensitive searches from signal strengths that would be routinely noticed by our society. Within this vast range of signal strengths, signal types and duty cycles, there is much room for unique observations, even at an early stage in the development of a complete survey. In addition to the sanity check always needed for intensive computer simulations, the possibility of detecting an ETI signal early on is another reason why a collection facility for real data must be incorporated into the OSS R&D effort.

Current Challenges

It is worthwhile to enumerate the problems that must be solved in order to develop a complete electromagnetic survey of the sky. It should be noted that several groups have understood the need to work on many of these issues. Two prominent groups are the Dutch astronomers and engineers at NFRA, and a group at Ohio State University (OSU), on a much smaller scale. Thus, in addition to R&D efforts focused on SETI issues and continuing exponential growth, we should collaborate with and support other related efforts whenever possible.

For the OSS, the list of issues that must be understood is daunting:

1. **Beamforming** For the 1hT, dishes will be large compared to a wavelength, and the number of beams formed for SETI within a primary field of view may reach 100. It is plausible that such beams could be formed directly as sums, using time delays to compensate for the different times of arrival of the wave front at different antennas. This is an excellent wideband solution. However, as the number of beams becomes large, the number of small, omnidirectional elements also becomes large. Thus, additive beamforming with time delays is less efficient than the $n \log_n$ FFT approach. Unfortunately, approaches based on the FFT method rely on wave phase rather than timing. Is it better to form time delayed beams and perform wideband, high data rate computations, or is it better to divide into bands where phase delays will give essentially correct beams? This is an important question for any omnidirectional antenna, and for antennas like the 1hT if they grow to become an SKA.

2. **Commonality** In the near future, how much of the OSS effort can be used for the 1hT? The obvious candidates are digitizing and timing strategies, beamforming, and data storage. It is worthwhile to investigate commonalities between SETI, general radio astronomy, and radar astronomy. What are the technical and networking difficulties in turning an omnidirectional receiving array into a generalized transmitting array? Theoretically, this is possible. Can it be done with current technology at a few dollars per square meter?

3. **Cryogenics** Clearly, as receivers become more numerous and smaller, bandwidth and cooling become issues. For the 1hT, the use of cooled receivers for optimum cost effectiveness is favored. For the OSS, it may be that the complexities of cooling should be abandoned in favor of large numbers of elements to achieve the best sensitivity.

4. **Sky Coverage** There are deep questions about how best to cover the sky. Should we use overlapping beams, or a regularly or irregularly spaced array, or closely or widely spaced elements? These questions become more complex when the relatively new area of interference excision is considered.

5. **Terrestrial Interference** How can terrestrial interference be removed? Should some generalized technique be developed? Is it better to use subtraction with similar arrays close together or anti-coincidence with two arrays that are widely separated? The current Project Phoenix system uses a pseudo-interferometer, with a signal elimination element much smaller

than the main detection element. Is this still an optimum strategy when one has more antenna flexibility?

The complementary and overlapping themes of the 1hT and OSS developments should be emphasized. While the 1hT strives to develop a significant, usable, collecting area in the next decade, the OSS will be developing processing and networking methodologies for the following decade. If interesting signals are detected, it will be a great bonus. We do not know exactly how follow-on generations of telescopes will be built because of our uncertainties about the development of detector elements and their most cost-effective size. The OSS activity will help ensure that we have a range of techniques to apply to the optimum hardware solution.

6. **Array Dispersal** We do not currently know how dispersed we can make arrays and still have the ability to combine their signals with precision. How should the timing information be entered into the data stream, and how accurate can we expect it to be? With only a few antennas, perhaps movable ones, and with all data being digitized and timed at each antenna, we can perform practical experiments in the context of the omnidirectional array to answer such questions. This same flexibility will allow a range of interference excision tests.

In addition to these questions about array construction and beamforming which are largely computing optimization issues, we must solve questions of data processing. These questions include: what platforms should be used, what fraction of the task requires specialized hardware, should the data processing be distributed near the collection elements, what are the networking requirements, is data recording technology useful, etc.?

7. **Computing** It would be desirable to have the selection of a particular observing mode be equivalent to the allocation of computing time. In other words, an astronomer desiring some particular array configuration brings his computer and forms his array. Except for a data spigot, the demands on the antenna facility are zero. The question for SETI is, how should computing be done cost-effectively? We wish to perform wideband spectral analysis. Is this function likely to continue its exponential growth using specialized chips, or will generalized computing eventually catch up? The commercial utility of spectral analysis may guarantee that specialized cost-effective chips for this purpose will always exist.

More important, is it necessary to have a whole set of computers, running a real-time operating system, carefully interlocked so that they can deliver precisely synchronized data to an observer upon demand? It may be that timing information in individual data streams will be sufficient. In that case, many low cost computers with UNIX-like and Windows-like operating systems could serve the needs of the system cost-effectively. One of the first requirements of the OSS Project is a study of this issue for beamforming and data processing.

8. **Local Processing** As systems grow larger, one needs to address the issue of where the data processing should be located. What data processing can be performed near the antenna elements? Any such capability decreases the load on the networks.

9. **Networking** What network architecture would be best? As new high-speed networks come into general use, it is not obvious which will be most cost-effective. However, with literally hundreds of thousands of elements, it is obvious that we should not design specialized networks. Such one-of-a-kind development is too expensive. The money would be better spent on computing and antenna elements.

10. **Data Storage** Last, but not least, is the issue of data recording. It is always better to identify an ETI signal in real time. The major reason for this is that, almost always, the classification of a signal as being from an ETI is wrong. For real time confirmation of an ETI signal, it is necessary to bring systems on-line that are not part of the initial detection process. The confirmation systems are less susceptible to errors of judgment or electronics than is the initial detection system, because they are dealing with a greatly reduced data rate.

Thus recorded data, even when complete, is less satisfying. Its status is likely always to remain ambiguous, if only because its reality is suspected by those not involved in its collection. The data might contain systematic errors, or be an outright hoax. However, recorded data could give interesting clues to real time observers.

There is another more important reason for looking into recording technologies, especially in the early stages. While many computationally intensive beamforming and data analysis techniques are being investigated, even a relatively narrow data stream cannot be processed in real time and tested in all interesting ways. It may be worthwhile to record data from an array in a snapshot mode. This could then be processed to the exhaustion of data and programmer alike before new data needs to be collected. After the system grows to the point that real time processing is possible, data can still be stored for a short time using the same facility. Thus, if a signal is detected in real time, its recording of the most recent history will permit additional testing of the candidate signal. Since the real time capacity of the system will certainly exceed the recording capacity by this date, this off-line process will not significantly change the computational load of the array.

Incremental Approach, or How to Get Moore's Law on Our Side

The budget for the OSS is strongly dominated by signal processing costs. These costs have been mitigated by Moore's Law for the past several decades. Our industrial colleagues assure us that this remarkable doubling of capability every 18 months for the same price will continue for at least the near and medium term. Money should be spent incrementally in a project where capability improves 2 dB per year.

The target year for getting the OSS on the air is approximately 2015. We have to wait that long to make the full project affordable in terms of signal processing. But just waiting is probably *not* beneficial to the project, since we need to ramp up our signal processing skills along with the hardware.

Table 8.3 shows an alternative. The single OSS project is broken down into a sequence of mini-projects, with an estimate of the computational costs for each, not including personnel. Each row in the table is a distinct project that lasts about four years, or about a graduate student generation. It starts small. The value in the last column giving the cost of computing in the first mini-project (2000 through 2003) is \$40,000. This should be enough to provide the 4×10^{10} operations per second needed to analyze a 4 MHz bandwidth for a four element array of wide-angle inexpensive elements.

Year	Array	Bandwidth	Computing (operations)	Cost
2000	2×2	4 MHz	4×10^{10}	\$0.04 M
2004	4×4	16 MHz	6.4×10^{11}	\$0.1 M
2008	8×8	64 MHz	1×10^{13}	\$0.25 M
2012	16×16	250 MHz	1.6×10^{14}	\$0.63 M
2016	32×32	1 GHz	2.6×10^{15}	\$1.58 M
2020	64×64	4 GHz	4.2×10^{16}	\$4.0 M
⋮	⋮	⋮	⋮	⋮
2036	$1,024 \times 1,024$	1,000 GHz	2.5×10^{21}	\$158 M
NOTE: Personnel costs are not included.				

Table 8.3: Cost of Computing for the Omnidirectional SETI Systems.

The next mini-project increases both the number of elements and the analyzed bandwidth by a factor of four. This step requires a factor of 16 (12 dB) more computing, but by this time Moore's 2 dB/year increase has provided 2/3 of this increment. That leaves 1/3, or 1 dB per year, as the required annual ramp-up in the project budget. Since a factor of 16 provides a qualitatively different result, the steps are large enough so that there will be a significant learning curve in each mini-project. This is just what we want: the project is enabling us to develop the skills that will be needed later. Mistakes can be corrected before they become serious. The 1 dB (25%) per year rise in the project budget is enough to teach us how to deal with change, but avoids the hyper-growth which would be needed if the project were to start from scratch in 2012. The project as planned would be virtually complete in 2016 with 1,024 elements. By that time, however, we may have become used to the idea of a constantly growing capability. If Moore's Law still applies, then the last three rows (2020 through 2036) show what might be possible.

This forecast, however, is predicated on the OSS budget being totally dominated by signal processing costs, and with aperture costs being nearly free. Once aperture costs become significant, the rules change dramatically:

- The optimal fraction of the budget devoted to Moore's Law components (signal processing) drops from 100% to as small a value as practicable – perhaps 20%.
- The disposable aperture concept is abandoned. It is replaced with an aperture that is carefully designed to last a very long time and require very low ongoing maintenance.

The reason is simple. Aperture value increases with inflation, whereas Moore's Law components *lose* value with time. It makes no sense to spend 80% of our budget installing signal processing components that we could install for 20% in three years time without losing capacity.

One final technical point should be considered. What is the trade-off between aperture and signal processing? In traditional antennas, signals from the whole aperture of a telescope are combined coherently at the focus. Thus, investing in additional coherent signal processing in either frequency (for pulses) or time (for CW) yields the same sensitivity improvement as does investing in aperture. If the bandwidth is constrained, incoherent integration is required for both pulses and CW. However, incoherent averaging improves sensitivity only as the square root of signal processing power. Hence, in the incoherent case, the sensitivity advantage of Moore's Law drops to 1 dB/year. Increasing the aperture at that rate is competitive with Moore's Law.

8.3.1   Budget and Plan for the Incremental Approach

The OSS should begin as an array of four antennas. The total collecting area of such an array is very small, of the order of 0.01 m^2, depending upon the wavelength and directionality of the elements. This system can be designed to

Figure 8.6: Dr. Michael M. Davis of the Arecibo Observatory, enumerates details of the Arecibo upgrade.

handle a bandwidth of approximately 4 MHz, tunable over the standard SETI range from 1 to 3 GHz. Data will be separately 'time-tagged' and digitized from each antenna. The data from each antenna will be fed into its own computer. The computers will be networked to a central processor where beams can be formed. The advantage of such a system is that each element, collector, and computer is small and portable.

Since timing is coordinated at the element level, a wide choice of networks can link the array. Thus, portability to test interference excision and various array configurations would be ensured. Since no special cooling is proposed, collecting elements can be inexpensive, for example, inexpensive dipoles could be used. The computers processing the data from the collecting elements would be PC class computers costing a couple of thousand dollars apiece. The central processor would be a workstation class machine costing about $20,000. The recording technology is certainly available as either a Redundant Array of Inexpensive Disks (RAID), or as a broadband recorder. This would permit both real time and exhaustive tests, at a cost of approximately $100,000.

The proposal is to work with this simple array through about 2003, with about four FTEs. This will give time for element technology to develop, and for the SETI Institute group to integrate its efforts with others working in this area. Other groups may be much further along in some areas, such as digital interference excision. However, the SETI Institute has significant expertise in SETI signal detection, and this first effort would optimize it and fit it into the larger astronomical picture.

Once digitization levels, networks, and computer architectures have been stabilized, it is time to invest in hardware. The computing industry is changing so rapidly now that the only surviving relic of this first stage may be algorithms for signal processing and networking, and no hardware at all.

Item	Estimate
8 FTE Years	$ 800,000
10 Elements	$ 40,000
Networking, etc.	$ 60,000
Data Processing (10 Workstations)	$ 200,000
Total	$1,100,000
NOTE: The 10 units (element + workstation) are to be used as twin clusters, each with four units and one spare.	

Table 8.4: Near Term Budget for OSS for the Period 2000 through 2003.

The estimated costs for the first few years of development are shown in Table 8.4. Extrapolation to further development stages seems so dependent on the first stage, that this is not the right time to develop a detailed long-term funding plan. For example, the detailed mid-term plan will depend upon whether the Dutch development of small, inexpensive antenna elements and experiments at Ohio State University succeed. However, in later stages, it is expected that about the same number of researchers will be required to keep the effort moving, about four FTEs. Meanwhile, the hardware efforts will be greatly increased.

8.4 Optical SETI

Since optical SETI is a new venture, the costs and timelines are not well defined. However, approximate costs and requirements are included in Table 8.5. These estimated costs assume that optical SETI will remain a research activity until 2005, primarily conducted at universities. A review of the technical status of this field is targeted for 2005, and a $10 million placeholder has been inserted into the budget table to accommodate a dedicated optical SETI facility, with operational funds through 2020.

Year	Phoenix Field Operations	Institute 1hT/SKA†	UCB RAL/NSF‡ Operations Support	OSS	Optical SETI	In-House Science/ Tech. Staff	Totals $M
2000	2.50	2.0		0.3	0.2§	2.0	7.00
2001	2.50	3.0		0.3	0.2§	2.0	8.00
2002	0.75	5.0		0.3	0.2§	2.0	8.25
2003	0.75	5.0		0.3	0.2§	2.0	8.25
2004	0.50	5.0		0.3	0.2§	2.0	8.00
2005*	0.50	5.0		0.3	0.3§	2.0	8.10
2006		3.1	2.0	0.3	5.0	2.0	12.40
2007		1.0	2.0	0.3	5.0	2.0	10.30
2008		1.0	2.0	0.4	1.0	2.0	6.40
2009		1.0	2.0	0.4	1.0	2.0	6.40
2010		1.0	2.0	0.4	1.0	2.0	6.40
2011		1.0	2.0	0.4	1.0	2.0	6.40
2012		2.0	2.0	0.6	1.0	2.0	7.60
2013		2.0	2.0	0.8	1.0	2.0	7.80
2014		1.0	2.0	1.0	1.0	2.0	7.00
2015		1.0	2.0	1.2	1.0	2.0	7.20
2016		1.0	2.0	1.6	1.0	2.0	7.60
2017		1.0	2.0	2.0	1.0	2.0	8.00
2018		1.0	2.0	2.5	1.0	2.0	8.50
2019		1.0	2.0	3.2	1.0	2.0	9.20
2020**		1.0	2.0	4.0	1.0	2.0	10.00
TOTAL	7.50	44.1‡	30.0	20.9	24.3	42.0	168.80

† In this time frame, additional fund-raising may be required to support SETI access to the SKA.
‡ RAL/NSF – UC Berkeley Radio Astronomy Laboratory/National Science Foundation.
* 1hT is completed in 2005. 1hT/SKA funds thereafter can provide a step function, such as expansion of the 1hT, or a contribution to SKA construction.
** OSS reaches 4,096 elements in approximately 2020.
§ These funds will be used primarily to support research outside the SETI Institute.

Table 8.5: Budget for Fiscal Years 2000 through 2020 ($Millions).

8.5   Resource Requirements

8.5.1   Total Funding Requirements by Project

Obviously, long range recommendations that cover decades can be addressed only in broad cost terms. For the period 2000 through 2020, the grand total funding requirement for the three SETI projects recommended by the SSTWG is $168 million, in FY 2000 dollars. Following are some comments about the budget.

1. $7.5 million are required for Project Phoenix and continuing field operations (2000 through 2005).

- The new SETI signal processing system (the NSS) is being implemented in phases beginning in 2000. The total bandwidth will be 100 MHz and the expected cost of $5 M has been allocated over 1999 through 2001.
- Telescope rental fees at Jodrell Bank and deployment costs for Arecibo and Jodrell Bank will be approximately $750,000 through 2003.
- As the 1hT begins to be operational, transfer of SETI operations to that site and development of remote observing capabilities will likely cost $500,000 per year for two years. Thereafter, observing expenses are subsumed under the annual budget for the 1hT/SKA, the operational costs are supplied by other agencies, and the personnel are line items.[5]

2. $44.1 million are required for 1hT and SKA (2000 through 2020).

- The 1hT budget totals $25 million through completion by the end of 2005. This is the current target for a capital fund-raising campaign at the SETI Institute. As an early next step, a detailed breakdown of annual budgets into their major components must be developed. Major components are labor, antennas, receivers, signal detection systems, system control systems, etc. Development milestones must also be set.
- The estimated costs for the 1hT do not include any funds for construction of SETI signal processing equipment. Using a figure of $5 M/100 MHz bandwidth derived from the current NSS development, and Moore's Law reduction in cost, $3.1 M has been added to the 2006 budget.
- After 2006, the funding in the *1hT/SKA* column is shown at a steady level. This permits funding a mixture of different activities, based upon the circumstances at those times. Possible, but not mutually exclusive,

---

[5]Since this text was prepared, the Paul G. Allen Foundation funded the R&D phase of the Allen Telescope Array, with the expectation that successful completion of the technology challenges will result in funding for constructing the array. Nathan Myhrvold provided funds for an electronics laboratory for the ATA.

activities include: maintenance, operations, constructing additional antennas and systems for the same array, duplicating the 1hT in another location, directly contributing to early phases of the SKA, etc.

- The maintenance and operations portions of these funds, for 2006 through 2020, is assumed to be provided by the Radio Astronomy Laboratory of the University of California at Berkeley, in conjunction with the NSF support of the Hat Creek Observatory. If the 1hT is built elsewhere, then these costs will become an addition to the required donor support for the project.

- The SETI Institute would like to be able to process 10 of the 100 beams from the SKA, which should begin to be available in 2012. An additional million dollars has been added for two years to cover the back-end costs of signal processing from the SKA. A larger and unanswered question is whether the SETI Institute will have to pay the then-current rate for the SKA, or whether its contribution to the prototype development of the SKA, via the 1hT, will guarantee access to the desired number of beams from the array. Another very significant capital fund-raising campaign may be necessitated from 2008 through 2009, when the SKA construction commences.

- However, throughout this entire time, the opportunity exists for a major donation step function. For example, a $25 million gift would enable us suddenly to build a second 1hT, or double the size of the first, or provide a huge incentive to the SKA effort.

3. $21 million are required for OSS (2000 through 2020).

- The OSS figures follow the master plan as presented by Mike Davis during the SSTWG meetings. The impact of Moore's Law results in the late investment, after early R&D into how this mission is accomplished. The next step is to associate each annual budget figure with array size, bandwidth, and the operations per second computation numbers, ending up with 4×10^{16} operations in 2020.

4. $24 million are required for Optical SETI (2000 through 2020).

- It assumes small costs for early experiments identified by the Optical SETI team, and a very gradual rise in expenditures over time.

- It is assumed that new developments, especially in infrared photon detection systems, will allow some expansion of this new venture.

- In Table 8.5 on page 265, the costs for the early years of optical SETI work are marked '§' to indicate that the bulk of these funds will be used to support research outside of the SETI Institute. If a larger scale endeavor is endorsed following the 2005 review, that effort will likely become an in-house program.

- An estimated $10 million has been added in 2006 through 2007 to reflect the potential need for a dedicated observing facility.

5. $42 million are required for in-house Science and Technical Staff (2000 through 2020).

- This is required to support the overall enterprise.

8.5.2   Budget by Year and Project

Table 8.5 on page 265 provides a detailed breakdown of the several projects and costs through the year 2020.

8.5.3   Additional Resources

In addition to the specific programs listed in the budget, Table 8.5, there are other activities that should be carried out within the SETI Institute and elsewhere to keep the field viable and vibrant. The desires of the SETI Institute will probably always outstrip available funds, but these activities are crucial to the future.

First, and foremost, is the cost of development and fund-raising, which makes everything else possible. Because of the brand-name recognition of the SETI Institute and its outstanding credibility, it should be possible to keep this cost to less than 10% of the funds raised. A related set of in-house activities involves public outreach and public education. If the SETI Institute wishes to continue to maintain its brand-name status, it must provide information and factual materials describing its projects to as large a segment of the population as possible. There is no clear demarcation between those outreach activities that inform and those that develop interests that eventually lead to gifts.

If the SETI Institute is to grow to encompass the many projects recommended by the Working Group, then a first class effort must be expended on all these activities. New scientists, engineers, and instrument builders are vital to the long-term health and success of the SETI Institute. Science education is therefore a vested interest for the SETI Institute. Specific grant requests will fund new projects in education, but a core education staff is required to generate those proposals and lead the efforts. In addition, there will be an ever increasing burden to serve the teachers and students in the classroom with updated materials on the Web. Funds for these services will need to be built into the costs of any educational products that are marketed, but the expertise and skills should reside at the SETI Institute, not with a publisher or distributor.

The SETI Institute also has a responsibility to work with social scientists, economists, and political figures worldwide to try and better understand the consequences of the successful detection of a signal, and to decide what preemptive actions might be taken to effect the most positive outcome.

Like any enterprise that wishes to remain viable in a rapidly changing technological environment, the SETI Institute must invest in basic research and look for ways to incorporate the results of corporate investments in research and development. Small grants should be made available to seed novel search efforts. These dollars are precious, and perhaps the most difficult to raise.

The costs for all these necessary, non project related activities will probably total about 25% of the budget in Table 8.5 on page 265, averaging about $2 million per year.

8.5.4   Scalable Design Philosophy

The Working Group identified a dilemma in establishing the budget targets. On the one hand, the highly experimental nature of the endeavor makes it risky to establish premature budgetary estimates when so many unknowns are still involved in the process. On the other hand, it seems entirely possible to fix the cost (or budget), work with these numbers, and adjust the pace and scale of the projects. This can be accomplished with a scalable design philosophy. Successes with early milestones allows one to carry forward that success and knowledge into the next design. In the case of the 1hT, the Working Group cautions against overdoing early precision in an attempt to have it exactly right at the inception, because the eventual array will include as many as 1,000 antennas. Once the design is known, the SETI Institute needs to work with production experts who know how to work with suppliers and get costs down for thousands of items.

It is recommended that the SETI Institute partner with other expert organizations whenever feasible. For instance, the Working Group notes that the SETI Institute team can act as a primary player in the back-end work, which is computing intensive, and apply significant in-house expertise to this. In contrast, antenna engineering is not an area of expertise within the SETI Institute, but ample opportunities for partnerships exist and should be maximized. [6]

In the infrared and optical, the SETI community initially needs to do many small experiments. Then, over a longer period, it should pursue opportunities to get more star coverage and better sensitivity. An exception to this might be 'Spaghetti SETI', a 10^6 star, targeted pulse search. Another exception might be a sky survey using technology derived from Cerenkov air shower gamma ray

---

[6]Since this text was written, the SETI Institute entered into a partnership with the Radio Astronomy Laboratory of the University of California, Berkeley to design and prototype the ATA.

telescopes. The budget provides a gradual increase in expenditures to reflect the increased investment over time.

With the OSS design effort, the Working Group expects that early R&D investment will have significant payoff in better understanding of how to do the job. The budget reflects this modest early investment, which is increased as the effects of Moore's Law and experience in active RFI excision, from 1hT and SKA work, make building the system more cost-effective.

The Working Group cautions that it is always hard to develop software, hard to predict what it will cost, and hard to predict how long it will take. It usually takes several tries to get it right. However, these are facts of life, and the SETI Institute needs to understand that it is just a complex kind of task. The concept of outsourced software development should be investigated. It could provide low cost, highly skilled, experienced resources for the software development effort. The key to making this approach successful is writing excellent specifications for the software.

8.6   Strategic Implementation

This book describes an ambitious plan for the continuation and extension of the SETI Institute projects over the next twenty years, including the funding requirements for the three major components of the program. It is now necessary to address how this plan will be implemented.

In this section, we are neither concerned with tactical approaches to implementation, nor with the fine details of everyday science and technology tasks. Instead, it is now critical to examine the *strategies* to be used for ensuring that the new SETI mission is actually accomplished in the most efficient and effective way possible over the twenty-year period.

This section divides the various facets of strategic implementation as follows:

- The philosophical approach to be taken by the SETI Institute itself is discussed: including those aspects of governance, management, structure, facilities, and operations which have the most direct bearing on strategy.
- Liaison with the scientific and engineering communities, public programs and education, and the acquisition of resources, including private and government support are considered.
- International collaborations and agreements, and the need for an overall strategic plan for the SETI Institute are explored.

Since the primary focus of the Working Group was on science and technology, it did not include experts in the fields of science education, mathematics education, and 'SETI and society'. Nevertheless, these areas are considered to be integral to and important for SETI. Some recommendations to the SETI Institute for future studies in these areas are given in the *Epilogue*, starting on page 277.

SETI is one of two major enterprises being conducted by the SETI Institute. The other is basic research in "Life in the Universe", which addresses the scientific foundations upon which SETI stands. The SETI Institute's future studies of "Life in the Universe" will be the subject of a plan comparable to this one, and will include its own strategic implementation section. Needless to say, the two plans must be complementary.

8.6.1   The SETI Institute

The SETI Institute should build on the strengths it has developed, and aspire to maintain its leadership in the field. It should slowly expand its niche at the frontier of science and technology, and encourage colleagues to collaborate in the SETI enterprise in a variety of ways. It must step up to the need to fund such collaboration. It should, at the same time, be both bold and persuasive in vision, and rigorous and efficient in execution. Effective external communication of the achievements and spirit of SETI to wide audiences is essential for continuing success. Some 5% to 10% of resources should be set aside for basic research in engineering and science. Project planning and implementation should combine clear objectives and procedures with the ability to respond flexibly to changes in scientific, technical, political, and institutional circumstances. Continuation of the existing interdisciplinary method of working is a top priority. Maintenance of close contacts with the scientific community is essential.

The SETI Institute should vigorously pursue the special skills and inventions which have already emerged. These include powerful algorithm development, radio frequency interference rejection, unique observational strategies, commensal projects with astronomy and other disciplines, new telescope and system designs based on Moore's Law, omnidirectional sky surveys, prototype tests for the SKA, and optical SETI. The SETI Institute should rely on initial collaboration with others for observatory development, management and operations. Then it should decide which of these skills it should develop in-house, in the future.

The SETI Institute should expand its studies of "Life in the Universe", and seek recognition that it is a significant player in this field. It should better define how this basic research is complementary to the underlying science of SETI, and to the astronomical and planetary science incorporated in SETI. It should also define how these factors influence the design, development, construction, testing, and operation of SETI telescopes and systems.

Last, but not least, and in spite of various desirable links with large institutions of academia, government, and industry, the SETI Institute should firmly retain its own unusual, stimulating, and innovative identity. It should also cherish and nurture all those philosophical qualities of governance and management that attract and keep good people and which inspire dedication, enthusiasm and productivity.

In summary, the SETI Institute must strive to enhance its record of achievements with firmly established credentials. A reputation for excellence and integrity is the top priority.

8.6.2   Liaison with the Broader Scientific Community

As part of the scientific community, the SETI Institute should vigorously maintain its determination to be a leader in the field, with a reputation for integrity and excellence. It is recommended that this vision be pursued for "Life in the Universe" science as well. A twenty-year plan should be developed for this area, with a clear indication of the sources of funds to support such work. The SETI Institute should endeavor to make more widely known that it is actually conducting significant studies on this topic. It would also be helpful to develop a clearer picture of the nature of the complementary relations between SETI studies and "Life in the Universe" studies.

The SETI Institute should continue a diversity of interactions with the greater scientific community. Peer review is essential. Collaborations are encouraged, especially in the field of astronomy. The 1hT, the OSS, and optical SETI have significant interactions with astronomical science, as described in various sections of this book. To some considerable extent, these telescopes and systems should continue to be deliberately designed to conduct both SETI and astronomy. It is also important to be aware of the possible serendipitous discovery of new astronomical phenomena during searches carried out with the new SETI telescopes and systems, as observations with them begin to open up new dimensions of astronomical search space.

Comparable joint endeavors should be pursued in the field of SETI engineering, and efforts made to interact with university and industrial research organizations. Further, SETI should encourage the participation of postdoctoral researchers in various aspects of the research. The infusion of new, young talent into the discipline is vital to the long-term future of the enterprise.

When adequate funding becomes available within the SETI Institute, small grants should be made available to outside organizations for innovative work in science and technology. A set of procedures for submitting and reviewing proposals should be established. Announcements of the grant opportunities should be widely disseminated.

SETI results should continue to be disseminated through the normal channels, and an increase in the number of papers published in refereed journals is desirable. The *Annual Report* of the SETI Institute should highlight these published results.

The interactions between the members of this Working Group have been so stimulating and productive that a mechanism should be found to continue them on a regular basis. Since plans are always changing in response to multiple variables, it is recommended that the SETI Institute form a standing committee to conduct a continuing review of the progress of implementing this SETI 2020 Plan. Perhaps some of the members of the current Working Group would be interested in playing a role. The committee could act, in part, as an ongoing peer review group, and ensure constant contact with the larger scientific and engineering communities.

8.6.3   Public Programs and Education

The public and educational programs should be continued and enhanced as resources allow. Both efforts are valuable for the dissemination of new knowledge in new ways, and they contribute to the wider understanding of science.

8.6.4   Funding and other Resources

The target figures for funding for the 1hT, the OSS, and optical SETI, as given in Section ?? on page ??, should be vigorously pursued. It is recommended that additional funds, shown as a percentage allocation (Section 8.5 starting on page 265), be set aside for basic SETI research, for the Development Office, and for education, public programs, and studies of SETI and society. Funding should be sought from private donors and from government sources. A combination of these two sources could result in a public/private partnership, which could have many advantages.

In thinking of resources for SETI, other types of contributions should not be neglected. Some examples are:

- 'in-kind' assistance from industrial research organizations, perhaps with reciprocal exchanges of people and technology;
- encouraging university faculty to take sabbatical leaves at the SETI Institute;
- volunteers who are willing to give their time to any aspects of SETI;
- the use of national and international data banks for archiving purposes;
- contributions from responsible elements of the media in disseminating the real SETI story;

Figure 8.7: The SETI Institute's Greg Klerkx makes a point.

- the inclusion of SETI studies in school and university curricula.

One of the purposes of preparing this SETI 2020 plan is to have it available for potential private donors, as a bold and resolute declaration of the overriding importance of discovering ETI life. Condensed versions of the book written in plain, attractive and easily understandable language should be prepared for a variety of audiences.

It soon may be appropriate to revisit the prospect of government support for SETI. It is recommended that an analysis be made of this possibility, starting with informal discussions at high levels, followed by formal proposals to NASA, NSF, or other agencies, at an appropriate time. There is now more awareness of the scientific challenge, and general excitement about extraterrestrial life, than existed five years ago. This stimulus comes from the continuing discoveries of extrasolar planets, from the possible discovery of extinct microbial life in the martian meteorites of Antarctica, from renewed speculation that life on Earth may have actually originated on Mars, and from the speculation about the interiors of Europa and Callisto as possible habitats for extraterrestrial life. Particular attention should be paid to how SETI complements the NASA Astrobiology program, to the development of new technologies which benefit both SETI and other disciplines, and to the use of new telescopes, like the 1hT and SKA, for a next generation DSN for communicating with interplanetary spacecraft. The SETI Institute will have the advantage of conducting its own privately funded SETI programs. Thus, there could be mutual benefits in a symbiotic public/private partnership.

8.6.5   International Cooperation and Agreements

The Working Group encourages the SETI Institute to continue its strong leadership in encouraging SETI worldwide, and in collaborating with SETI scientists in other nations. The role of the SETI Institute in the international SKA consortium should be given high priority. It should be proactive and under continuous review as the 1hT program emerges. Formal agreements should be developed among observatories around the world for the tracking of candidate signals, with the SETI Institute perhaps providing receivers and signal detection systems where appropriate. Studies of SETI and society should continue at an international level.

Finally, the SETI Institute should continue its pattern of devoting some small efforts to supporting international professional societies, such as the International Astronomical Union (IAU) Bioastronomy Commission 51, and the International Academy of Astronautics (IAA), by providing travel grants, meeting support, and establishing prizes for outstanding papers or reviews.

8.6.6   Strategic Plan for the SETI Institute

A Strategic Plan should be developed by the SETI Institute to cover all its activities. This obviously should incorporate the SETI material in this book, the forthcoming long range plan, and the implementation strategy for its "Life in the Universe" studies. Finally, it should thoroughly discuss the governance, management, organization, and operations of the SETI Institute.

Chapter 9

Epilogue

This epilogue, by John Billingham, looks out beyond the plans described in this book and considers the consequences of success in our search. The implications of successfully detecting an alien signal will depend on the nature of the signal and on factors well outside the physical sciences, including international agreements, space law, social science, etc.

The nature of a signal from an ETI will fall between two extremes. At one extreme are simple, information poor signals, while at the other extreme are complex and information rich signals. The first extreme poses questions such as: Shall we respond? If so, then what should we say? Who decides the answers? The second extreme requires that the message be interpreted before appropriate questions can be posed, especially because of the significance of opening a dialog with an advanced civilization.

A signal received from one civilization will imply the existence of others. The immediate response to the detection of a signal will be demands for more information, which can only be obtained by continuing and expanding searches.

The discovery of ETI will have societal implications at least as significant as the discoveries of Galileo and Darwin. They include historical analogs of contact, sociological and psychological implications, legal, political and institutional issues, international policy, the impact on individual philosophy and organized religion, and interactions with the educational process and the media.

The SETI Institute should continue to work with appropriate disciplines and interested organizations in an attempt to find some answers to these questions, to develop a post detection plan, and to seek resources for these tasks.

9.1   Background

This book is, in general, a plan for the science and technology of SETI over the years 2000 through 2020. Briefly, its scope embraces the underlying logic of SETI, new opportunities for searches and more powerful search systems, recommendations for specific technical approaches, and strategies for implementing these new enterprises. The potential for making great advances in SETI capabilities is evident from the discussions of these opportunities.

The SSTWG did not devote significant effort to another major question: "What would be the next steps after the actual detection of a signal that provided incontrovertible evidence for the existence of ETI?"

The question was not addressed for several reasons. The challenge of developing this book was complex and difficult enough. The next steps after a signal is detected will depend to a considerable degree on the nature of the detected signal, so answers to the question would require much speculation. While some of the post detection steps might involve expertise in the physical sciences and technology, many do not. For example, decisions may involve international policy and decision making, sociology and social psychology, institutional affairs and space law, and the media and education. These disciplines were not represented within the SSTWG.

In spite of these caveats, the SSTWG felt that the implications of detecting another civilization are so profound that they deserve some mention in this book, if only to point out that there is a formidable array of issues to be addressed in the years to come. As examples, three such issues are considered briefly.

9.2   Responding to an Extraterrestrial Signal

The type of signal will lie between two extremes. One is a simple continuous wave or pulsed transmission, the other a complex message rich in information and taking years to acquire. Intermediate situations of all types can be envisioned.

The first extreme can be dealt with now, before detection occurs. All we have to do is postulate some sensible candidate signals within the search envelopes being explored now and in the near future. The second extreme must wait until after detection, and indeed until the complex message has been completely deciphered.

In studying the first case of a simple signal with no message, the following issues have to be addressed:

- Should we respond?
- If so, why?
- If not, why not?
- If we respond, what should we say?
- Who decides the answers to these questions?
- Who sends the response?
- What processes might be put in place for studying these issues?

Today, there are no agreed-upon answers to any of these questions, and they deserve to be addressed in some depth. After all, a signal could be discovered at any time.

The importance of reaching some conclusions arises from the great significance of opening a dialog with an advanced technological civilization, about which we would initially know very little. This is not a matter of physical science, but of international policy relating to interstellar communication.

Some initial considerations of these issues have been made by the SETI Committee of the International Academy of Astronautics. Its subcommittee on *Questions of Policy Regarding Transmissions from Earth* prepared an Academy White Paper which was presented to the United Nations Committee on the Peaceful Uses of Outer Space on June 8, 2000.

The SSTWG recommends that the SETI Institute continue to work with appropriate disciplines and bodies in an attempt to find some answers to these questions. It should also seek funding from government and private sources to achieve this end.

9.3 Post Detection Searches

If a response were sent to the civilization whose signal we had detected, it could take decades, centuries or even a millennium to arrive at its destination. Any subsequent dialog would be very protracted, on the timescale of human lifetimes, although fast in geological time. After signal detection, there will be a clamor for more information to be provided as soon as possible and, no doubt, substantial funds will be provided for further searches and studies.

There is a natural way to seek additional information, and that is to carry out an enlarged SETI endeavor. This is because signals from one civilization imply that there will likely be a good number of other civilizations which are also transmitting, and whose signals are now falling on the Earth. These would be the targets of the expanded SETI searches, and might have a decent chance of being detected comparatively quickly.

In a very real sense, these larger enterprises would grow out of the new searches and systems described in this book. As support is sought for these new searches and systems, it seems likely that the SETI Institute, along with other organizations, might be called upon to answer questions about the next steps to be taken after the first detection.

9.3.1   A Possible Answer

What would an expanded SETI program be? Little effort has been addressed to this challenge, but there are some obvious immediate answers. For simplicity, assume that the detected signal is simple in character. Since the first signal will be detected by the new generation of searches and systems, an obvious step would be to dramatically scale up the number of telescopes, their sensitivities, locations, and signal processing systems that would use next generation computing systems. Scaling laws, however, are notorious for being nonlinear.

Assuming that funding levels would be high enough to warrant major planning on an international level, the responsible approach would be to conduct a series of conceptual design studies to define the significant parameters, boundary conditions, options, technical approaches, costs, and schedules. Then, at least some thought would be applied to the response to a very specific question: "How would you put together the post detection SETI program?"

Presumably, those who have developed some plans will be better situated to play significant roles in their implementation than those who have not. International bodies that routinely carry out advanced studies, such as the International Academy of Astronautics, might become involved. Such studies would have aspects of science, technology and international policy.

It is possible that such studies will be linked to the major question of how to respond to the first signal detected. It has been suggested that Earth should not respond until there is an understanding of the nature and mores of the ETI civilization, or indeed of the possible galactic community. There will be advocates for waiting at least until signals bearing complex messages have been detected and understood, before we start transmitting. If we have an inclination to join the galactic club, perhaps we should know all about it before signing on! Such a policy would add further momentum to the need for a Post-Detection SETI Plan.

The SSTWG recommends that the SETI Institute, together with interested other organizations, develop and refine a Post-Detection SETI Plan, and seek government and private resources to undertake the task.

9.4   Societal Implications of Signal Detection

It is generally agreed that the discovery of ETI would be a significant milestone for our civilization. The reason is simple – inevitably it would change the way in which our society evolves far into the future. This would be true whether the discovery period were rapid or protracted, and regardless of the nature of the information. We would come to view ourselves and our place in the Universe in a very different light. The changes would be as profound as those resulting from the revolutionary discoveries of Galileo and Darwin, which changed our understanding of our place in the solar system and of our biological evolutionary heritage.

The prospect of being on the threshold of amazing new insights should stimulate us to explore the implications of a SETI discovery on our society. Again, this is not a question just for scientists and engineers, but for everyone. At the very least, it is of deep academic interest. At most, it is of enormous practical interest for all human beings.

The disciplines involved in questions of the *Search and Society* range over a wide spectrum. They include studies of historical analogs of contact, sociological and psychological implications, legal, political and institutional issues, international policy, the impact on individual philosophy and organized religion, and interactions with the educational process and the media. There is a small but growing literature on these topics. (See Section 9.5 *Further References*.)

Currently, the resources available for continuing in-depth studies on the *Search and Society* are extremely small. The SSTWG recommends continuation, and indeed expansion, of these studies. It further suggests that the SETI Institute and related organizations take a lead in encouraging active interest among scholars in the societal disciplines, and in seeking greater resources to support them.

9.5   Further References

Almar, I., 1995, "The Consequences of Discovery: Different Scenarios", *Progress in the Search for Extraterrestrial Life*, ASP Conference Series 74, 499.

Billingham, J. et al., 1999, *Societal Implications of the Detection of an Extraterrestrial Civilization*, SETI Press, Mountain View, California.

Dick, S., 1996, *The Biological Universe: The Twentieth Century Extraterrestrial Life Debate and the Limits of Science*, Cambridge University Press.

Harrison, A., 1997, *After Contact: The Human Response to Extraterrestrial*

Life, Plenum Press, New York.

M. Michaud & J. Tarter, eds., 1989, Special Issue of *Acta Astronautica: SETI Post Detection Protocol*, Pergamon Press.

A. Tough, ed., 2000, *When SETI Succeeds: The Impact of High-Information Contact*, Foundation for the Future, Bellvue, Washington..

Vakoch, D. A., 1998, "The Dialogic Model: Representing Human Diversity in Messages to Extraterrestrials", *Acta Astronautica* 42.10-42.12, 705-710.

John Billingham

Executive Secretary,
SETI Science and Technology
Working Group

Appendix A

Cyclops Revisited

Excerpts from the *Project Cyclops* report of 1972 are reproduced in this appendix. This narrative, by Jill C. Tarter, describes subsequent developments and the current view. The *Project Cyclops* report was the original blueprint for SETI searches.

The *Project Cyclops* report [Oli71], edited by Dr. Bernard Oliver and Dr. John Billingham, has largely governed the observational work on SETI during the past 27 years. While Chapters 5, 6, and Appendix D of the *Project Cyclops* report still serve as a superb tutorial on signal detection theory, other aspects of the report have been overtaken by events and two decades of continued progress in many areas of science and technology. It is, therefore, worthwhile to revisit the conclusions and recommendations from that report to see where major shifts in strategies are warranted, and where our current understanding of the Universe and our place therein have changed. In particular, discoveries in optical astronomy, radio astronomy, and biology have changed our views in the following areas:

- the formation of planets;
- the development of Habitable Zones and the evolution of life;
- signal propagation through the interstellar medium, solar corona, and ionosphere.

Furthermore, our own technical evolution has changed our views of how ETIs might generate signals and at what wavelengths. This applies both to their domestic uses and deliberate beacons. In particular:

- A much wider range of wavelengths is now considered practical for interstellar communication. We now know that it is possible to build high-power sources, and aim them accurately enough to utilize frequencies up through the optical.

- Our understanding of signal processing and computing has improved, and the amount of computation considered 'practical' has changed dramatically.

- ETIs will probably be using more efficient coding schemes for domestic communications than we are, and they will be harder to detect as leakage. On the other hand, beacons are easier to build than we originally thought. Both these assumptions derive from the fact that computing continues to become less expensive.

The Cyclops array path through the SETI Decision Tree is shown in Figure A.1.

A.1   Cyclops Premises Revisited

In this section, premises from the *Project Cyclops* report are indented and italicized. The text that follows each premise provides an updated and current view of the natural world.

> *1. Planetary systems are the rule, not the exception. Most stars are now believed to possess planetary systems. These are a by-product of the mechanism of stellar formation whenever the gas cloud out of which the star condenses is slowly rotating. Since the Galaxy as a whole is rotating, most interstellar clouds are also.*

Surely, one of the most profound changes in our state of knowledge in the past three decades has to do with the very first premise. In addition to the nine planets that comprise our own solar system, we now know something about many planets orbiting nearby stars and pulsars [Mar00]. (See Table 2.1 on page 46.) This is an active field of research, and at least five methods are used for extrasolar planet detection. These are:

- pulsar timing studies;
- radial velocity surveys;
- astrometric studies;
- microlensing observations;
- transit/occultation studies.

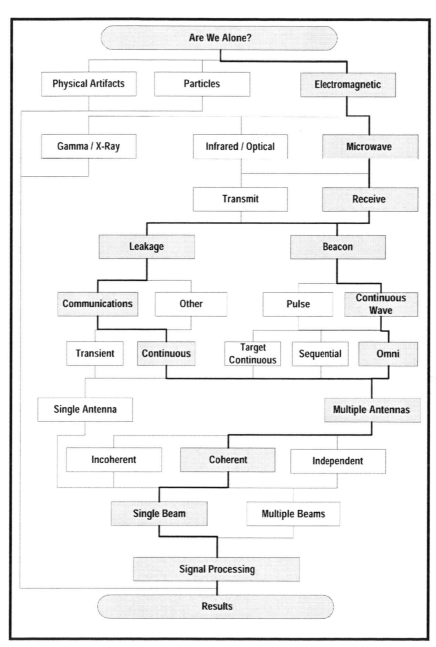

Figure A.1: SETI Decision Tree – Cyclops Path.

All of these techniques suffer from serious limitations and selection biases. Taken as a whole, however, we are beginning to understand the formation of planetary systems.

On a longer timescale, both NASA and the ESA have outlined a series of space-based missions that include astrometric precursors. They lead up to large imaging-interferometers with the capability of making single-pixel images of any orbiting planets around nearby stars and conducting crude chemical assays of their atmospheres [Leg00]. Within the first few decades of the 21st century, we will surely have a much more detailed understanding of the likelihood of planetary systems and we may have inferential evidence about the existence of some form of life on planets in orbit around nearby stars.

> *2. Many planetary systems will contain at least one planet in the stellar ecosphere, where it will be kept warm by its star, but not too hot, and where its atmosphere can evolve from the primitive reducing one to an oxidizing Earth-like atmosphere.*

Our thinking on this has changed since the *Project Cyclops* report as we have understood the Earth's history better and studied other planets and moons in our solar system. The past three decades have witnessed an elaboration of the definition of the planetary ecosphere, or Habitable Zone, to now encompass giant moons, as well as the recognition of temporal evolution. This has led to the concept of the Continuously Habitable Zone around solar-type and smaller stars [Kas97]. The Viking and Pathfinder missions to Mars have convinced us that our neighboring planet was relatively wet and warm at the same time as life was originating on Earth.

Since the *Project Cyclops* report, our best estimates of the early atmospheric constituents on Earth have shifted away from the highly reducing chemistry which worked so well in the Miller-Urey experiment, toward a more redox-neutral state [Des98], [Kas93]. The eventual transition to an oxygen-rich atmosphere appears to have depended not only on the evolutionary invention of photosynthesis, but also on the subduction and burial of large amounts of carbonates [Des94]. This introduces the necessity of adding plate tectonics to the list of items required in order to get a planet that is 'just right', at least for advanced, oxygen-breathing life. Furthermore, we now guess that during the first billion years of planetary existence, the atmosphere would have been steam-heated many times by major ocean-boiling impacts [Sle98]. During this period, the Habitable Zone might have consisted of large regions deep in the ocean or deep underground.

The necessity for geological activity may be placing a new effective limit on the minimum mass a planet needs for the emergence of intelligent life. However, since the *Project Cyclops* report, it does appear that the scientific view of potential habitats for unknown kinds of life has expanded rather than contracted.

This topic is discussed in more detail in Section 2.5 on page 51, but overall, the conclusions still stand.

> *3. Organic precursors of life are formed in abundance from the in-gredients of a primordial or a tectonic atmosphere, but in any case are also found in space and on certain meteorites (the carbonaceous chondrites).*

More than 100 different interstellar molecules have now been observed, possibly including glycine, the simplest amino acid [Sny97]. Organic chemistry is indeed ubiquitous. It is possible that at least some of these interstellar organic mole-cules survived the process of inclusion into the protosolar nebula without their chemistry being reset. On the primitive Earth, the resource pool of the first organics on which chemical evolution elaborated to produce life as we know it may have had contributions from incoming comets and meteorites as well as from volcanic outgassing [Chy92]. Current measurements of cometary deu-terium/hydrogen ratios suggest an upper limit of 10% to the water contributed by infalling sources [Lau99]. That would still be a substantial organic reservoir derived from interstellar and protoplanetary sources.

Non-racemic mixtures of amino acids have been verified within the Murcheson meteorite [Cro97] and several new sources of circularly polarized light have been identified in star formation regions [Bai98]. It is, however, unclear whether this is telling us anything about the origin of chirality favored by terrestrial life.

> *4. Main sequence stars cooler than F5 stars have lifetimes suf-ficiently long for Darwin-Wallace evolution to be effective. The ecospheres of stars cooler than spectral class K7, or perhaps K5, are so close to the star that tidal friction will either stop or greatly diminish the planet's rotation.*

We have recently begun to appreciate how episodic and random was the process of evolution that led to intelligent life on Earth [Gou89], and how intimately it was connected with our astrophysical environment [Alv80]. On Earth at least, complex life had to wait out the early period of heavy bombardment [Sle98], and the later glacial frustration [Kir00], until the Cambrian period of rapid spe-ciation. Subsequently, major extinction events have repeatedly eliminated large fractions of extant species, providing room for the emergence of opportunistic new species. This current understanding contrasts with the more or less gradual pace of Darwin-Wallace evolution assumed by the *Project Cyclops* report. One can, therefore, argue that we were particularly slow in our evolutionary history. Elsewhere life may have evolved to intelligence on a shorter timescale, or it may not have evolved at all.

At the other end of the stellar spectrum, as M type dwarf stars evolve, they become hotter, and the location of the Habitable Zone moves farther from the star. This allows for the possibility of clement planets existing beyond the distance where tidal locking of planetary rotation occurs [Doy93]. Optical and radio observations have identified M dwarfs as a frequent source of energetic flaring [Lan94]. Such flares would probably make the origin and continued existence of planetary life problematic but it is also conceivable that they might serve as stimuli for accelerated evolutionary change.

We are biased towards terrestrial models of life; but we have no robust way of deciding if there is a more universally valid model using our current knowledge. Thus, it seems reasonable to relax the F5 to K7 constraints that the *Project Cyclops* report placed on stellar spectral type, and to suggest that all nearby stars out to 5 pc be observed. At larger distances, selectivity should increase; and since measurements of photospheric activity of main sequence stars permit a fairly clean discrimination of stars younger than three billion years [Hen96]; distant targets should include F0 to K9 stars older than three billion years [Lat93].

> 5. *Intelligent life will have existed at some time during the life of the star on favorably situated planets circling a substantial fraction of the middle class main sequence stars in the Universe.*

The probable existence of intelligent life on suitable planets around solar-type stars remains the premise of SETI. There is a stronger case today for the prevalence of life elsewhere than there was three decades ago. The scientific community, however, is strongly divided over whether intelligence will always, or even often, accompany the origin of life [War00]. Intelligence may be the inevitable evolutionary result of predator/prey relationships [Daw88], or it may be a complete fluke [Myr94]. SETI may be the only method by which this question can be answered.

By 2020, we shall have a much clearer census of planetary systems in the Milky Way Galaxy, and, perhaps, an indication of whether life is abundant. While it is gratifying to see that the major space agencies have now embraced the search for our own cosmic origins and for life elsewhere it is particularly disappointing that they have failed to push their search plans another step forward to encompass the search for intelligent life.

> 6. *The longevity of technologically advanced civilizations is unknown within orders of magnitudes. Other than assessing (while struggling to improve) our own chances for long-term survival, we know of no way to resolve this question, short of actually making contact with other races.*

While we still do not know the average longevity of a technological civilization, we can use the Copernican Principle to estimate our own technical life span on the basis of the current age of our technology [Got93]. The Copernican Principle states that there is nothing special about us or our location in time and space; we are not the center of anything. Therefore, our probable longevity as a radio-transmitting technology can be used to estimate an upper limit to L, the average longevity of any radio-transmitting technology in the Milky Way. However, since we have a sample size of one, and do not know the lifetime for the one sample we do have, we need to make assumptions about the distribution to get any answer. Naturally, different assumptions give different answers. Different assumptions and their resulting average lifetime estimates are discussed in Section 5.2 on page 115.

In general, the results from these calculations, which are uncertain, are not as optimistic as those from the *Project Cyclops* report. It proposed a search of the nearest 10^6 stars, out of the 10^{11} stars in our Galaxy, corresponding to a distance of approximately 1,000 ly. For this strategy to succeed, there must be at least 10^5 civilizations in the Galaxy (statistically speaking). This in turn implies that an average communicative lifetime of a civilization must be at least 10^5 years (assuming that about one new planetary system forms per year, and under the optimistic assumption that all planetary systems have good planets and develop intelligent life).

> 7. *The establishment of interstellar contact may greatly prolong the life expectancy of the race that does so. Those races that have solved their ecological and sociological problems and are therefore very long lived may already be in mutual contact sharing an inconceivably vast pool of knowledge. Access to this "galactic heritage" may well prove to be the salvation of any race whose technological prowess qualifies it.*

The speculation that contact between intelligent species may provide a stabilizing influence that fosters increased longevity must have been very comforting in the midst of the tensions of the Cold War. It is still comforting at a time when we understand ourselves to be such a young technology in an old Galaxy [Got93], a situation that Carter [Car83] uses to argue against the prevalence of intelligent species in the Milky Way. However, it remains just a speculation.

A.2 Cyclops Conclusions Revisited

The *Project Cyclops* report concluded that the best search strategy was to build an expandable search system, working at the low end of the microwave band,

and looking for powerful, omnidirectional beacons set up by other civilizations. How did they decide this? How have these conclusions held up?

An advanced technology might be expected to have requirements for:

- energy production;
- additional real estate;
- interplanetary travel;
- information processing;
- strategies for long-term survival.

In the years since the *Project Cyclops* report, a few searches have been conducted that tried to find evidence for large-scale fission or fusion energy production activities in the vicinity of nearby stars [Fre85], [Val86]. Seyfert galaxies have jokingly been referred to as the 'industrial accidents' of the cosmos. Asteroids with companions were sought as possible examples of mining activities in our solar system [Pap95]; but subsequent detection of moons orbiting Eugenia and Ida have been interpreted as natural objects. Unlike information processing, these other plausible activities of an advanced technology do not allow one to calculate a meaningful threshold for sensitivity. Also, since these activities are not something for which our current technology can provide a model, there is no basis for interpreting a negative result. It is impossible to decide whether the activity is not taking place, or whether it is being pursued below the detection threshold.

In contrast, we do know how much power we devote to information transfer, and we can make educated extrapolations into the future. Searching for activities that represent reasonable or plausible extrapolations of *nonexistent* terrestrial technology should not be the basis for large-scale systematic searches. Rather, these offer attractive possibilities for searches conducted in parallel with more traditional astronomical observations.

The *Project Cyclops* report considered some of these same arguments and concluded that the most sensible search strategy, with the greatest potential for a significant positive result, was to look for evidence of information transfer. The report summarizes the properties of the ideal information carrier as:

- The energy per bit should be minimized, all other things being equal.
- The velocity should be as high as possible.
- The particles should be easy to generate, launch, and capture.
- The particles should not be appreciably absorbed or deflected by the interstellar medium.

From these properties, photons were selected as the carriers of choice, and in the opinion of the Working Group, this continues to be the best possibility.

However, the 'minimum energy per quantum' argument has been replaced by 'energy per bit' in the realization that a few powerful infrared/optical quanta may be more detectable than many radio quanta.

The strategy of how and where to look for photons was presented as a detailed list of 15 conclusions in the *Project Cyclops* report [Oli71]. The evolution of our own technology, including our conceptual picture of the Universe, suggests several alternative or additional search strategies. Before reaching our own conclusions, it is useful to begin by considering the specific conclusions of the *Project Cyclops* report.

> *1. It is vastly less expensive to look for and to send signals than to attempt contact by spaceship or by probes. This conclusion is based not on the present state of our technological prowess but on our present knowledge of physical law.*

The *Project Cyclops* report never suggested that interstellar travel was impossible, just uneconomical. Subsequent debate in the literature [Oli90b] clarified what special relativity permitted as the most efficient type of interstellar spacecraft (one that leaves its exhaust at rest with respect to the accelerated frame), but did little to achieve consensus about what was *too* expensive. Today, just as in 1971, any discussion of this topic ends with the same caveat about 'our present knowledge of physics'.

We still cannot discount interstellar travel based upon some 'unknown' physics, but there appears to be no way to craft a systematic search strategy that exploits it. One published effort [Har90] attempted to line up the known gamma ray burst sources in four-dimensional, space-time to see if they might reveal the trajectory of an antimatter annihilation rocket. Keeping Occam's razor in mind, this sort of speculative inquiry should be supported, but a systematic search needs to be based on the physics we do understand.

> *2. The order of magnitude uncertainty in the average distance between communicative civilizations in the Galaxy strongly argues for an expandable search system. The search can be begun with the minimum system that would be effective for nearby stars. The system is then expanded and the search carried farther into space until success is achieved or a new search strategy is initiated.*

The concept of an expanding search continues to make the most sense, particularly in light of the exponential growth of several key technologies over the past few decades.

A microwave search of 1,000 nearby stars (out to 155 ly) over the frequency range from 1 to 3 GHz is within the grasp of today's technology, and is being carried out by Project Phoenix [Tar96]. Its limiting sensitivity is sufficient

to detect transmitters comparable to our own strong radars (10^{12} W, EIRP).
Extending the sensitivity of a search down to leakage signals, with power comparable to our broadcast TV (10^7 W, EIRP), or extending the search range out to several thousand light years may be feasible within the next twenty years. This, however, will require ways to improve the efficiency of the search and a much larger, dedicated collecting area than is now possible. Searches at other frequencies have only recently begun, and these also may grow over time.

> *3. Of all the communication means at our disposal, microwaves are the best. They are also the best for other races and for the same reasons. The energy required at these wavelengths is least and the necessary stabilities and collecting areas are fundamentally easier to realize and less expensive than at shorter wavelengths.*

Conclusions regarding the correct frequency at which to conduct a search must be revisited in light of technological advances since the *Project Cyclops* report. In particular, it is necessary to revisit the arguments favoring microwave searches over optical/infrared searches. In the crucial tables (Tables 5-3 and 6-2) [Oli71] comparing optical, infrared, and microwave transmit and receive strategies, the *Project Cyclops* report made three assumptions when comparing microwaves to optical which now seem suspect:

- The power levels available in the optical were orders of magnitude less than those available in the microwave. Progress in lasers has removed this discrepancy.

- Omnidirectional beacons and narrowband continuous wave signals were required. Thus, narrowband heterodyne detection would be used in the infrared/optical, as in the microwave. Reducing the bandwidth in this manner maximizes the S/N for microwaves, and the *Project Cyclops* report assumed the same strategy would be used in the optical. But short pulses in the optical, coupled with direct photon detection, can offer the same S/N improvements, without the additional quantum noise introduced by heterodyning. These pulses will be broadband, thus easing the search in frequency space, and would be as obviously artificial as the coherent signals favored by the *Project Cyclops* report. Modern, high-speed, and high quantum efficiency detectors make this approach practical.

- Omnidirectional beacons were required: since it was claimed that the high theoretical gains of infrared/optical transmitters could not be used because of seeing problems in the atmosphere and inadequate knowledge of star positions, proper motions, and distances. The seeing problem now seems solvable with adaptive optics or space-based systems, and the pointing problem is being addressed by space-based astrometry missions. Signal propagation from nearby stars looks very good in the infrared, and is acceptable in the optical. Section 2.8 on page 61 shows that extinction and scattering may limit the detectability of optical pulses to distances of a few thousand light years, while in the near-infrared they may be detected from the galactic center.

4. The best part of the microwave region is the low frequency end of the 'microwave window' – frequencies from about 1 to 3 GHz. Again, this is because greater absolute frequency stability is possible there, the Doppler rates are lower, beamwidths are broader for a given gain, and collecting area is less expensive than at the high end of the window.

Arguments about Doppler rates apply only to the omnidirectional beacons envisioned by the *Project Cyclops* report. High-gain beacons from phased arrays can Doppler compensate for their targets. While it is true that higher frequencies require higher surface accuracies for the receiver, they also permit greater gains for the transmitter for a given antenna diameter.

Telescope construction mechanisms for advanced technologies may be dominated by the cost of the computation and electronics, rather than by surface accuracy. Furthermore, interstellar signal propagation is degraded less at higher frequencies. At best this is a weak conclusion, and to the extent that current antennas are being upgraded to work at higher frequencies, our SETI searches should extend upwards as well.

5. Nature has provided us with a rather narrow quiet band in this best part of the spectrum that seems especially marked for interstellar contact. It lies between the spectral lines of hydrogen (1,420 MHz) and the hydroxyl radical (1,720 MHz). Standing like the Om and the Um on either side of a gate, these two emissions of the disassociation products of water beckon all water-based life to search for its kind at the age-old meeting place of all species: the water hole.

The significance of water to life as we know it remains unchanged, and it heavily influences observational plans to search for evidence of life elsewhere. Today, we know of many more different interstellar molecules than we did at the time of the *Project Cyclops* report. This somewhat dilutes the 'magic' nature of any molecular, line-based frequency and argues for a systematic search through all available frequencies. On the Earth, it is particularly the low end of the microwave spectrum that is being saturated by terrestrial communication needs. There is no place to hide from satellite transmitters, except the Shielded Zone of the Moon (SZM) [Pan81], and we have not yet succeeded in active excision of these sources from our observations. Therefore, practical considerations mean that low frequency coverage will not be as complete as higher frequency coverage for our near-term searches. It is also arguable that a desire for higher information transfer rates for internal communication purposes will cause an advanced technology to utilize higher frequencies.

> *6. It is technologically feasible today to build phased antenna arrays operable in the 1 to 3 GHz region with total collecting areas of 100 or more square kilometers. The Cyclops system is not nearly this large, but we see no technological limits that would prevent its expansion to such a size.*

This remains true today; cost is the main impediment. Modern phasing technology can extend the frequency range of large arrays to considerably higher frequencies (at least 10 GHz) for even large arrays.

> *7. With antenna arrays equivalent to a single antenna a few kilometers in diameter at both the transmitting and receiving end, microwave communication is possible over intergalactic distances, and high-speed communication is possible over large interstellar distances. Thus rapid information transmission can occur once contact has been confirmed between two civilizations.*

This remains true today.

> *8. In the search phase we cannot count on receiving signals beamed at us by directive antennas. Neither can we afford to overlook this possibility. Beamed signals may be radiated at relatively low powers by communicative races to as many as a thousand nearby likely stars and for very long times. Long range beacons, intended to be detectable at any of the million or so likely stars within 1,000 light years, will probably be omnidirectional and very high powered ($> 10^9$ W).*

This conclusion has been challenged throughout the course of the Working Group meetings. Phased arrays of many individual transmitters could be used to illuminate even the million, or so, targets within the 1,000 ly horizon assumed in the *Project Cyclops* report. They could do so continuously or sequentially, with a power delivered to the receiver that differs by a factor of a million. The sensitivity of terrestrial radio astronomical observatories has been improving 10 dB per decade for the past five decades. At that rate, only 60 years of technological advances would be required for reception capabilities to progress from the stronger, sequential to the weaker, continuous beacon signals. This is such a short cosmic interval that this conclusion does not remain strong. Indeed, a joint consideration of transmission and reception strategies produces no compelling optimum.

What if high-powered beacons, of any duration, do not exist at any frequency? Does it make sense to search for leakage radiation from an advanced technology?

In the vicinity of a planet, an advanced technology may generate less waste emission (leakage) than does our current terrestrial technology. Broadcast communications may be replaced by point-to-point systems, on fiber or cable, or beamed from orbit. Multiple uses of the limited available spectrum may result in various spread-spectrum modulation techniques, with the suppression of coherent components and concentration to higher frequencies.

At optical frequencies, the local flux of communication beams may exceed the stellar background, but the integrated power cannot do so without producing undesirable thermal consequences. At microwave frequencies, the total power radiated will not grow indefinitely, as spatial, temporal, and frequency reuse of the spectrum causes the emission to resemble Gaussian noise. This presents a difficult signal processing challenge for the receiver. At present, only the (KL) Transform [Mac94] shows any potential for recognizing the difference between such 'pink' and white noise. While too computationally intensive for the present generation of systems, this capability should be added to future systems as soon as the computing requirements become affordable.

Although astrophysical studies may take place far from the planetary surface of the home world of an advanced technology, its spectral leakage may still retain historical artifacts from an epoch of ground-based observations. The vestiges of the infrastructure for spectrum management may still be discernible. Certain frequency bands might be under-utilized where it was important to keep open windows on the observable Universe, e.g., for radio astronomy, etc.

An advanced technology may occupy real estate on multiple planets within its planetary system or it may have engaged in vast astroengineering projects to reconstruct its planetary system and provide more livable real estate. This latter possibility can be investigated by infrared searches for Dyson spheres, if spectral index discriminants can be developed to distinguish these from infrared excesses due to natural dust in the system. Kardashev [Kar73] has suggested that a Planckian spectral index of -2 will characterize large-scale engineering projects, and be steeper for dust. If an advanced civilization lives on multiple planets, then, once again there arises a reason to expect broadcast leakage signals for navigational purposes, the equivalent of terrestrial VHF Omnidirectional Receiver (VOR) beacons.

Microwaves are a versatile technology that can be used for a wide variety of purposes. In addition to communication, they can be used for power transfer, navigation, and radar detection, ranging, and investigation of distant objects. Therefore, with our limited ability to predict the behavior of a technology far in advance of the Earth, there seems to be some rationale for leakage searches at microwave frequencies. Shannon's Theorem suggests that we should not expect leakage to be narrowband.

> *9. Beacons will very likely be circularly polarized and will surely be highly monochromatic. Spectral widths of 1 Hz or less are probable. They will convey information at a slow rate and in a manner that does not seriously degrade their detectability. How best to respond will be contained in this information.*

Circularly polarized, narrowband signals continue to represent an easily recognizable, and arguably technological, choice for microwave beacons. However, intentional beacons at optical frequencies may well be broadband, with rapid pulses permitting very high data transfer rates.

> *10. The efficient detection of beacons involves searching in the frequency domain with very high resolution (1 Hz or less). One of the major contributions of the Cyclops study is a data processing method that permits a 100 MHz frequency band to be searched simultaneously with a resolution of 0.1 Hz. The Cyclops system provides a receiver with a billion simultaneous narrow channel outputs. Although the Cyclops system bandwidth is 100 MHz, no very great technological barriers prevent widening it to 200 MHz. This would permit searching the entire 'water hole' simultaneously. If our conclusion as to the appropriateness of this band is correct, the problem posed by the frequency dimension of the search can be considered solved.*

The first statement, that it is feasible to design a billion channel search system, is now rather conservative. Today, we would use digital signal processing to accomplish this, rather than the optical lensing, analog processing, of the Cyclops system. Today, a 10 GHz span with 0.01 Hz resolution is certainly possible, although expensive, and will become easier in the future.

The second statement, that the frequency search problem can be considered solved, is much more debatable. As we have learned more about the atomic and molecular constituents of our Galaxy, and as our instrumentation has opened up higher radio frequency regions, the 'water hole' of the *Project Cyclops* report has lost most of its apparent uniqueness. The appropriate limits for ground-based microwave searches are set by increased noise from galactic synchrotron emission below 1 GHz, and by contributions from atmospheric O_2 and H_2O above 10 GHz. Current state-of-the-art detectors at microwave frequencies can now provide simultaneous access to several GHz of the spectrum. We can already anticipate advances in digital signal processing that can accommodate analysis for narrowband signals across the terrestrial microwave window.

In the infrared/optical region of the spectrum, very short broad pulses remove the frequency search domain problem. CW signals can be sought with subthermal resolutions (resolving powers of 5×10^4) covering a wavelength range of about an octave by using echelle spectrographs. Covering all frequencies in the infrared/optical region is best done with multiple telescopes.

If the value of L is small, and the distance between civilizations is great, signals at microwave frequencies can have their amplitudes modulated by interstellar scintillation, rising above a detection threshold only occasionally [Cor91a], [Cor91b]. Transmission strategies that produce sequential beacons with low duty cycles will be easier to miss. For both these reasons, it is desirable to be able to search all-targets, at all-frequencies, all-the-time. Covering all targets all of the time with high sensitivity has no foreseeable solution in the infrared/optical portion of the spectrum, but this task may be approachable at longer microwave frequencies.

> *11. The cost of a system capable of making an effective search, using the techniques we have considered, is on the order of six to ten billion dollars, and this sum would be spent over a period of 10 to 15 years. If contact were achieved early in this period, we might either stop expanding the system or be encouraged to go on to make further contacts. The principal cost in the Cyclops design is in the antenna structures. Adopting an upper frequency limit of 3 GHz rather than 10 GHz could reduce the antenna cost by a factor of two.*

The *Project Cyclops* report assumed that antennas would be built the way they had been historically, but it did not correctly anticipate the digital signal processing revolution with the cost savings and capacity increase that would result. As long as the upper frequency of operation does not drive the costs of a receiving system, the upper frequency of operation should be kept as high as possible.

> *12. The search will almost certainly take years, perhaps decades and possibly centuries. To undertake so enduring a program requires not only that the search is highly automated, it requires a long-term funding commitment. This in turn requires faith. Faith that the quest is worth the effort, faith that man will survive to reap the benefits of success, and faith that other races are, and have been, equally curious and determined to expand their horizons. We are almost certainly not the first intelligent species to undertake the search. The first races to do so undoubtedly followed their listening phase with long transmission epochs, and so have later races to enter the search. Their perseverance will be our greatest asset in our beginning listening phase.*

The *Project Cyclops* report assumed that the search effort would take decades and expand to cost billions of dollars; and it asserted that funding would be possible. In the USA, SETI now has a 38-year observational history, with modest government funding spanning less than a decade before being terminated. The *Project Cyclops* report contemplated sharing time on the facility with traditional radio astronomy, but the cost and performance of the Cyclops system were far in excess of what the radio astronomy community would support for its own purposes. Unless the current searches find an ETI signal detection in the very near future, funding for SETI will be counted in the millions of dollars, not billions. The potential for making the search an international project holds out the possibility that the millions might become hundreds of millions, but only if it can be linked to other international astronomical projects.

> *13. The search for extraterrestrial intelligent life is a legitimate scientific undertaking and should be included as part of a comprehensive and balanced space program. We believe that the exploration of the solar system was and is a proper initial step in the space program but should not be considered its only ultimate goal. The quest for other intelligent life fires the popular imagination and might receive support from those critics who now question the value of landings on 'dead' planets and moons.*

Today, astronomers are again wondering if all the other planets and moons of our solar system are really so 'dead'. Interest is high to return to Mars seeking extinct fossils or extant life deep under its surface. Jupiter's moons Europa and Callisto certainly appear to harbor liquid water oceans beneath their ice-covered surfaces and, perhaps, may even harbor primitive life. While SETI remains a legitimate scientific undertaking, it is not the only focus for the study of life beyond Earth. NASA's recently completed *Astrobiology Roadmap* [Mrd99] asks three questions:

- How did life begin and evolve?
- Is there life elsewhere in the Universe?
- What is the future of life beyond Earth?

That makes SETI sound like a mainstream enterprise, but, unfortunately, its checkered political past reduces it to a barely visible effort. The *Astrobiology Roadmap* identifies radio searches for intelligent life elsewhere as a long-term goal. Even though SETI has been 'politically incorrect' in the recent past, there is significant public interest in it. SETI logically fits within the major new NASA initiatives of *Origins* and *Astrobiology*. It continues to be a legitimate scientific exploration and its political status will change over time.

> *14. A great deal more study of the problem and of the optimum system design should precede the commitment to fund the search program. However, it is not too early to fund these studies. Out of such studies would undoubtedly emerge a system with a greater capability-to-cost ratio than the first Cyclops design we have proposed.*

The increase in digital signal processing capacity in the decades since the *Project Cyclops* report has profoundly altered the capability-to-cost ratios which it may be possible to achieve. In particular, continued growth in processing capacity for the same cost (doubling every 18 months, according to Moore's Law) leads to the conclusion that concrete, steel, and software should be considered to be capital costs, but the actual signal processing hardware should be considered an operating expense. Any properly designed processing system will be flooded with too much data to process at its inception. It will reach parity with the data rates during its middle age, and have excess data processing capacity towards the end of its life cycle.

> *15. The existence of more than one Cyclops-like system has such great value in providing complete sky coverage, continuous reception of detected signals, and in long base-line studies, that international cooperation should be solicited and encouraged by complete dissemination of information. The search should, after all, represent an effort of all mankind, not just of one country.*

Any very large microwave array project requires significant amounts of land, far from population settlements, in areas where the satellite coverage is sparse, and where political stability encourages the placement of an international project. These requirements are not easily met within the developed nations of the northern hemisphere. Bilateral and multinational agreements now exist to enable the construction of the largest optical telescopes and large arrays at millimeter wavelengths. Ground-based astronomy is being driven to international collaborations by its need for access to high-altitude sites, and by the cost of these increasingly expensive research tools. SETI should find a way of joining these efforts.

An international consortium now exists to encourage collaboration in the construction of a radio astronomical research instrument having one million square meters of collecting area (10% of the full-blown Cyclops system) somewhere in the world. Recent interest in such large collecting areas by the NASA Deep Space Network might naturally lead to multiple instruments in the northern and southern hemispheres. Designs that permit multiple, independently steerable beams could accommodate joint usage in ways not envisioned by the *Project Cyclops* report.

For the SETI Institute, the challenge for the coming decades will be to creatively couple private sector philanthropy with the international collaborations among governments to advance both SETI and astronomical exploration. Patronage is a long-standing tradition for astronomical research, continuing to the present with the remarkable contributions of the Keck Foundation. Within the next twenty years, contributions from very wealthy individuals who are eager to support SETI could be comparable to the contributions that some individual governments are able to make in support of astronomical research.

A.3   Things Not Considered

The *Project Cyclops* report recommended construction of a large array using elements approximately 100 m in diameter. The antenna size was derived from the assumption that the cost per antenna scaled approximately as a construction term; proportional to d^2, plus a fixed cost. The construction cost of an array of any given area was then independent of the element diameter d. The fixed costs for the array were minimized by making the fewest possible elements. Since the *Project Cyclops* report concluded that large diameter antennas were optimum, all subsequent calculations were performed assuming that the array was phased to create only a single beam. For a search of a million stars, this is a reasonable assumption since there will be, on average, only one target within the field of view of antennas greater than 30 m in diameter. No trade-off study was made to consider the benefits of using a focal plane array, nor was a study made of forming m beams simultaneously with smaller elements whose field of view would contain at least m targets.

If a very large array is to be built for an affordable cost in the relatively near future, new methods of construction will need to be exploited to overcome the historical cost models. Construction of arrays that permit multiple simultaneous beams, which examine multiple target stars may turn out to be the most cost-effective approach. The multiple beam approach is one way to shorten the search process; are there others?

The *Project Cyclops* report recommended only a targeted search approach. For up to several million targets out to several thousand light years it makes sense to follow the target-by-target approach of the *Project Cyclops* report. As long as the gain of the system (about the number of synthesized beams needed to tessellate the sky) remains greater than the number of targets, the most efficient search strategy will not waste time surveying the empty regions of the sky between targets.

If we abandon the proximity constraint (and thereby inflate the required transmitter power for a given receiver sensitivity) and the direction of individual, suitable targets is no longer known, and their number exceeds the gain of the system, then a sky survey becomes an efficient way to conduct a search.

The choice between an untargeted sky survey or a targeted search is just one of the many strategic choices that must be made for a SETI receiver. The correct decision (i.e., the one with the highest probability of a successful detection) must be based upon both the luminosity function of ETI transmitters and their temporal characteristics. Neither of these functions is known. If transmitters are continuous in time, then a power law luminosity function for transmitters of power P, with $dN/dP \sim P^{-\alpha}$ favors a sky survey for values of $\alpha < 5/2$ [Kar73]. On the other hand, transmitters may have very low duty cycles, perhaps because they are being cycled among many targets. In such cases, search strategies that require dwelling for long times on a limited set of targets have a greater probability of success than do sky surveys which rapidly pass over any area of the sky (provided that the target sets of the transmitter and receiver overlap). Of course, in either case, transmitter and receiver must be using the same frequency.

Appendix B

Beacons, Beacons and More Beacons

This appendix, by Louis K. Scheffer, describes the designs for six kinds of beacons that we could build, and which we expect ETIs to be able to build. It also discusses how we might search for such beacons.

The beacons discussed in this appendix are only paper designs, sketched out to see how practical beacons might be constructed. At present, there are no plans to construct or operate any of these beacons. Figure B.1 on page 304 shows the pathway for beacon searches through the SETI Decision Tree.

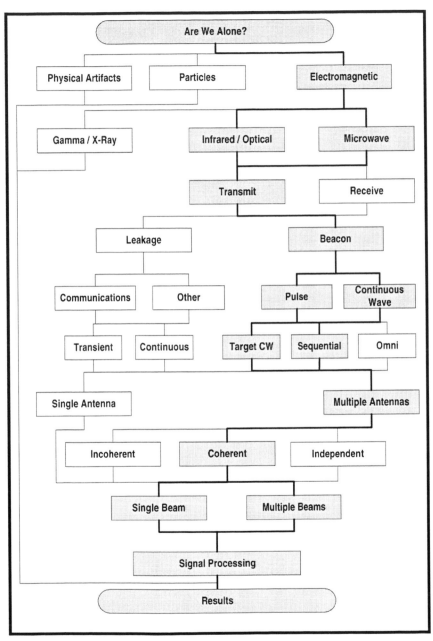

Figure B.1: SETI Decision Tree – Beacon Possibilities.

B.1 A Big Beacon

Building a radio beacon with phased array technology is straightforward. A beacon with M modules (1 W each) might cost \$1 M to \$10 M and could irradiate N selected stars at a time with an EIRP of (M^2/N) W each. Even with simple control systems, the trade-offs between power on target, duty cycle, and number of targets simultaneously irradiated can be controlled by software and changed at will. This makes it equally easy to build either low duty cycle, high-power beacons, or high duty cycle, low-power beacons. For example, a system with 10^8 1 W modules, costing \$10^8$ to \$10^9$, could do any of the following:

- Hit one target continuously with a 10^{16} W EIRP signal.
- Hit 1,000 targets continuously with a 10^{13} W EIRP signal.
- Hit 1,000 targets with a 10^{15} W EIRP signal and a 1% duty cycle.

In order to reach a particular target system, we need to take into account both the sensitivity and duty cycle of the receiver when designing the beacon. With this, we find that reaching a Phoenix-style system from 1,000 ly away requires $M = 6.5 \times 10^6$. Reaching a BETA or Argus style system from the same distance requires approximately $M = 4 \times 10^8$.

Argus or BETA systems are about equally difficult to contact, but require completely different strategies. Argus requires high power, but low duty cycles are sufficient; BETA requires much less power, but the duty cycle must be nearly 100% to have any chance of success.

Given that some computations for beamforming are required for any phased array system, the additional computation needed for Doppler correction is inexpensive. Therefore, any beacon will probably be Doppler compensated.

As listeners, this gives us hope that we will see some relatively high-power beacons. Among other things, we should listen for Doppler compensated narrowband beacons. We probably have enough sensitivity; what we need most is the ability to listen at all-frequencies, all-the-time. Phoenix-style systems should make follow-up as fast as possible, to lessen duty cycle problems for the beacon transmitters.

B.1.1 An Anti-Cyclops

A good starting point is an 'anti-Cyclops' – a beacon to irradiate the nearest 1,000 (plausible) stars with a circularly polarized beacon at a 'water hole' frequency. However, there has been one major change since the *Project Cyclops* report; the technology for detecting and analyzing other planetary systems has improved considerably. If we assume that some fraction α of the original target stars from the *Project Cyclops* report will be considered likely candidates, then the range will need to be extended by $\alpha^{1/3}$ to encompass the same number of target stars. Of course the value of α is not clear yet, since only a few results are in from the study of extrasolar planets. For the purposes of this design, we will assume $\alpha = 0.001$, corresponding to a range of approximately 1,000 ly to include 1,000 likely targets.

Note that α is really $\alpha(t)$. As a SETI search goes on, we can be more and more selective about which stars we either look at or irradiate.

One way to build a beacon is as a phased array with many, small, nearly omnidirectional elements. By adjusting the phase of each element, we can produce either a single beam in the direction of one of the targets, or many beams covering many targets simultaneously but at reduced power. For example, assuming 10^8 elements of 1 W each, we can produce one beam with an EIRP of 10^{16} W, or 1,000 simultaneous beams with an EIRP of 10^{13} W as seen by each target, or anything in between.

B.1.2   What Receivers should we Target?

Assume that silicon technology is good enough so that the signal analysis is limited only by antennas, feeds, and receivers. Assume also that all received frequencies are analyzed. (This is not true of current systems, all of which analyze the full bandwidth in chunks of 20 MHz or so, but this is the most obvious area for improvement.) Assume that each receiver has, at minimum, a detector for narrowband, source-compensated, CW signals.

There are several possible types of receiving systems which differ considerably in antenna size, noise figure, bandwidth, and duty cycle. The general characteristics of different receiving systems are shown in Table B.1.

System Name	Antenna Size	Noise Temp.	BW	Margin	Flux	Duty Cycle
Phoenix	100.0	20	0.01	10	3.5×10^{-27}	1/N
BETA	25.0	50	0.01	20	2.8×10^{-25}	10^{-5}
Argus	3.7	100	0.01	35	4.5×10^{-23}	1.0
None	0.3	100	1.0	10	2.0×10^{-19}	1.0

Table B.1: Characteristics of Receiving Systems.

Note that the *Margin* column is the signal-to-noise ratio, in the given bandwidth, that is necessary to record a detection without too many false alarms. Phoenix has a low margin, because it has active follow-up. BETA has a medium margin, because the target must be seen in multiple beams. Argus has a high margin, because it looks at many channels all-the-time with no beam or separate follow-up.

Targeted searches, such as Phoenix, achieve good sensitivity; but they have very low duty cycles. (1 in N, where N stars are considered.) The required margin is relatively low, because it has a geographically distant backup receiver.

The all-sky searches such as META, BETA, and SERENDIP have extremely low duty cycles. For example, a 30 m antenna with a 30 cm wavelength gives a duty cycle of one part in 10^5; so, if used as a transit instrument, a source is in the beam for only four minutes. A bigger dish at the same frequency will have better sensitivity but a smaller duty cycle. Higher frequencies cause the duty cycle to drop even further.

'All-sky', 'all-the-time' systems try to look at the whole-sky all-the-time (OSS), or 'all targets', 'all-the-time' (Argus). These systems have low sensitivity, but a high duty cycle. For an omnidirectional sky survey system, there is a question of how much Doppler correction can be achieved. This is limited by the relatively large beam width and the variation of Doppler shift with position.

B.1.3   Basics of Phased Arrays

The principle behind a phased array beacon is simple. Electromagnetic signals add linearly in voltage, whereas the power varies as the square of the voltage. For example, for a single isotropic radiator of power P, an observer will receive a signal power that is proportional to P. However, for two such radiators, geographically separated but adjusted, so that their signals are exactly in phase at the observer, the signals add linearly in voltage, and the observer will see twice the voltage of the single transmitter. This is the same result that would be achieved by a single transmitter of power $4P$.

Similarly, three transmitters in phase result in a received power of $9P$. In general, N transmitters of power P can generate the same received signal strength as an isotropic transmitter of power $N^2 P$. Naturally, this power gain in the desired direction is offset by cancellation in other directions; so that, the total radiated power, integrated in all directions is conserved (NP).

If the transmitters are moving with respect to each other, perhaps because they are located on the surface of a rotating planet, then the phase of each transmitter must be actively manipulated to achieve the desired superposition in a given direction. This constant manipulation of phases is well within the reach of modern electronics.

This is exactly the reverse of the process used by radio telescopes made up of multiple antennas. By computing the desired phase shifts and then summing the signals, the desired signal voltage adds up linearly; whereas, the uncorrelated receiver noise only adds up as $\sqrt{N}$. Thus, the S/N (in power) increases as $N/(\sqrt{N^2})$, or N, as expected.

B.1.4   Phased Array Beacon Design

Assume that a budget for a phased array beacon is equal to that of a major astronomical facility – on the order of $\$10^8$ to $\$10^9$. For constant total power, the EIRP increases as the square of the number of modules; so the largest feasible number of modules is preferred. Using 1999 technology, the optimum might be 10^8 modules, each built with a single chip and costing $\$1$ to $\$10$. Since each is a single chip and extensive heat sinking is difficult at this cost, each might have an output power of about 1 W.

These modules might be arranged on a flat area, 10 km on a side, with four overlooking viewpoints that are used to measure the three-dimensional coordinates of each module. This gives a density of one transmitter per square meter, on average. The RF power of 1 W/m^2 is a small fraction of the solar incidence of 1 kW/m^2; thus, provided that the beam is focused at infinity, it should not hurt machines or animals that fly through it. The antenna would not block a very large fraction of sunlight; so the land beneath it could still be used for agriculture or ranching.

Phase Adjustment

The primary technical problem is to correctly adjust the phases of each module. One possible solution is for a transmitter with a clear view of the array to send a master signal at 700 MHz to all modules. Each module receives 700 MHz, doubles the frequency, adjusts the phase according to an individual schedule, and then transmits it at 1,400 MHz. In this scheme, only the master transmitter needs stability.

To correctly phase the beams, we need the three-dimensional location of each transmitter module. We could control this by accurate physical construction; but it seems easier to measure the location of each module instead. This can be done by a technique borrowed from laser range finders as follows:

1. Select a module which will be the only one retransmitting the master carrier.

2. Turn all adjoining transmitters on, but not to transmit, in order to make the interaction effects as correct as possible.

3. Modulate the master carrier (FM) with 10 kHz.

4. Measure the phase of received modulation at four known locations. This will give the location of the module to within a small fraction of the wavelength of the 10 kHz signal.

5. Switch the master carrier modulation to 100 kHz, then 1 MHz, etc., each time estimating a more accurate location.

6. Carry out the final reading with carrier phase.

As a 1 degree accuracy in phase gives 1 mm precision, it is more than enough for this application.

Power

The modules could be powered from the normal commercial power grid or by solar power. In many ways, this is an ideal application for solar power, as the required power density is small, and the failure of any module is of minor importance. It also reduces the operating budget, which may be important for a potentially long-term operation. This is especially true in a long-term scientific experiment where, historically, initial capital is much easier to come by than operating funds. Solar power would enable modules to be sealed (with no need for connectors), which should improve reliability and ease of construction.

An RF power output of one watt implies that three watts of power may be required. At 10% efficiency and 1,000 W/m^2, this requires an area of 0.03 m^2 (about 17 cm on a side). This blocks only 3% of the available light.

Polarization

Each module might have crossed dipole radiators, to be able to radiate in any polarization. It might have a separate receiver antenna tuned for 700 MHz. Superimposed on the 700 MHz command carrier would be 50-bit frames, one per microsecond. Each module decodes this data stream to get its own commands. Each module could adjust its own phase according to a linear or quadratic schedule. Linear phase tracking is good for 100 seconds; quadratic phase for about 500 seconds.

With a much more intelligent module, Doppler compensation is possible. If a CMOS chip is used, this may not be prohibitive for 500 to 1,000 targets. Because each module can set the phase of X and Y antennas independently, it is possible to transmit either circular or linear polarization (LP) signals.

B.1.5 Usage

A beacon is limited by the power budget. Within this budget, the energy may be spread across a single beam or many beams, and in many patterns over time. For example, it could:

1. Hit each target for two hours each year with a 10^{16} watt beam.

2. Hit each target more or less continuously with a 10^{13} watt beam.

3. For pulsed operation, for example, hit each target with a one second, 10^{15} watt pulse, once every 100 seconds for one-tenth of the year.

Which of these appears to be best?

A system such as Phoenix watches 1,000 stars over several years. The system spends a total of about one day on each star, but observes each frequency for only 300 seconds. To correspond with this regime, we want to make sure to send at least one strong signal to each star every 100 seconds.

One second pulses repeated every 100 s might be a good choice. This is a 1% duty cycle, so if we are trying to signal 1,000 stars, we need to hit 10 of them simultaneously. This means that from 1,000 ly, we need a beacon of 4×10^{12} W EIRP for each target, or 4×10^{13} W EIRP total, split 10 ways. This implies 6.5×10^6 modules for a cost of between \$6.5 M and \$65 M. This is easily feasible today.

To reach a BETA type system we need a higher EIRP; approximately 3×10^{14} W from 1,000 ly. Because the duty cycle for the receiver is so low, we need to hit all the targets all the time; so we need a total power of 3×10^{17} W EIRP, or about 5.5×10^8 modules. This is about 100 times more expensive than reaching a Phoenix system.

To reach an Argus type system, we need high peak power but only a very low duty cycle; so we could just hit each star in turn. The peak power required from 1,000 ly is 5×10^{16} watts, requiring 2.5×10^8 modules; this is easier than reaching a BETA system by a factor of about two.

Note that if we target Argus, then Phoenix and BETA will not see it, because their duty cycle is too low. If we target Phoenix, then BETA and Argus will not see it, because their sensitivity is too low. If we target BETA, both power and duty cycle will be unnecessarily large for Phoenix; but Argus will not see it, because it is too weak.

Maybe we could alternate between them, perhaps every 10 years, or even alternate days!

B.1.6   Aiming

There are several limitations on how narrow a beam should be. To be useful, it must be wide enough to encompass the position of the target when the beam gets there, accounting for all uncertainties. Since the target stars are moving, the beam has to lead them to hit the spot where the target will be when the beam arrives. The basic strategy is to extrapolate the proper motion of the target by twice the round-trip light time, and broadcast to that position. A 10 km antenna with a 20 cm wavelength produces a beam of about 4 arc seconds in size. This sets the minimum limits of accuracy needed.

The first limit concerns inaccuracies in initial location and proper motion estimation. However, Hipparcos data is adequate here – about 0.001 arc second.

The second limit is distance. Since the proper motion must be extrapolated by twice the round-trip travel time, distance measurements are crucial. For example, a target star at 1,000 ly may have a proper motion of about 0.03 arc second/year. Hipparcos distances (the best available now) have about a 300 ly uncertainty at this distance; leading to a 20 arc second uncertainty in location which is too big. However, assuming that GAIA or SIM (the successors to Hipparcos) are built, they will measure proper motions and parallaxes to 0.01×10^{-3} arc second per year, which will give sufficiently accurate distances.

The final limit is accelerations. Unmodelled accelerations will result in motions that are not accounted for by extrapolation of proper motions. However, this appears too small to be significant: as 4.0×10^{-11} m/s^2 unmodelled acceleration implies only a 1.6×10^{-3} arc second error at 1,000 ly. We need to be sure of hitting the planet while aiming at the star; but this is easy since 1 AU = 0.288 arc second at 10 ly.

B.1.7   Alternatives

The beacon could run off solar power during daylight, and commercial power at night, to provide a higher duty cycle on each target. We could build three stations, spaced around the world, and run them on commercial power. The duty cycle could then be continuous except when a target is occulted by the Moon or Sun.

B.1.8   Conclusions

A deliberate beacon will be very bright, between 10^{13} W and 10^{16} W, even with our current technology. So we probably have enough sensitivity. We just need better duty cycles, (i.e., more frequency coverage more of the time). Stability will be high, because it is easy to Doppler compensate the beacon while the beam is forming.

It is about 100 times easier to reach a Phoenix type system than a BETA or Argus system, which are about equally difficult but require different strategies. The hardware can trade-off evenly between peak power, number of stars illuminated at one time, and duty cycle; this is subject only to the total power budget. So there is little to choose between these options. Fast follow-up makes the job of the beacon easier since it can then use lower duty cycles.

B.1.9   Other Uses

A large transmitter could be used for other applications as well. Some possibilities are:

- deep space and planetary radar;
- commanding spacecraft;
- ionosphere experiments.

B.2   Mid-range Beacon

This section discusses the possibility of two types of less expensive microwave beacon designs. The first is a beacon made of very large numbers of very small transmitters on one printed circuit (PC) board with hundreds of drivers on one chip. The second is a beacon made of an array of small dishes (about 3 m across).

Both types of beacons could be run in a mode where the total radiated power is low although the EIRP is still quite substantial, leading to beacons that could be run with an amateur radio operators' license, but still contribute noticeably to the Earth's EIRP. An interesting point is that both designs strongly favor higher frequencies.

B.2.1   Phased Arrays of Very Small Elements

Phased arrays are promising as beacons since the EIRP scales as N^2, where N is the number of elements and each element adds power and area. Two 'printing' technologies help us to build as many elements as possible, as inexpensively as possible. We can print several antennas on a PC board, and we can print several drivers on a single chip. So a 'tile' in this system consists of a single PC board with a single chip on it.

How many elements can we get in a single low cost tile? The shorter the wavelength, the more elements per tile and the more cost-effective the system. For example, suppose we use 10 GHz, or 3 cm. At this wavelength, the required $\lambda/2$ spacing is 1.5 cm. To allow for non-zenith pointing, we might use a slightly larger spacing, e.g., 2 cm, which would allow for 41 degrees off-bore sight before efficiency suffers. At 2 cm spacing, a PC board 33 cm on a side (about the size of a PC motherboard) would contain 277 antennas with a single chip to contain 277 programmable phase shifters. If each transmitter uses a milliwatt, then the total power output is only 277 mW, so heat dissipation is manageable.

How big an array could be built this way? 10^9 elements seems practical, requiring 3.6 million tiles. If a tile could be built for $10, which should be possible at that volume, then the total cost would be $36 M.

Such an array forms a powerful beacon. Although each element radiates only 10^{-3} W, the antenna gain is 10^{18}, giving an EIRP of 10^{15} W. This is about 50 times the EIRP of the Arecibo and JPL planetary radars! Since each element radiates only into a cone around the zenith, each element only requires a small gain. If each element is a dipole above a ground plane, then the gain is doubled.

Assuming these tiles can be built, then the infrastructure could be quite simple. Using the technology developed for tile floors in shopping centers, pour a flat concrete slab and glue the tiles onto the slab. This requires a slab 630 m on a side. In practice, we would probably want smaller slabs separated by access roads. As a guess, we could distribute the pilot signal though copper water pipes embedded in the slab.

The chips would each need to contain a 16 element phase shifter, generating all phase shifts in increments of $\pi/8$ radians. Each of the 277 outputs would then be connected to one of these phase taps. Each chip would be programmed with an X and Y phase gradient that defines the pointing, by controlling the taps for each output. 10 GHz is fast for today's CMOS, but might be easy to achieve in a few years. If so, the chip would be very inexpensive. The simplest chip gives up the ability to hit N targets with 1/N of the EIRP, but a more complex chip design could allow this, and might not be much more expensive.

Amateur Radio Operation

The basic idea is simple: An amateur operator is allowed to transmit about 1,000 W in a few bands devoted to communication between amateurs. In general, no limit is placed upon antenna gain, and, in fact, high gains are used for amateur pursuits such as Moonbounce (or EME, for Earth-Moon-Earth). In general, this pursuit is unremarkable and only followed in a few specialist hobby magazines. No permits and no public discussion are needed – just a relatively easy to obtain license based on simple technical competency.

The chip could be built with a 'ham' mode, where each output is 1 μW instead of 1 mW. If this was used for all chips (for the total of 1,000 W permitted for amateur operation); then the EIRP would be only 10^{12} W, which would still be the third strongest transmitter on the planet!

How does this compare to the EIRP obtained from an amateur pointing the largest legal transmitter through the largest available dish? For a fixed size antenna, the greatest EIRP is obtained at the highest frequency. Some clubs have built, or obtained the use of, 30 m class antennas. If these could be used at a wavelength of 3 cm, which provides an antenna gain of 10^7, then with 1,000 W out, an EIRP of 10^{10} W results. This is not bad, but the 'ham' mode of the professional facility is (not surprisingly) better.

B.2.2   Phased Arrays of Small Dishes

Suppose it costs $3,600 each for a 3 m dish, with mount, and a 100 W transmitter installed. Then, for the same $36 M (suggested for the small element design), we get 10,000 dishes. At 3 meters and 3 GHz, each dish has a gain of 10^5, so each dish has an EIRP of 10^7 W, for 10^{15} W total. In 'ham' mode, each is 100 mW, for an EIRP of 10^{12} W. (Feed inefficiencies would reduce this somewhat, but the 100 W is just a guess anyway.) We would give up electronic steering over long angles, and could only hit multiple targets within a primary beam, but perhaps, this could share infrastructure with a 1hT design. We could gain horizon to horizon tracking.

Can we do better than this with big dishes? Probably not. Transmitters do not share many of the features that push receivers to larger dishes. Spillover is not nearly as important, and a single frequency is adequate; so feeds are not as complex and costly, and no cryogenic cooling is needed.

Even one tenth of either full array would have an output of 10^{13} W EIRP, which is about the same as the Arecibo or JPL planetary radars. Construction could be incremental and continue after operation has begun.

B.2.3   Conclusions

There are at least two comparatively inexpensive approaches that look promising for very high EIRP beacons. Both types of beacons work better at higher frequencies, with the EIRP scaling as the frequency squared. These straw man designs were calculated at 10 GHz, the upper end of the Earth surface microwave window. If the beacons were operated at 'water hole' frequencies, then each beacon would be only 1/36 as powerful for the same cost. Frequencies above 10 GHz would work even better until other limitations are reached. Such limitations are the limits of pin counts for ICs, surface tolerances and pointing for inexpensive dishes, and the atmosphere itself.

Searches that are looking for beacons should go as high in frequency as possible, since bright beacons are easiest to build at high frequencies.

B.3   A Low Cost Personal Beacon

This section considers an easy and self-consistent, transmit-and-listen strategy. Suppose we want to implement a strategy that will work, provided both civilizations follow the same strategy. What could we do along these lines? Here is such a strategy for the 1,000 closest stars.

Build two beacons, one in the Northern and one in the Southern Hemisphere. Each is low power (1,000 watts) to allow use under an amateur license. (This does not resolve the ethical issues of transmitting, but does resolve the legal issues. Individuals are allowed to radiate about 1,000 W of CW power in certain specified bands as a hobby.) Each beacon would broadcast at 3 cm (10 GHz) with a 10 meter dish, giving an EIRP of 10^9 watts; and each would hit 500 stars for 100 s each, once every day. (This allows about 60 s to re-point the antenna between broadcasts.) The signal will be Doppler compensated.

Assume that each civilization receives with a telescope with a square kilometer aperture and a 20 K noise temperature, with the appropriate pulse detector. This gives a S/N of about 30. Real time interferometry and a continuous day of listening will be used for confirmation.

The interesting point here is that a very low cost beacon, well within the range of an individual's finances (and very easy for an organization like the SETI Institute), makes the job of the receiver much easier. Compared to leakage, it has a higher EIRP, uniform repetition, and Doppler compensation. It is inexpensive to run (8,000 hours/year at $0.08/kWh = $640 per year for power), and so could be kept up for a long time. No new technology is needed. Except for the ethical issues of "Should we transmit?", it would be easy to do and would give us a consistent stand in front of donors. We would not be relying on the ETIs to do anything that we are not doing ourselves.

B.4   Adding a Beacon Capability to the SKA

By adding a small (about 1 W) transmitter to each element of a large array, we can achieve rather high EIRPs. It is not at all clear whether we can do this without damaging the receiver, but, if we can, then we could use it as a planetary radar, beacon, spacecraft command transmitter, etc. Also, it is much easier to make a frequency-agile 1 W transmitter than a 100 kW transmitter, so we potentially have the possibility of a much broader selection of frequencies.

The calculations differ somewhat from those given in Section B.1, on page 305. Each element, by itself, has an appreciable gain, and a fairly narrow beam. The individual EIRP is multiplied by N^2 to get the total EIRP, where N is the number of elements. This gives much higher EIRPs for the same number of elements. However, it also means, we can only form multiple beams within the relatively small beams of the individual antennas.

Table B.2 shows the total EIRPs achievable for an array of 1 km^2 total area, dishes of 4.4 and 7.3 meter diameters, and an aperture efficiency of 50%.

Size (m)	Number of Elements	1 GHz	3.1 GHz	10 GHz
4.4	65,000	4.5×10^{12}	4.5×10^{13}	4.5×10^{14}
7.3	24,000	1.6×10^{12}	1.6×10^{13}	1.6×10^{14}

Table B.2: Achievable EIRPs.

We can compare this to the world's current EIRP champion, the Arecibo planetary radar. After the upgrade, this has an output of about 133 dBW, or an EIRP of 2×10^{13} W. So, if we express the results as a ratio of the EIRP to that of Arecibo, then we get the figures in Table B.3.

Size (m)	Number of Elements	1 GHz	3.1 GHz	10 GHz
4.4	65,000	0.22	2.2	22.0
7.3	24,000	0.08	0.8	8.0

Table B.3: EIRP and Arecibo Compared.

If we can add the transmitter without disturbing the receiver unduly, it would add quite a powerful transmitter capability. This might be straightforward with an off-axis design where blockage is not important: just put the transmitter feed next to the receiver feed, and accept the lower performance from not being at the optimum focus. Since the power is low, we can probably use a single chip, and cooling is not required; so it could be very inexpensive.

If the chip is simply a broadband power amplifier and copies the signal from the fiber-optic feed cable, then all the phase compensation could be done at the central station. This allows many phase adjusters on one chip at the central station, and might be combined with the receiving phase adjusters (assuming simultaneous operation is not attempted, as it would probably not be on a simple system). This could give a very powerful, frequency agile transmitter for a relatively low, incremental price.

B.5   A Simple Optical Beacon

A straw man design by Horowitz showed the feasibility of building a simple optical beacon. Just take a big laser and aim it backwards through a big telescope. You have a beacon!

B.6   Phased Array Optical Beacons

This section looks at building an optical beacon from a phased array of semiconductor lasers. Possible advantages include riding the semiconductor learning curve, high efficiency, a design with no moving parts, and fast beam steering. The main disadvantage of such a system is that (at least with our current technology) the optimum pulse length is longer than the detector response time, and so the simple coincidence schemes of Horowitz and Werthimer do not work well. However, a somewhat more complex receiving strategy that looks at the time span of the last N photons avoids this problem, and would be more sensitive to many astrophysical phenomena as well.

Howard et al. [How00] has shown that building an optical beacon is possible with near-term technology. This design uses a large laser feeding and some fast beam-steering optics. Such a design can generate easily detectable pulses even at interstellar distances (tens of photons/square meter/pulse at 1,000 ly).

However, the revisit time is slow at 1,000 ly (1,000,000 target stars, 0.1 seconds settling, implies only one revisit each 100,000 seconds or about 28 hours). Can we improve the revisit rate? In particular, it would be nice to get reconfirmation within one observing session, e.g., 1,000 seconds. Furthermore, this design involves large, precise and fast moving mechanical parts which, at least with Earth technology, can be a problem for facilities that must run for very long periods of time.

First, consider a beacon design for the Werthimer/Horowitz detectors. Assume a 1 ns pulse at 1,000 ly distance. The observer looks for two photons of the same color (frequency), for 1 ns, in a 1 m^2 aperture. The star puts out 10^6 photons/m^2/s at this distance. Assume 10^{19} infrared photons/J for computational convenience.

Existing laser diodes emit 2 W continuously. Current lasers can produce 20 W pulses with low duty cycle. Assume this output can be improved to 100 W pulses. Vertical Cavity Surface Emitting Lasers (VCSELs) emit perpendicular to the wafer surface. They are relatively small and can be grown on wafers at 50 μm spacing.

A 100 W pulse at 1 ns = 10^{-7} J = 10^{12} photons. Assume the beacon can cover

1/10 of the sky $= 10^{38}$ m^2 at 1,000 ly. So one diode gives 10^{-26} photons/m^2/pulse. We need a gain of about 10^{27} to make up for this, and get 10 photons/pulse/m^2. With phased arrays, this implies 3×10^{13} phased emitters.

This is a sizeable number, but these lasers are printed on a wafer surface with normal semiconductor manufacturing technology. Today, we can put these lasers on 50 μm spacing, and we can make 300 mm wafers (although of silicon, not the laser material). Combining these technologies, we get 3.6×10^7 lasers per wafer. So, we need 10^6 wafers – a large number, but not impossible. These form a square about 10^3 wafers on a side, or about 300 meters.

Can we phase the lasers properly? We could illuminate them from the back, with a laser with a long coherence length. Each laser would need a phase shifter, but this can probably be built with conventional photonics located between the lasers.

B.6.1   Longer Pulses – Transmitters

Can we use fewer lasers? We can increase the received energy by making longer pulses. With semiconductor lasers, this is much easier than increasing the peak power. How does this trade-off look?

For example, at 10^6 photons/s, the following photon counts occur naturally with a probability of 10^{-12}:

- 4 photons in < 1 ns;
- 6 photons in < 10 ns;
- 12 photons in < 100 ns;
- 40 photons in $< 1,000$ ns;
- 0 photons for 40,000 ns.

Thus, for equal detectability, longer pulses require more energy, but less peak power. For the example above, the 1,000 ns pulse requires 10 times the total energy per pulse, but the peak power is 100 times less. So, if we are peak power limited, longer pulses look very attractive.

Suppose we try for 40 photons in 1,000 ns, or 0.04 photons/m^2/ns. This is a factor of 250 less peak brightness, but 4 times the total energy per pulse. However, if we are peak brightness limited, it is a good trade-off, enabling the use of a transmitter that is $\sqrt{250}$ or about 16 times smaller.

The details are as follows:

- A 100 W pulse 1,000 ns long $= 10^{-4}$ J $= 10^{15}$ photons.

- Assume we cover $1/10$ of the sky $= 10^{38}$ m^2 at 1,000 ly.
- So one diode gives 10^{-23} photons/m^2/pulse.
- We need a gain of about 4×10^{24} to make up for this, and get 40 photons/pulse.

With phased arrays, this implies 2×10^{12} phased emitters. This is about 60,000 wafers, or about 250 wafers on a side, or about a 75 meter square. This is only $1/16$ the size of the short pulse system, as expected.

What about power density? We have 4×10^8 emitters per m^2. So when we are emitting, dissipation is about 40 GW/m^2, assuming 50% efficiency, which has already been demonstrated in semiconductor lasers. Since we are using this array to cover $1/10$ of the sky, we need to visit 100,000 stars in 1,000 seconds, or one pulse every 10^{-2} seconds. So our duty cycle is 10^{-4}, and the average power is 4 MW/m^2, or 400 W/cm^2. Power transistors easily handle this power density today, so it is possible; but a powerful cooling system will be required.

What about total power? 2×10^{12} lasers at 100 W output each and 50% efficiency require 4×10^{14} W when they are turned on. The duty cycle though, is 10^{-4}, so the power required is about 40 GW.

B.6.2 Longer Pulses – Receiver

How could we detect this longer pulse?

Suppose we 'time-tag' each incoming photon, and put them in a shift register. Then, by looking at the difference between the Nth entry and the first we see how long it took to get N photons. For each incoming photon, and each N up to some number (e.g., 100), we compare this to the minimum and maximum numbers expected for that many photons and generate an alert if the number falls outside the expected bounds. The computer controlling the observation can generate the minimum and maximum bounds for each photon count, given the average counting rate and the desired false alarm rate.

The computation rate for this is reasonable. Since we need to redo the calculation for each incoming photon, it is easiest for faint sources. If we build the whole receiver in hardware (quite practical in one chip) for 100 photon memory, we need 100 stages of shift register (perhaps 16-bits wide) 100 subtractors, 200 'min-max' registers, and 200 comparators. If built in modern CMOS, and since any possible pipelining is suitable, it could easily cycle in 2 ns. This would be good for incoming rates of up to 500×10^6 photons per second. With a 1 m^2 aperture, any star at least 50 ly away is suitable. Closer stars could use a smaller telescope or a filter.

If done in software we need to do a store, a hundred subtracts, and 200 comparisons. At a wild guess, with a 250 MHz DSP chip, 500 instructions required, and a parallelism of 2, we would be able to keep up with one photon every microsecond. So this would only work on distant stars (more than 1,000 ly) and is not quite practical yet. Exactly how this would fit into the DSP instruction set would have a big impact on the actual throughput. Each DSP chip would need to be considered individually to get accurate estimates.

There is a side benefit. This system should be much more sensitive to interesting astrophysical phenomena than the simple coincidence detector. Eclipsing stars, pulsars, etc. which exhibit many variations in time will generate nonstandard statistics and trigger the detector.

Appendix C

Noise Temperature Estimates for Next Generation Amplifiers

This appendix, by Sander Weinreb, examines amplifiers designed to cover the range from 1 to 10 GHz, that use state-of-the-art High Electron Mobility Transistor (HEMTs) operating at temperatures of 300 K, 80 K, and 20 K. Estimates of the noise temperature, as a function of frequency, are provided.

C.1 Introduction

At present the microwave antenna with the largest collecting area on Earth is in Arecibo, Puerto Rico, a 305 m diameter spherical reflector with an effective area of approximately 30,000 m^2. For comparison, the VLA in New Mexico has an effective area of approximately 10,000 m^2.

A larger collecting area would greatly increase the sensitivity of the search for ETI, for various radio astronomical observations, and for wide bandwidth communications with deep space probes. Initial investigations have started for the conceptual design of an antenna with an area on the order of 10^6 m^2. A wide variety of formats, ranging from 10 Arecibo-type antennas to 10^6 very small fixed antennas are being considered. In the middle of this range is an array of about 10,000 antennas, each of which has a diameter of 9 m. This latter approach can exploit the technology and cost basis of highly developed and low-cost direct broadcast satellite receiving systems. The frequency range being considered for this array is continuous coverage of 0.3 to 10 GHz.

The following sections discuss how to estimate the noise temperature and bandwidth of low-noise amplifiers using current state-of-the-art low-noise HEMT transistors operating at temperatures of 300 K, 80 K, and 20 K. This information, together with cost estimates of the receivers, is required to minimize the total cost for antennas, receivers, and processing equipment for a given figure-of-merit such as effective area or gain divided by system temperature, G/T_{sys}.

C.2   Methodology and Assumptions

Fortunately, the noise-theory and device data needed to solve this problem are well in hand. The field-effect transistor model published by Pospieszalski [Pos89] gives the dependence of noise temperature as a function of physical temperature, frequency, and generator impedance given the equivalent circuit of the transistor as determined by measurements of S-parameters. The transistor noise is described by assigning thermal noise at ambient temperature to all resistors in the transistor equivalent circuit except the drain to source resistance, R_{ds}, which is assigned a measured noise temperature, T_{drain}.

The transistor models for several GaAs and InP HEMT transistors with different width and manufacture (Lockheed-Martin, TRW, and Hughes) have been measured at the University of Massachusetts. The differences between the best transistors of the manufacturers are relatively small, and the temperature dependence of the model is also small as observed from the small changes in frequency dependence of several MMIC LNAs that have been cooled from 300 to 20 K.

At frequencies below 10 GHz, the gate leakage current of an InP HEMT can contribute significantly to the noise temperature. As suggested by Pospieszalski [Pos93], the noise due to gate leakage current can be modeled by two shot-noise current sources across the gate-to-source and gate-to-drain junctions of the transistor using the relation, $in2 = 2e|I|$, for each noise generator where $e = 1.6 \times 10^{-19}$, and I is the Direct Current (DC) I_{gs} or I_{gd} leakage current. I_{gs} and I_{gd} must be determined separately by measuring $I_{gs} + I_{gd}$ as a function of V_g with $V_d = 0$ and I_{gd} as a function of V_d with $V_g = 0$.

These measurements have been made on a typical InP HEMT with the result that I_{gs} and I_{gd} = 2.5 and -2.0 mA at 300 K, 0.032 and -0.030 mA at 80 K, and 0.0001 and -0.006 mA at 20 K, all evaluated at the low-noise bias point of $V_g = +0.1V$ and $V_d = +0.6V$. These values would increase the minimum noise temperature at 300 K by 27 K at 1 GHz and 14 K at 10 GHz; the increase at cryogenic temperatures is negligible. In conclusion, unless gate leakage is reduced the noise of InP transistors will be higher at 300 K than GaAs HEMTs which normally have much lower leakage. A model of a GaAs HEMT will thus be used to predict 300 K noise performance and the InP models will be used at 80 K and 20 K.

A typical model for a state-of-the-art, generic, 0.1 μm InP HEMT, with width 50*M1 μm, is described in Figures C.1 below, and C.2 on page 324.

Figure C.1 shows the circuit file of a transistor-equivalent circuit including noise sources of a typical InP 0.1 μm HEMT transistor with width equal to 50*M1 μm. The drain noise temperature is a function of temperature and can be determined by one noise figure measurement of the transistor at any frequency. Based upon previously published noise figures and cryogenic results [Smi96], and our own measurements of the noise of wideband InP MMIC LNAs, a value of $T_{drain} = 300$ K $+ 6T_{amb}$ has been used in our analysis, where T_{amb} is the ambient temperature. This information is sufficient to determine the minimum possible noise temperature, T_{min}, of the transistor given an ideal lossless input-matching network. T_{min} is plotted as a function of frequency at $T_{amb} = 300$ K, 80 K, and 20 K in Figures C.3, C.4, and C.5 on pages 324 - 326, respectively.

```
! DEFINE GENERIC 50*M1 um InP HEMT
! MODEL REF PLANE +/- 30uM FROM CENTER
NCS 2 5 I={0.32*MAG(IGD)*M1} R=1E10
NCS 2 3 I={0.32*MAG(IGS)*M1} R=1E10
IND 12 1 L={6+10/M1} ! LG
CAP 12 0 C=7.5 !CPG
RES 1 2 R={4/M1} ! RG
CAP 2 3 C={48*M1} ! CGS
RES 3 4 R=0 ! RI
RES 4 10 R={7.2/M1} ! RS
IND 10 0 L=3.9 ! LS
CAP 2 5 C={7*M1} ! CFI
CAP 5 4 C={2} ! CDS
RES 5 6 R={10/M1} ! RD
CAP 1 6 C=1 ! CFO
CAP 8 0 C={8+10*M1} !CPD
IND 6 8 L=23 ! LD
VDCS 2 5 3 4 GM={60*M1} TAU=0.15 R1=1E6 R2=1E6 & F=1E6 A=0
REST 5 4 R={170/M1} T={300+6*TEMP}
DEF2P 12 8 HEMTGS1 !THE FET MODEL
```

Figure C.1: Transistor Equivalent Circuit for Typical InP HEMT Transistor.

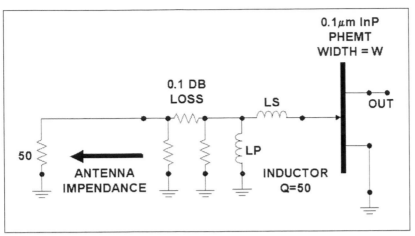

Figure C.2: Circuit Model for Estimating Feasible Noise Temperature.

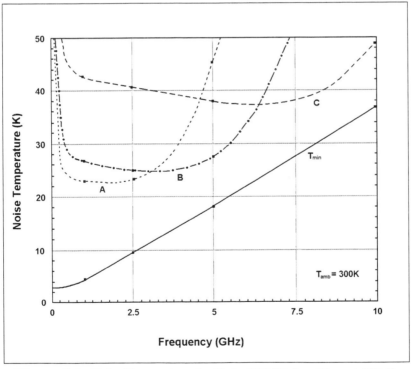

Figure C.3: Noise Temperature for GaAs HEMT Amplifiers at 300 K.

Figure C.3 shows the noise temperature, T_n, for three designs of GaAs HEMT amplifiers operating at 300 K with the model of C.2 optimized for the frequency ranges A (1 to 2.5 GHz), B (2.5 to 5 GHz), and C (5 to 10 GHz). The optimum circuit parameters for the three ranges are (width in μm, Ls in nH): A (1,000, 1.5), B (750, 1.2), and C (400, 0.8). T_{min} is the minimum noise temperature for the transistor with an ideal lossless input-matching network optimized for each frequency.

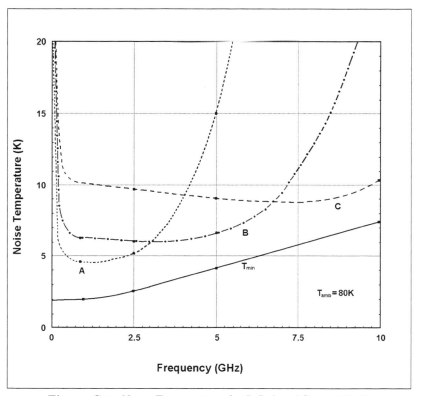

Figure C.4: Noise Temperature for InP Amplifiers at 80 K.

Figure C.4 is the same as Figure C.3 except for InP amplifiers operating at 80 K. The optimum circuit parameters (width in μm, Ls in nH) are: A (1,500, 2), B (600, 1.2) and C (300, 0.8).

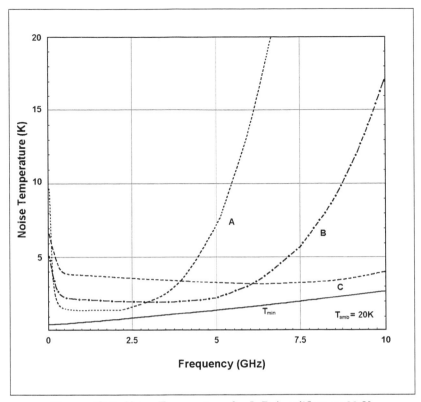

Figure C.5: Noise Temperature for InP Amplifiers at 20 K.

Figure C.5 is also the same as Figure C.3 except for InP amplifiers operating at 20 K. The optimum circuit parameters (width in μm, Ls in nH) are: A (1,250, 2.5), B (600, 1.5), and C (300, 0.8).

The remaining problem is to determine the noise introduced by a real, lossy, input network and additional transistor noise that occurs because the input-matching network cannot produce the minimum-noise generator impedance, Z_{opt}, over a wide range of frequencies. A simple model of the input-matching network, shown in Figure C.2 on page 324, has been selected for this study.

The model allows the transistor width to be varied to match the real part of Z_{opt} to 50 ohms at the middle of the desired frequency range and allows a series inductor, Ls, to be selected to provide reactance equal to the imaginary part of Z_{opt} at the center frequency. A shunt inductor, Lp, is also optimized in the model to attempt to reduce the noise temperature variation with frequency. The parameters Ls, Lp, and W were adjusted at each temperature to provide a minimum average noise over the desired frequency range.

A quality value (Q) of 50 is assigned to both inductors as an estimate of the loss of small inductors in this frequency range and a loss of 0.1 dB is added to the input of the amplifier to account for losses in the transmission line and connector or transition to waveguide. An off-chip series inductor may be desirable to achieve high Q. A more complex network with multiple poles could reduce the frequency variation of noise temperature in a specified band but would have more loss and is not considered in this appendix. Only the first stage of the low-noise amplifier is considered because in the frequency range under 10 GHz the transistor has gain > 10 dB and the second stage will add < 10% to the first stage noise temperature.

C.3   Results

The parameters, Ls, Lp, and M (transistor width) were varied while observing the noise temperature, T_n, of the amplifier as a function of frequency as computed by the MMICAD$^T M$ circuit simulation program. Gate leakage current was scaled with width. The shunt inductor, Lp, had little beneficial effect above 1 GHz and a value of 100 nH was used for all cases. Three sets of series inductance and the transistor width were chosen to minimize the noise temperature in the 1 to 2.5, 2.5 to 5, and 5 to 10 GHz ranges with results shown in Figures C.3 - C.5 on pages 324 - 326 at respective ambient temperatures of 300 K, 80 K, and 20 K. The optimum widths, such as 1,500 μm for the 1 to 2.5 GHz band, are larger than usually used for low-noise amplifiers (but certainly manufacturable and used for power applications). Smaller widths could give the same result with an input network which transforms the 50 Ω generator to a higher impedance, but this will tend to reduce bandwidth. The results of this simple model indicate that wider transistors should be used for wide bandwidth low-noise amplifiers at the low microwave frequencies.

C.4   Summary and Conclusions

The results are summarized in Table C.1 on page 328 which shows both the average amplifier noise temperature, T_n, from Figures C.3, C.4, and C.5, and a first estimate of the total system noise temperature, T_{sys}, which adds the noise temperatures due to feed loss, ground pickup from spillover, atmospheric noise, and sky temperature. For the cooled receivers it is assumed that the feed transition is also cooled and less noise is added from this source. The results are shown for octave frequency ranges and also for the full decade, 1 to 10 GHz range. The feed loss and spillover contributions need careful study. Efficient octave-band feeds are not easily designed and an efficient decade band feed may not be feasible.

The results show approximately a factor of three reduction in system noise temperature by cooling the low-noise amplifier from 300 K to 80 K. In quantities of 10,000 elements, if we assume costs of \$300 for 300 K receivers and \$3,000 for 80 K receivers including refrigerator and vacuum dewar, the improvement of G/T_{sys} costs \$81 M for a 10,000 element array with 3 octave band receivers. This needs to be compared with the cost of increasing the total antenna area by a factor of about three.

Ambient Temp.		1 - 2.5 GHz T_n/T_{sys}	2.5 - 5 GHz T_n/T_{sys}	5 - 10 GHz T_n/T_{sys}
300 K	Rec.	23/38 K	25/40 K	45/65 K
	Sys.	45/65 K	45/65 K	45/65 K
80 K	Rec.	5/17 K	6/18 K	10/22 K
	Sys.	10/22 K	10/22 K	10/22 K
20 K	Rec.	2/14 K	3/15 K	4/16 K
	Sys.	5/17 K	5/17 K	5/17 K

Table C.1: Estimated Receiver and System Noise.

Appendix D

A Wideband Reflector Feed for the SKA

This appendix, by William J. Welch, describes a wideband version of the SKA. It consists of a large number of small reflectors of approximately 5 m diameter, with prime-focus feeds and a bandwidth extending between 0.3 and 10 GHz. If the amplifier is combined with a log-periodic dipole array feed of the same bandwidth, the entire frequency range can be received at once, offering a substantial multiplex advantage. Two planar arrays mounted at right angles would provide dual polarization.

In this appendix we examine a wideband version of the SKA, with a bandwidth extending over 0.3 to 10 GHz. This is largely motivated by the fact that it is now possible to build an HEMT-based low-noise amplifier that operates over that entire range (Weinreb, 1997, private communication). The simplest antenna that has wideband capabilities is the parabolic reflector, and a possible design for the SKA consists of a large number of small reflectors of approximately five meters in diameter with prime focus feeds. If the amplifier is combined with a feed of the same bandwidth, the entire frequency range can be received at once, offering a substantial multiplex advantage. The 30:1 frequency range suggests the use of a log-periodic dipole array for the feed. Two planar arrays mounted at right angles will then provide dual polarization.

D.1   The Log-Periodic Array

The properties of such arrays are reviewed by Rumsey [Rum66]. Figures D.1 and D.2 show schematics of a five meter reflector with a log-periodic dipole feed, realized as traces on two orthogonal PC boards. The HEMT preamplifier is a very small compact integrated circuit, which can be located at the vertex of the feed, and integrated with the transmission line of the feed to minimize line losses. The transmission line can then be excited by simply extending the amplifier input terminal across the gap, and no further balun is needed. The vertex of the feed is at the reflector focus. An offset feed option is also shown in Figure D.2. This might be particularly useful at low frequencies where the aperture blockage of the feed may be substantial.

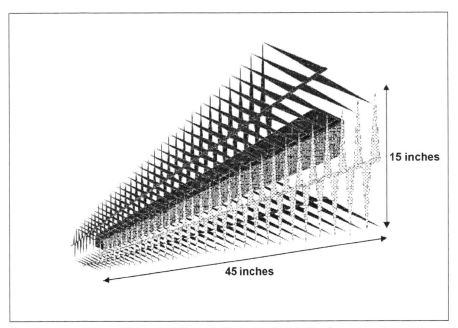

15 inches

45 inches

Figure D.1: Log-Periodic Dipole Feed.

One characteristic of the log-periodic dipole array is that its phase center is located about 0.8 wavelengths from the vertex. For wideband operation, such as we are considering here, the only sensible arrangement is to put the vertex of the array at the focus of the parabola. Then the reflector is fed 0.8 wavelengths out of focus at every wavelength. Figure D.3 shows the gain loss with focus error. For the usual focal ratio $F = f/D = 0.4$, the loss is about 40% with a focus error of 0.8 wavelengths. Increasing F to 0.6 reduces this loss to 15%, and at $F = 0.8$ this loss is only 6%. The effect of focus error is a broadening of the base of the main beam on the sky, and so there is no ground pickup associated with it.

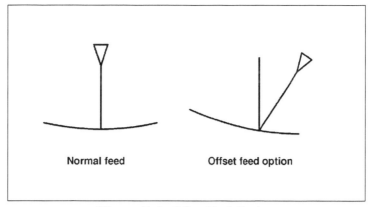

Figure D.2: Offset Feed Option.

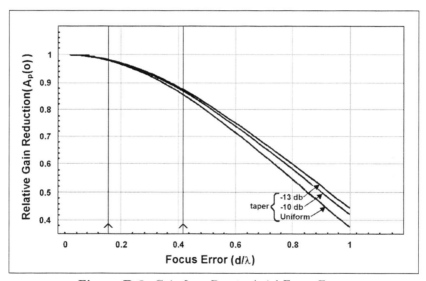

Figure D.3: Gain Loss Due to Axial Focus Error.

$A_p(o)$	Relative Gain Reduction from the in-focus case;
$x = d/\lambda$	Focus Error in wavelengths (d/λ); $x = d/\lambda = (17)(s/\lambda)(1 + 16(F/0.41)^2)$;
taper	strength of the field at the edge of the dish relative to the dish center.

The larger value of F requires more directivity for the feed, and a feed gain of 10 to 11 dB is necessary to ensure low spillover. Such gains have been achieved for log-periodic dipole arrays [Isb60]. However, the most completely characterized high gain antenna described by Isbell [Isb60] had a gain of about 9 dB, and we have worked out a feed design based on that case. This required some compromise with the focal ratio F, in order that the spillover past the parabola is not too high. Figures D.4 thru D.7[1] summarize the properties of this feed which has an angle of 34 degrees and log-period of $1.05 = 1/0.95$.

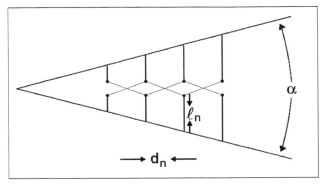

Figure D.4: Switched Dipole Array.

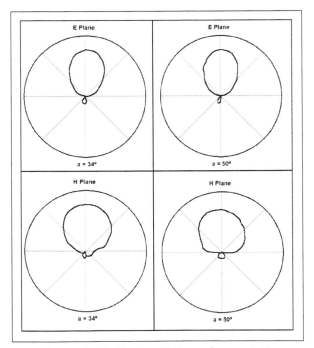

Figure D.5: Typical Patterns for $r = 0.95$.

[1]These figures are from Victor H. Rumsey, 1966, "Frequency Independent Antennas", Academic Press, New York, Figures 5.15, 5.17, 5.20, 5.23, pages 70-75.

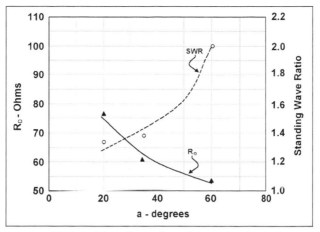

Figure D.6: Input Impedance $r = 0.95$.

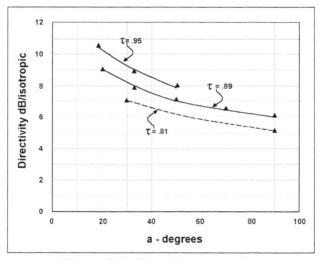

Figure D.7: Directivity vs. r and a.

Table D.1 on page 334 is a summary of the overall antenna/feed properties. We assume a reflector diameter of five meters, which is a minimum diameter for operation down to 300 MHz, or 1 m wavelength. The illumination efficiency and spillover are based on the patterns shown in Figure D.5. To estimate the expected system temperature, we have used the calculations by Weinreb (Appendix C on page 321). He found that a room temperature amplifier could be designed to work over most of the wide bandwidth with a noise temperature of about 25 K. If the amplifier could be cooled to 80 K, its temperature could be 15 K (or lower) over the 0.3 to 10 GHz band. At a physical temperature of 20 K, an amplifier noise temperature of 10 K results. In the table we give the result for the compromise feed with its 34 degree angle. In Table D.1, in the *Goals* column, we give the results expected from a higher gain feed.

Parameter	Value	Goals
D	5 m	
F	0.6	0.8
f	3.0 m	4.0 m
BW	0.3 to 10.0 GHz	
Feed Parameters:		
Length	1 m	
Angle	34 degrees	20 degrees
Period	1.05	
Input VSWR	1.4	1.2
Illumination Efficiency	0.87	
Forward Spillover	10%	
Backward Spillover	1% to 3%	
Defocus Factor	0.85	0.94
Feed Ohmic Losses	0.05 dB	
T_{sys}	60 K †	36 K §
† Uncooled receiver, sky brightness included.		
§ With improvements, and 80 K cooled receiver.		

Table D.1: Summary of Overall Antenna Feed Properties.

D.2   Conclusions

The results shown in Table D.1, are sufficiently encouraging that we plan further study. The goal will be the design of a 10 to 11 dB gain feed with reasonably symmetric E and H plane patterns. This should permit a larger focal ratio with less focus loss and less spill over. The possibility of cooling both the preamplifier and the input end of the feed will also be investigated. With no cooling, a system temperature of about 60 K is expected. Lowering that by a factor of two would be valuable.

This type of feed has several advantages. One, of course, is the capability of observing over a large number of wavelengths at once. Further, the collecting area and input impedance are frequency independent. Feed and spillover losses are modest, and the construction should be fairly inexpensive.

Appendix E

Optimum Small-Parabola for the SKA

This appendix, by John W. Dreher, discusses a very preliminary feasibility analysis for a small-parabola (3 to 30 m) option for the SKA. It is expected to significantly underestimate the actual costs when it is built. It assumes a general use of the array, rather than exclusive SETI use, and adopts the simple (A_e/T_{sys})/Cost figure-of-merit. It deviates from the SKA straw man specification, by having a frequency range from 1 to 10 GHz. Going to lower frequencies would have only a small effect on the results, but going to higher frequencies might take us out of the accuracy range available for commercial paraboloids, escalating costs significantly. The discussion considers only the array costs. It ignores beamforming costs, because the figure-of-merit does not include the number of beams.

In this appendix we attempt to refine the small-parabola (3 to 30 m) option for the SKA in a little more detail. We use a model that is expected to scale with, but that will significantly underestimate, the actual cost of the array when built. We will take the perspective of the general use of the array rather than the SETI use, and adopt the simple (A_e/T_{sys})/Cost figure-of-merit. In particular, we will set $(A_e/T_{sys}) = 2 \times 10^4$ m^2/K following the SKA straw man specification. We will deviate from this specification by taking the frequency range to be from 1 to 10 GHz. Going down to 500 MHz or so would have only a small effect on the results, but going up might take us out of the accuracy range available for commercial paraboloids, escalating costs significantly. Finally, we only include the array costs – in particular we ignore beamforming costs, since our figure-of-merit does not include the number of beams, furthermore the array costs dominate the total costs in this diameter range for all cases. This is a very preliminary feasibility analysis, and is certainly no replacement for a real 'Phase A' type of study.

E.1   Cost Model

Small dishes are an attractive option for the SKA because the existing market for direct broadcast TV reception, both domestic and at CATV head-end ground stations has produced units that are made on larger scales than encountered in research applications, and hence at much lower cost.

Figure E.1 shows some typical costs per unit area for such small dishes. The cost figures are based on retail prices for single items, adjusted downward to remove the contribution of the mounts. Actual prices would be lower for purchase in quantity, but probably not by too much given the extremely competitive market for these devices, particularly at the smaller sizes used in the home application. All these dishes are supposed to work up to the TVRO K_u band (12 GHz), but there are concerns that some of the less expensive units may suffer from some performance degradation at K_u band.

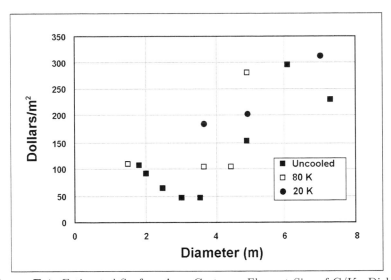

Figure E.1: Estimated Surface Area Costs vs. Element Size of C/K_u Dishes.

There is the appearance of a trend in cost per unit area that is approximately linear against diameter. Such a dependence is, of course, just what one would expect from the square-cube law; however, it may have more to do with the volume in which units are produced than physical principles.

For large paraboloids, past experience suggests that the cost is proportional to diameter to the 2.7 power and cutoff frequency to the 0.7 power. For our preliminary analysis, we adopt a cost per unit area of $50(D/3)^{0.7}$ where D is the diameter in meters. This cost is one tenth the area cost relation found for large parabolas. Part of this difference is that no costs for the mounts are included, but the larger part must be attributed to economies of scale.

The design of the mounts is a vexing problem. Many options need to be investigated, such as standard azimuthal mounts optimized for mass production and assembly, mounts based on linear actuators (no gears), and common mounts for multiple dishes. We adopt as a goal a mount cost just equal to the cost of the dish itself. As a sanity check, our cost model predicts $38,000 for a 7 m dish plus mount, while a price check with one manufacturer yielded $48,000 for an azimuthal motorized antenna of this size (in quantity one) – about 1 dB higher.

For the LNAs we use the estimate given in Appendix C on page 321: $600 for a dual polarization InP HEMT, plus post-amplification, power supply, and casing, in quantities of 1,000. We double the cost for cryogenic units. We further reduce the costs a bit for greater quantities, down to $500 at quantities of 10,000. Note that this LNA cost is rather higher than the $10 assumed for the 'tile' option.

For cryogenics, we adopted $2,000 for an 80 K system, based on large quantity pricing. We double that for a two-stage, 20 K system.

The same industry that gave us these economical dishes also produces fairly sophisticated feed horns for about $100 to $200 (dual polarity). Of course these units are narrowband, totally unsuited for our application, but they are mechanically similar to the type of wideband feeds developed by Kildal for Arecibo [Kil89], so we estimate that we could employ the same manufacturing processes to get feeds for about $300. Another option is the ultra wideband log-periodic feed described in Appendix D on page 329. Based on the cost of the typical roof top TV antenna, a weatherproof gadget of similar complexity, we estimate the log-periodic would cost $50 in large quantity production.

E.2 Optimization

Equipped with this cost model, we investigated the cost of the array with (A_e/T_{sys}) held constant at 2×10^4 m^2/K, with three feed/receiver options. Figure E.2 on page 338 shows the cost of the array (only) as a function of element diameter for these three options, and the options are described below.

The least expensive receiver option is a single, 1 to 10 GHz LNA at room temperature, which thus favors the smaller antennas. For such a wide frequency span, the log-periodic feed is appropriate. An aperture efficiency of 50% and a T_{sys} of 60 K were assumed. Since this option results in large numbers of small antennas, some downward adjustments were made in costs to reflect the huge quantity purchased, in particular the LNA cost used was approximately $250, based on the cost of commodity units. The array cost is minimized at $423 M for $D \approx 4$ m, with $\approx 1.9 \times 10^5$ elements in the array. The total cost of the SKA will be higher, as shown in Table 6.6 on page 185.

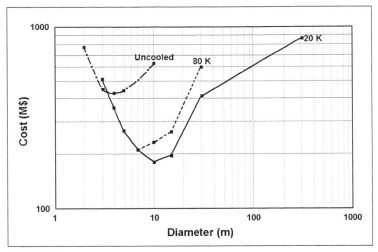

Figure E.2: Array Cost vs. Element Size.

Going to an 80 K cryogenic system reduces T_{sys} dramatically. We looked at the option of a single 1 to 10 GHz LNA versus using three 'octave' LNAs, which have better performance, and the latter option always won, so that is what is plotted in figure E.2. The three dual-polarization LNAs share a common Dewar. We assumed that there are three horn feeds when calculating the costs, although how and where to mount them remains uncertain so far. A representative system temperature of 18 K was adopted, based on Appendix C on page 321. The aperture efficiency was taken as 50% for $D \leq 5$m, since for very small dishes it will become difficult to get the feeds, mounts, etc., all squeezed in. For larger dishes, up to 30 m, we used 60%, while at 30 m we assumed a shaped Cassegrain would be used yielding 70% aperture efficiency. Again, some guesses were made as to cost variation with quantity. The array cost is minimal at \$208 million for $D \approx 7$ m and approximately 1.6×10^4 elements. The cryogenic maintenance issue might push us toward somewhat larger dishes for best overall economy.

Finally, we considered using a 20 K cryogenic system, and were surprised that it actually yielded the lowest cost of all. The minimum was \$180 million at $D \approx 10$m with 5.5×10^3 elements. We fear that 20 K cold stations may require much more maintenance than 80 K cold stations, so the approximately 10% cost reduction for the 20 K option does not seem worth it.

The best candidate, then, seems to be a system with parameters similar to a 7 m dish equipped with an 80 K cryogenic system and three 'octave' LNA feeds. Tempting as it is to have no cryogenics, that option seems to double the array cost for fixed A_e/T_{sys}.

Appendix F

SETI Figure-of-Merit

This appendix, by John W. Dreher and D. Kent Cullers, explains the process of defining figures-of-merit for different search options. It was originally published in: C. Cosmovici, S. Bowyer & D. Werthimer, eds., 1997, *Astronomical and Biochemical Origins and the Search for Life in the Universe*, Proc. IAU Colloquium 161, eds. C. Cosmovici, S. Bowyer, & D. Werthimer, (Editrice Compositari, Bologna.)

A figure of merit for SETI observations, which is an *explicit* function of the EIRP of transmitters, is developed. It allows us to treat sky surveys and targeted searches on the same footing. The product of terms measuring the number of stars within detection range, the range of frequencies searched, and the number of independent observations for each star, is calculated for each EIRP. For a given set of SETI observations, the result is a graph of merit versus transmitter EIRP. This technique is applied to several completed and ongoing SETI programs. The results provide a quantitative confirmation of the expected qualitative difference between sky surveys and targeted searches: the Project Phoenix targeted search is good for finding transmitters in the 10^9 to 10^{14} W range, while the sky surveys do their best at higher powers. Current generation optical SETI is not yet competitive with microwave SETI.

In early years of SETI the search domain was entirely uncharted, and technological innovations presented obvious opportunities to explore new realms. In this decade, we have begun systematically examining microwaves with forays into other regions of the electromagnetic spectrum. Clearly, future searches will be enhanced by improved technology.

As a community, we need a metric to measure search merit. Commonly used metrics agree that good searches have large bandwidth, look at many stars, and detect weak signals. A careful analysis shows that the number of stars visible to a given search strongly depends on assumed transmitter power and is approximately proportional to computing capacity. Any particular figure-of-merit must incorporate several assumptions and will suffer from a degree of unavoidable over simplification.

A significant problem with previously proposed figures-of-merit, e.g., [Dra83], [Gul85], is that they treat all solid angles as equal, making no allowance for selection of the directions in which to observe. As a result these methods are not suitable for evaluating targeted searches. In addition, the vexing problem of the luminosity function of the transmitters has often been treated in an implicit fashion that does not clearly express the underlying assumptions. Our new technique solves both of these problems.

F.1   Figure-of-Merit

Our figure-of-merit is an explicit function of P_ξ, the EIRP of the transmitters; this choice allows us to treat sky surveys and targeted searches on the same footing. This function, f_{merit} is the product of three terms designed to provide a plausible (but not unique) representation of the parameter-space 'volume' searched for transmitters with given EIRP:

$$f_{merit}(P_\xi) = N_{star}(P_\xi)\ F_{freq}\ N_{looks} \qquad\qquad (\text{F.1})$$

In equation F.1:

f_{merit} = Ignores two very important features, search-interference rejection techniques and detectability of different signal types, because we were unable to develop any simple and convincing metric for them.

P_ξ = Is taken to be that portion of the total transmitted EIRP that the search detectors under consideration will respond to. For example, in the case of a narrowband detector looking at a TV signal, we would take the power to be that of the carrier alone, disregarding the power in the side bands.

N_{star} = Is the approximate number of stars within the detection range of a given search at the specified value of P_ξ. We use the simple approximation of the number of 'good' stars within a given distance given in Section 2.3 on page 42.

N_{looks} = Is the number of independent observations made of each star. We assume that looking at a star more than once is an improvement; that is, we think that the duty cycle may be considerably less than, unity for the type of very powerful transmitters that current generation SETI systems can hope to detect. We considered using a logarithmic measure for the number of observations, reflecting the plausible assumption that if one has looked at a star many times, looking once more will not help as much. However, given that the duty cycle for beacons of high EIRP may well be small, and that N_{looks} is typically not large for current SETI observing plans, we chose the simpler linear metric because it is easier to actually implement. Some SETI programs have protocols that require more than one observation on a target for an ET signal to be recognized. In this case we count only the number of independent tries that are made. For example, Project Phoenix demands a chain of observations and re-observations, we count this as one independent look.

F_{freq} = Is a measure of how much of the available signal space is examined by each search:

$$F_{freq} = \eta_{pol} \int_{f_{low}}^{f^{high}} df/f \qquad (F.2)$$

which reduces to:

$$\eta_{pol} ln(f_{high}/f_{low}) \qquad (F.3)$$

The first factor η_{pol} allows for the polarization coverage of the search. We assume that the most-detectable signals will be strongly circularly polarized. A search that includes both circular polarizations is deemed to be complete and η_{pol} is unity. If only one circular polarization is searched, we set η_{pol} to 0.5. A search with linear polarization will respond to both circular polarizations, hence $\eta_{pol} = 1$, but only half the power will be received, which increases the detection threshold and thus reduces N_{star}.

The second factor is a measure of the frequency span searched. We have adopted a fractional measure for two reasons. First, it provides a frequency-independent measure of how well a given band has been searched. Second, in many cases the minimum detector bandwidth is set by Doppler effects that scale with frequency, so that this measure corresponds, approximately, to the number of possible 'channels' searched.

F.2   Results

For a given SETI program, the figure-of-merit will be a graph of $f_{merit}(P_\xi)$. We have applied this technique to several SETI programs to demonstrate the method. Where explicit star lists were not available, we used a uniform flattened disk model with a scale height of 1,000 ly and the local luminosity function. In all cases, for stars within 20 ly, actual star counts were used rather than the luminosity function.

F.2.1   Completed SETI Programs

Figure F.1 shows the figure-of-merit, f_{merit}, on a log scale, plotted as a function of the EIRP of the transmitter, P_ξ, also on a log scale. Three recently completed SETI microwave programs are shown:

- the subset of the SETI Institute Project Phoenix program completed at the Parkes antenna of the ATNF [Tar96];
- the recently completed UC Berkeley SERENDIP III program using the Arecibo antenna concurrent with other research programs [Bow00], [Wer96];
- the Harvard/Planetary Society META program [Hor93].

The curves begin at the point where just one star is within range. As might be expected, the two sky surveys show basically a power law slope with index 1.5, with some fine structure due to the actual star counts at the low power end, and with a gradual roll-off at high powers caused by the flattened star distribution. The shape of the Phoenix curve is quite different. At low powers it rises along a similar power law, but then plateaus when all the targeted stars (209 in this case) are within range, and finally starts rising again due to contributions from untargeted stars in the background. It is instructive to note that the merit functions previously given by Dreher and Cullers [Dre97], and Gulkis [Gul85] apply only to this last background component, essentially ignoring all the actual targeted stars that are the point of this type of search!

F.2.2   Current SETI Programs

Figure F.2 on page 344 shows the results for four projects that were expected to be in progress by the end of 1996. The figure-of-merit, f_{merit}, is plotted on a log scale as a function of the EIRP of the transmitter, P_ξ, also on a log scale. The anticipated results for four current (as of 1999) SETI programs are shown:

- SERENDIP IV;

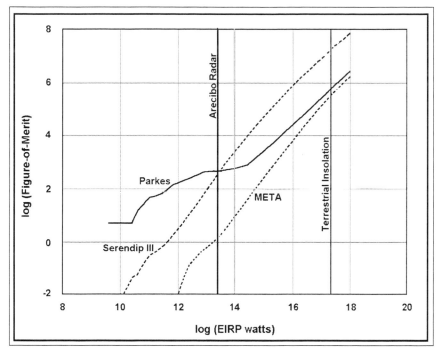

Figure F.1: Figure-of-Merit vs. Transmitter EIRP – Completed SETI Programs.

- BETA;
- the SETI League's Project Argus;
- Project Phoenix, the continuing baseline program.

The parameters of these future searches are not as well determined as those of the completed work. Also shown is a curve for the SETI League Argus project, an ambitious amateur effort. We have used the nominal parameters of Argus; it is shown dotted since, although this effort was inaugurated in 1996, it will probably be years before it reaches anything like its nominal design goals.

On both Figures F.1 and F.2 we have for reference also marked the (post-upgrade) EIRP of the Arecibo S-band planetary radar, which is the most powerful transmitter on Earth, and the total power of sunlight falling on Earth.

F.2.3 Infrared and Optical SETI Programs

Figure F.3 on page 345, shows the results for:

- the ongoing infrared SETI project of Betz [Bet86];

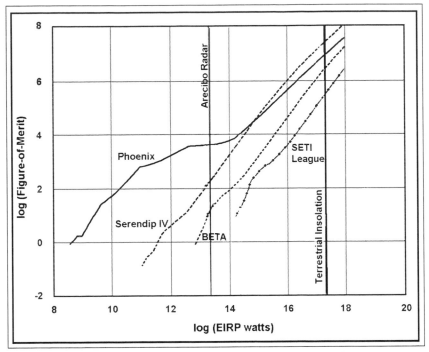

Figure F.2: Figure-of-Merit vs. Transmitter EIRP – Current SETI Programs.

- the planned program of the Columbus Optical SETI Observatory (COSETI) [Kin96], an amateur effort.

Figure-of-merit, f_{merit}, is plotted on a log scale as a function of the EIRP of the transmitter, P_ξ, also on a log scale.

The 10 μm search of Betz is targeted on the nearest stars and, unlike Phoenix, the contribution of background component is negligible because of the small solid angle seen by infrared/optical systems. The detection scheme proposed for COSETI requires the transmitter's EIRP to substantially outshine the parent star, if only for a nanosecond or so.

We see that to be detectable, infrared/optical transmitters must have much higher EIRPs than microwave transmitters. But since transmitter gain is proportional to λ^{-2} much higher transmitter gains can be realized at shorter wavelengths.

Consider two large antennas of about the same cost. In the microwave, the Arecibo antenna has an effective diameter of 225 m and a gain of 73 dB at $\lambda = 13$ cm. In the infrared, the Keck telescope has an effective diameter of 10 m and a gain of 127 dB at 10 μm. For reception, on the other hand, the

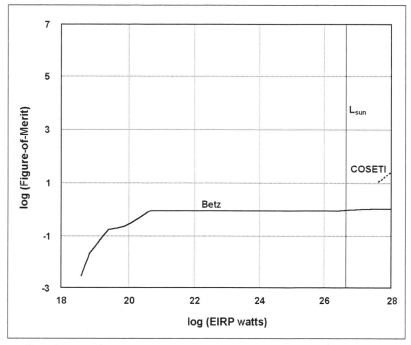

Figure F.3: Figure-of-Merit vs. Transmitter EIRP – Infrared/Optical SETI Programs.

microwave region offers larger effective areas, lower noise, and a more mature technology. As a concrete example, the detection threshold of Project Phoenix at Arecibo will be 8×10^{-27} W/m^2, while Betz's threshold at 10 μm is 7×10^{-17} W/m^2. Thus, on the transmitting side the infrared has a potential advantage of the order of 10^5, while on our receiving side it has a disadvantage of the order of 10^{10}. At our present technological level, then, infrared/optical SETI is not yet competitive with microwave SETI, although it holds some promise as our technology develops.

F.3 Conclusions

Our new figure-of-merit has allowed us to investigate the relative merit of various SETI programs. As expected, the high sensitivities that can be achieved with a targeted search such as Project Phoenix result in the highest merit for transmitters with EIRP $\leq 10^{14}$ W, while the best of the current generation of sky surveys, SERENDIP, has greater merit for more powerful transmitters. Infrared and optical SETI currently offer much lower merit. Our figure-of-merit, however, does not incorporate metrics for two extremely important features of every survey – the types of signals to which the survey is sensitive, and the techniques used to verify that signals are of extraterrestrial origin.

In the future, we suggest that new techniques will be needed to provide vastly improved time coverage, corresponding to high values of η_{looks}, so that the programs can have a reasonable chance of detecting ultra-high-powered, highly-directive beacons that can reasonably be expected to have correspondingly small duty cycles. Targeted searches, on the other hand, need to improve their sensitivity by a factor of about 10^2 so that they become sensitive to low-directivity 'leakage' emissions at terrestrial levels. In both cases, we expect that improvements in DSP technology will play a big role.

Finally, all microwave SETI will be severely challenged in the near future by ever-worsening interference. To some extent, improved signal processing and antenna design may mitigate this interference. This problem may be a factor pushing SETI into the infrared/optical regions in the next decades [Kin96b].

F.4   Further References

Betz, A. L., 1986, "A Direct Search for Extraterrestrial Laser Signals", *Acta Astronautica*, 13.10, 623-629.

Drake, F. D., 1981, "Quantitative Estimates of the Probability of Success of SETI Programs", *NAIC Report*, 155.

Gulkis, S., 1985, "Optimum Search Strategy for Randomly Distributed CW Transmitters", *The Search for Extraterrestrial Life: Recent Developments*, IAU Symposium 112, ed. M. D. Papagiannis, 411-417.

Horowitz, P., & Sagan, C., 1993, "Five Years of Project META: An All-Sky Narrow-Band Radio Search for Extraterrestrial Signals", *ApJ*, 415, 218-235.

Kingsley, S. A., 1996, "Prototype Optical SETI Observatory", *The Search for Extraterrestrial Intelligence, (SETI) in the Optical Spectrum II*, Proceedings of SPIE, eds., S. A., Kingsley & G. A., Lemarchand, 2704, 102.
See http://www.coseti.org

Werthimer, D., Donnelly, C., & Cobb, J., 1996, "SERENDIP SETI Project: New Instrumentation for SETI and Results of the Recent Arecibo Sky Survey", *The Search for Extraterrestrial Intelligence (SETI) in the Optical Spectrum II*, Proceedings of SPIE, eds., S. A., Kingsley & G. A., Lemarchand, 2704, 9.

Appendix G

Thinking Big

This appendix, by D. Kent Cullers, considers the long-term possibilities of a search unbounded by costs, and relying solely on foreseeable technological developments.

Those of us whose work is intimately connected with computer technology have experienced the heady phenomenon of exponential growth. The phenomenon has persisted for three decades. If sustained, it can cure any problem of computational demand existing today, in amazingly short times. Many desirable conditions, such as a stable population, can be swept away by the same kind of growth. Whatever we think about the likelihood of exponential growth persisting, working in an industry that has this characteristic of rapid change encourages us to consider possible futures that are very different from our present. Thus, serious people have asked the question, "Can we continuously survey the whole galaxy over all interesting wavelengths for all interesting signals in the next century, and should we plan for this eventuality?"

What is required to fully survey this galaxy for signals of a technology like ours? The most important difficulty in giving an answer is that the electromagnetic signature of our technology is rapidly evolving. It seems probable, given the direction of our own evolution in communications, that leakage signatures will increasingly resemble white noise. We can assume, not implausibly, that advanced technologies will radiate into the atmosphere with powers of terawatts, and that such signals will be radiated over frequency bands of gigahertz. This yields power densities in the microwave region of kilowatts per hertz. Coincidentally, this is approximately the density radiated in this frequency range by a solar-type star. Thus, the presence of noise-like technological radiation, with plausible power outputs based on our own civilization, yields "stars" with an apparent microwave flux about twice as high as that expected from a natural source.

This is a very difficult case for detection. To see this, first consider the best case one could expect. In such a case, detection is performed by a matched filter, with a shape exactly the same as that of the transmitted signal. Detection sensitivity then depends only on the energy of the signal divided by the thermal noise energy (kT_{sys}), where T_{sys} is the system noise temperature. This suggests that unlimited sensitivity could be attained in free space reception; if the beacon signal of a transmitter were sufficiently simple to match a pure sinusoid, for example. Given sufficient integration time, enough energy could be accumulated to detect a signal, however weak it might be. Theoretically, a pure sinusoid is perfectly defined in frequency and can be matched by a channel of indefinitely narrow bandwidth, if the integration time is long enough.

However, such a sinusoid is spread in frequency as it travels through the interstellar medium which causes scattering and multipath. Its minimum bandwidth is approximately 0.01 Hz. Given an integration time, consistent with this finite bandwidth, an Arecibo sized antenna, and a transmitter in free space, simple calculation shows that the necessary transmitted power in megawatts is given by the square of the transmitter distance in light-years. Depending on precise estimates of antenna temperatures and tolerable false alarm rates, sensitivities can vary by a factor of two, but this is unimportant for the conclusions that follow. As a rule of thumb, using a matched filter at Arecibo, a one megawatt transmitter can be received at one light-year; while a transmitter at 1,000 ly distance requires a terawatt.

When detecting a CW signal coherently,

$$P = L^2 \tag{G.1}$$

where P is the power in megawatts, and L is the distance in light years. Obviously, if one had enough time, incoherent detection could be used to increase the range, but in this case, range grows only as $(\text{time})^{\frac{1}{4}}$.

A megawatt is, incidentally, about the effective power of our horizon-beaming TV transmitters. Simple application of the inverse square law leads to a remarkable conclusion. To receive this typical Earth-like transmission from across our 100,000 ly diameter galaxy, we would need an antenna 100,000 times the diameter or ten billion times the area of Arecibo. This antenna would be a little more than twice the diameter of the Earth!

Computational capacities associated with radio astronomy are growing exponentially, with a doubling time of about 1.5 years in speed and cost-effectiveness. If the antenna area were to share this same exponential curve, antenna diameters would be growing by a factor of 100,000 every 51 years. However, it is virtually certain that antennas will not have collecting areas larger than the Earth, even in the next century.

Thus, barring a new technology just on the edge of physical feasibility, such as space-going, nanotech, antenna-building robots, searching the galaxy for Earth-like leakage is not possible in the near future. Let us, instead, be guided by limits imposed on us by resource constraints, and let us consider searches with the SKA.

What signals can such a system be expected to find? Arecibo could detect a terawatt beacon from 1,000 ly, within which distance lie about one millions solar-type stars. The SKA will be about 25 times better in sensitivity, and able to see about 100 times more stars. Thus, a significant search for beacon signals of large, but not prohibitive, powers can be made over perhaps 1% of the galaxy in the next two decades.

The situation is much less promising for wideband, noise-like signals. To see this, consider averaging a gigahertz input band filled with Gaussian noise, for a thousand seconds (ten times longer than for our matched filter case). This example may seem rather arbitrary. However, a thousand seconds is a reasonable time to track a star, and coherence is no longer an issue, since we are considering the detection of arbitrary signals spread over significant bandwidths. Matched filtering cannot be used. Thus, a consequent advantage is that it is possible to integrate power longer than the time allowed for coherent detection. As is well known, with independent physical measurements, averaging the 10^{12} uncorrelated power samples in the one gigahertz band yields a quantity with fluctuations one millionth as large as those of a single sample. Thus, single power samples, such as those used in a matched filter, have fluctuations equal to the power itself.

Now consider what constitutes a tolerable false alarm rate, when there is no signal. Continuing to use our one gigahertz band as an example: there are within this region a billion independent samples per second; so in a thousand seconds, this yields a trillion samples. In other words, any matched filtering operation that loses no information, and that allows reconstitution of the input stream from the filter outputs, has this same data volume.

It seems reasonable that a human operator could tolerate one false alarm every thousand seconds (twenty minutes), under conditions of noise alone. In any case, sensitivity depends only weakly on false alarm rates, when the false alarm rate is low. The significant issue is not the precise false alarm rate, but the very different character of the curves, depending on whether matched filtering or power averaging is used. The curve for a matched filter is that of a single power sample distribution, an exponential with standard deviation equal to the mean and to the e-folding length. For large numbers of averaged samples, on the other hand, the curve is Gaussian with, in our example, a standard deviation equal to one millionth the mean of the samples.

Thus, to gain the same false alarm rates in both the matched filter and averaged cases, one needs a 27σ threshold for the matched filter, and a 7σ event for the averaged power. Given these thresholds, signals approximately equal to their respective levels will be equally detectable. Comparing a 7σ event in a trillion averaged gigahertz samples with a 27σ event in a single sample of 0.01 Hz gives a power ratio of $(7 \times 10^9/10^6)/(27 \times 0.01) = 25,925$ in sensitivity favoring, of course, the matched filter.

This great discrepancy in sensitivity implies that, given the same power, a beacon for which a matched filter can be constructed is detectable at powers 25,000 times less than a similar amount of noise-like, broadband transmission. Thus, a 40 MW beacon is as visible as a terawatt of broadband noise. Viewed another way, given equal powers, the broadband signal would have to be 160 times closer than the beacon to be equally detectable.

Thus, our terawatt noise signal, distributed over a gigahertz, could be seen by Arecibo at a distance of about 6 ly. Within this distance there is one solar-type star, Alpha Centauri. The SKA could detect this same signal at about 30 ly, a range which includes about 25 solar-type stars. This is about the same capability as for beacons, or even leakage transmitters, having about 40 MW of EIRP. Since such powers are almost typical of terrestrial transmitters, it appears that the case for detecting leakage transmissions is only slightly worse than that for broadband super civilizations.

The problem with leakage is that signals drift due to the Doppler effect. However, integration for timescales of a thousand, rather than a hundred seconds, makes up for the loss of coherence in the narrow bandwidth. Thus, it seems that we should continue to search for simple leakage signals, assuming the computational demands are manageable. We should also add a search for broadband technological noise, because looking for simple megawatt signals or noise-like terawatt signals yields approximately equivalent sensitivities. If terawatt EIRP signals are plausible, as discussed in the next section, over the next twenty years, we may find at least one powerful technology.

G.1   Beacons and Beaming

As discussed elsewhere, a beacon, if detected with a matched filter receiver, may use an arbitrary transmission scheme. To illustrate this, consider both a CW beacon and a pulsed beacon emitting low duty cycle pulses, and assume that both beacons are received for an equal amount of time in free space. We have already noted that free space is probably too simple a model for the interstellar medium. The real characteristics of the interstellar medium place a limit on what we can do in a practical sense. However, free space transmission certainly gives a zero order approximation for discussion. If the CW and pulsed beacon deliver equal energies in equal times, then they have the same average power.

Use two Fourier Transforms as the detectors of our two beacons: one for the CW signal with a transform length equal to the pulse repetition period; and the other for the pulsed signal with a length equal to the pulse duration. Both FFTs are assumed to analyze the same bandwidth. Since the FFT conserves information, the output rates of both transforms are equal to their input rates, and hence to one another.

In both cases, one FFT output will contain all the energy of the incoming signal. For the CW signal, this will be a single channel of the long FFT used to detect continuous beacons. For the pulsed beacon, one channel, its width determined by the pulse duration in one particular spectrum, will contain the signal energy. Now the energy in an FFT channel is just $kT_{sys}bt$, where b is the bandwidth and t is the integration time. Since, for FFTs, bt is unity, the energy per channel is kT_{sys} in both detectors. Furthermore, since all channel outputs are single sample powers, the statistical distributions are identical. Thus, setting equal thresholds in both detectors gives equal false alarm rates per unit time.

To be detectable, signals must be approximately equal to the detector threshold. Thus both detectors, in terms of E/kT, are identically sensitive to their respective beacon types. The great advantage of the CW detector is that no guessing is required as to the pulse length. Of course, the beacons should be corrected to an inertial frame, so that no guessing is necessary about planetary-induced drift.

As a corollary, note that beams deliver signals that, like CW beacons, depend in their detectability, only on the average power delivered to the part of the sky they beam. They are not better beacons, only different. Their naturally broader frequency spectrum makes removal of drift easier and, they may be modulated. However, modulation is also possible with CW signals, either in polarization or phase. For either signal type, modulation compromises sensitivity slightly.

It is possible to imagine using structures like the SKA to beam the sky. The SKA has a gain that is approximately the square of circumference divided by transmitted wavelength, and equal to approximately a billion. Thus, while beaming only a thousandth of the sky, such an antenna could target a million stars. This would require carefully phased elements to achieve the pattern (perhaps digitally controlled). It would also require some care, to guarantee that stars with large proper motions, at large distances, are in the right places when the signal arrives. However, it would mean that gigawatt CW beacons, equal to about one nuclear power station output, could illuminate all solar-type stars within 1,000 ly with signals having terawatt EIRPs. This, in turn, is consistent with the ability of our proposed SKA to receive terawatt signals from perhaps 1% of our galaxy.

A civilization with a slightly larger telescope, or wishing to radiate more power, could illuminate a hundred million solar-type stars with terawatt EIRPs. It may be possible, within a couple of centuries, to use dedicated arrays, part time for transmitting and part time for receiving This would increase both our visibility and duty cycle for receiving. Conceivably, at least in the radio regime, we will soon be able to develop programs that would allow exploration of 1% of our galaxy, if reciprocated by other civilizations. Given the million-star goals for the next twenty years, and the plans in Section 8.5.1 on page 266 for achieving them, observing the galaxy within a century seems more feasible than at first glance.

G.2 The Optical Regime

In the optical, the sources of background noise are not the same as for microwaves. For microwaves, the noise sources are primarily the cosmic background, but, as frequency increases toward the optical regions, there is much less cosmic background and much more stellar interference. In addition, the techniques used to phase together elements in radio begin to fail. If we want omnidirectional coverage of the sky, the number of elements increases as the square of the frequency. This causes a catastrophic increase in the computational beamforming requirement, because of increases in the number of elements and the sampling bandwidth.

To create phased arrays, we may still need heterodyne techniques, which introduce their own quantum noise, as discussed elsewhere. Hence, we can envision many optical telescopes targeting individual stars with very high gain, rather than a vast phased array of antennas. In addition, because photon counting is used instead of matched filtering, gates need not match the pulse or observation length. It is only necessary to accumulate power and compare the result against the expected fluctuations in the background count.

G.2.1 Calculation of Photon Count

In this photon counting regime, the dominant background comes from the star near the signaling civilization. solar-type stars emit about 10^{12} photons/s/m^2 at one light year. Thus, if we are to optically search our closest million stars, most of these will be at about 1,000 ly distance. At such a distance, the typical star emits 10^6 photons/s/m^2 in the optical, much more than the 3 K blackbody, background of the Universe.

A transmitter is certainly capable of emitting 100 J in a nanosecond pulse. In the optical, a 10 m transmitting telescope has a gain of about 10^{16}, giving the pulses an effective power (while on) of $10^{16} \times 10^2 \times 10^9 = 10^{27}$ J/s. This is several times the total output of a typical solar-type star. When one further considers that the laser signal sent through the transmitting telescope is monochromatic, a set of filters, of modest resolving power, can cause the signal to outshine the star in the laser band by many factors of 10. Thus, in principle, it is only necessary to crudely gate stars in nanosecond blocks of time over rather wide bands. In such blocks, the expected number of photons from a background star is much less than one, while a highly directional pulse in a particular nanosecond could deliver tens of photons. Thus, optical searches are easy, assuming that directionality of the transmitter is high and that the receiver can easily eliminate background light.

Appendix H

Time and Frequency in SETI

This appendix, by Leonard S. Cutler, discusses and compares various methods of recording time and temporal coordination across both space and time. This is also done in terms of the requirements of astronomy, in general, and VLBI and SETI, in particular.

H.1   Some General Aspects

Time is important as the fourth dimension. Along with three spatial dimensions, it is used to specify an event in the Einstein sense. It is particularly important to radio astronomy, VLBI, and SETI. Like the spatial dimensions, time must have an origin and a unit.

Timescales are generated by clocks. A good clock consists of two major parts: a frequency or rate generator, that ideally produces an absolutely constant frequency; and an integrator. In a mechanical clock or watch, the rate generator is the pendulum or balance wheel; and the integrator is the gear train combined with the hands. Moderate quality electronic clocks or watches use a quartz crystal controlled oscillator as the rate generator. In today's best clocks, the rate generator is based on the frequency associated with transitions between a pair of energy levels in isolated atoms, molecules, or ions; and the integrator is, essentially, a frequency divider and counter combination. The constant of integration for the integrator is the time at which the clock is initialized, and the time to which the hands or counter are set.

Presently, the basic unit of time, the SI second, is defined by the frequency of the transition between the hyperfine levels in the ground state of the cesium 133 atom. The SI second is the time interval during which exactly 9,192,631,770 cycles of the hyperfine transition frequency occur. This was set to agree, as well as possible, with the second based on the older astronomical definition, the ephemeris second. The present accuracy of the realization of the SI second, by averaging several of the best laboratory cesium beam standards, is a few parts in 10^{14}. New cesium fountain standards are being evaluated and the accuracy of these will be a few parts in 10^{15}.

There are many timescales in use today. The subject is quite complex and will only be touched on here. For practical reasons most of them are related to the rotation of the Earth: the solar day, or the sidereal day. The scale based on the SI second is the International Atomic Time scale (TAI). The most commonly used scale is Coordinated Universal Time (UTC), which is maintained close to the mean solar day. The rate of the UTC scale is exactly equal to the rate of TAI. Corrections to UTC time required by the slowing and irregularities of the Earth's rotation rate are applied as 'leap seconds' added or subtracted when needed, usually at the end of June or December. UTC is thus always within a fraction of a second of UT1, mean solar time. Sidereal time, based on the rotation of the Earth with respect to the fixed stars, is the time used by astronomers and, of course, SETI. The difference between sidereal time and UTC is provided by the Earth Orientation Department of the US Naval Observatory, and the International Earth Rotation Service.

The timescales mentioned thus far are presently based on atomic clocks running on the rotating geoid. The clocks are synchronized to an imaginary set of clocks running in a coordinate system with the same effective gravitational potential as on the geoid, but not rotating with respect to the fixed stars. This is necessary because clocks on the rotating geoid cannot be synchronized by electromagnetic signals directly because of a relativistic effect.

Another general relativistic effect should be mentioned. The rate of ideal clocks depends on the effective gravitational potential. The effective gravitational potential is due to the combined effects of mass/energy and the rotation with respect to the fixed stars. If the Earth were a fluid, or completely covered with a fluid, the surface would be an equipotential of the effective gravitational potential, because if it were not an equipotential fluid would flow until it was. Ideal clocks anywhere on an effective gravitational equipotential surface run at the same rate. This is approximately the case for the oceans of the Earth because there are dynamical effects caused by the tides that produce small departures of the ocean surface from a true equipotential. Corrections to clock rate are routinely applied for clocks running at different elevations on the Earth and thus at different potentials. The effect is about 1.1 parts in 10^{16} per meter of height at the Earth's surface.

Two more timescales are of interest mainly to astronomers. Geocentric coordinate time is a coordinate time in the general relativistic sense with its spatial origin, nonrotating, at the center of mass of the Earth. Its rate is that of ideal clocks in the absence of the effective gravitational potential of the Earth, but in the presence of the gravitational potential of the solar system.

Barycentric coordinate time is a coordinate time in the general relativistic sense with its spatial origin, nonrotating, at the center of mass of the solar system. Its rate is that of ideal clocks in the absence of the effective gravitational potential of the solar system.

H.2 Frequency and Frequency Stability

High quality oscillators generally are designed to produce, ideally, a pure sinusoidal waveform, usually electrical. When frequency-locked to an atomic/ionic transition or phase-locked to a maser based on an atomic/ionic transition, these can be used as rate generators in good clocks. However, noise and other effects degrade the signal. The real output as a function of time t can be represented as:

$$V_{out}(t) = V_0(1 + m(t))\cos(2\pi f_0 t + \phi(t)) \qquad \text{(H.1)}$$

where V_0 is the nominal amplitude, f_0 is the constant average frequency, $m(t)$ is the amplitude modulation, assumed $\ll 1$, and $\phi(t)$ is the phase modulation assumed $\ll 1$ radian. $m(t)$ and $\phi(t)$ are, in the ideal case, random variables with zero mean. A perfect oscillator or rate generator would have $m(t) = 0$ and $\phi(t) = constant$. However, even ideal atomic clocks are perturbed by shot noise, thermal noise, and other types of noise.

The total instantaneous angular frequency is the time derivative of the argument of the cosine and is thus:

$$2\pi f_0 + d/dt(\phi(t)) \qquad \text{(H.2)}$$

The instantaneous angular frequency difference from $2\pi f_0$ is $\omega(t)$:

$$\omega(t) = d/dt(\phi(t)) \qquad \text{(H.3)}$$

There is always some conversion of phase modulation to amplitude modulation and amplitude modulation to phase modulation even in linear dispersive circuits, but with careful design and optimization this can be minimized.

Frequency drift is another common behavior of oscillators. Linear drift can be modeled as an additional, systematic phase angle $at^2/2$ where a is the constant drift rate in radians/s^2. For this case, $\omega(t) = at$, a linear drift of frequency with time.

The power spectral density of phase is the Fourier Transform of the autocorrelation function of $\phi(t)$ if the integral converges. Likewise, the power spectral density of angular frequency is the Fourier Transform of the autocorrelation function of $\omega(t)$ if the integral converges. Unfortunately, most oscillators have divergent low frequency processes, such as flicker of frequency noise, which must be handled by suitable filtering or analysis techniques. Any pure sinusoidal components in $\phi(t)$ will show up as bright lines (delta functions) in the power spectral densities of angular frequency or phase.

Often the spectral densities are expressed in Fourier frequency, rather than Fourier angular frequency. Also, one-sided spectral densities are frequently used.

The spectral densities of frequency and phase, when the total RMS phase is small, are good measures of frequency stability. Modern spectrum analyzers use the digital techniques of sampling and discrete Fourier Transform to achieve high quality results in the low to moderate frequency range.

Direct measurements of the spectral density of the signal are also useful, but contain contributions from amplitude modulation. High frequency signals can be analyzed directly with suitable spectrum analyzers, but these will generally not have very high resolution. The signal can be heterodyned down to a frequency suitable for high-resolution, digital analysis, provided the oscillator used as the reference for the heterodyning has noise approximately equal to or lower than that of the signal being analyzed.

For low modulation frequencies, corresponding to long observation times, digital, time-domain measurements are very useful. There are many time-domain techniques, but one that is often used and gives convergent, stable results for flicker of frequency modulation is the Allan variance or its square root, the Allan deviation. This measure is based on the mean of the square of differences between successive measurements of normalized average frequency. The normalized frequency $y(t)$ is:

$$y(t) = \omega(t)/2\pi f_0 \qquad (H.4)$$

The normalized frequency averaged over various time intervals T is:

$$<y(t)>(T) = (\phi(t+T/2) - \phi(t-T/2))/2\pi f_0 T \qquad (H.5)$$

This is true since $\phi(t)$ is the time integral of $\omega(t)$.

A log-log plot of the Allan variance or the Allan deviation versus the averaging time T gives a good picture of the frequency stability of an oscillator. The unavoidable shot noise and thermal noise can have two effects. Inside the bandwidth of the frequency-lock loop or the bandwidth of the oscillator resonator they perturb the frequency directly and produce a constant or white spectral density of frequency. This leads to a dependence of $T^{-1/2}$ in the Allan deviation in the range where this noise is dominant. White noise that is just added to the signal produces a white spectral density of phase. This leads to approximately a $1/T$ dependence where it is dominant. Flicker of frequency noise, which has a $1/f$ spectral density of frequency, produces a constant value where it is dominant, and linear frequency drift produces where it dominates a linear dependence on T.

H.3   Frequency Standards and Clocks

H.3.1   Hydrogen Maser

There are two versions of the hydrogen maser, active and passive. The active maser uses atomic hydrogen in its upper hyperfine level to produce maser oscillation. State selection takes place in a beam of atoms passed through a quadrupole or hexapole magnet producing a beam of atoms in the upper hyperfine level. This beam is injected into a Teflon coated bulb inside a cavity tuned to the hyperfine transition frequency, about 1,420 MHz. Oscillation takes place via stimulated emission, provided the total cavity losses are low enough, and the inflow rate of hydrogen is high enough. The Teflon coated bulb stores the hydrogen with a low, wall-collision, relaxation rate, but this does introduce a frequency shift that depends on the nature of the coating and on the bulb size. The power level available directly from the maser oscillation is very low. Consequently, a high quality oscillator/frequency-synthesizer combination is phase-locked to the maser signal with the appropriate time constant to produce the actual output signal at a useful level and frequency.

The major characteristics of the active maser are:

- very good short-term and medium-term frequency stability;
- flicker of frequency level at about 1 part in 10^{15};
- drift rates of a few parts in 10^{16} per day;
- accuracy a few parts in 10^{12} (limited by bulb wall frequency shift).

The low drift rate is only achieved in masers, with automatic cavity tuning since frequency pulling by changes in cavity resonance frequency is considerable. The cavities are temperature sensitive, and do drift in frequency with time. Low drift rates also require very high quality, stable well-aged Teflon wall coatings.

The active maser is an excellent frequency and timing source for VLBI, and for SETI when separated sites are used coherently mainly, because of its very good short-term and medium-term stability. This allows the longest periods of time for the uncertainty in relative phase between sites to get to one radian, approximately the limit for coherent processing.

The passive maser is similar to the active maser but the operating conditions, one of these being much lower cavity Q, are such that oscillation does not take place. Instead, the hyperfine resonance is detected in any of several ways, including the rapid changes in cavity susceptance at the atomic resonance, detected by using a probe signal from an oscillator/frequency-synthesizer combination. The oscillator/frequency-synthesizer is frequency-locked to the resonance and provides the useful output signal. The short-term stability is not nearly as good as that in the active maser due to limitations in the signal-to-noise ratio in the resonance detection process. Accuracy, drift rate, and flicker of frequency level are comparable.

H.3.2   Cesium Atomic Beam Standard

The cesium atomic beam standard uses a beam of Cs^{133} atoms selected in one of the hyperfine levels that is passed through a microwave cavity excited by a probe frequency from an oscillator/frequency-synthesizer combination. If the microwave signal is at the hyperfine resonance frequency, about 9,192.6 MHz, and at the right amplitude, the atoms make a transition to the other hyperfine level and are detected. The oscillator/frequency-synthesizer combination is frequency-locked via the signal produced by the detected atoms to the hyperfine transition frequency, and provides the useful output signal. Hyperfine state preparation and detection is done either by the use of inhomogeneous magnetic fields and a hot wire detector or by optical pumping with laser radiation and optical detection.

Because the atoms in the beam are essentially free, the frequency of cesium beam standards is quite predictable and very reproducible. It was mainly for this reason that, as mentioned above, cesium was chosen as the basis for the SI second. The main source of error is probably uncertainty in the distribution of phase in the microwave cavity.

The major characteristics of the best laboratory cesium beam standards are:

- resonance linewidths of a few tens of Hz;
- very good accuracy of about one part in 10^{14}, probably limited by knowledge of cavity phase distribution;
- reproducibility of < 1 part in 10^{14};
- moderately, good, short-term stability – limited by shot noise;
- very low flicker frequency noise, probably < 1 part in 10^{16};
- very low drift rate, presently unmeasured.

The major characteristics of the highest quality commercial cesium beam standards are:

- resonance linewidths of about 400 Hz;
- accuracy of 7 to 8 parts in 10^{13};
- reproducibility of about 1 part in 10^{13};
- fair short-term stability, of about 8 parts in 10^{12} at one second – limited by shot noise;
- flicker noise of a few parts in 10^{15};
- drift rates of a few parts in 10^{16} per day.

High quality commercial cesium standards are fairly good candidates for SETI work. Their high accuracy, low drift, and low flicker noise make them very good clocks. Their short-term stability levels permit coherent processing times between independent spatially separated sites of about 100 seconds, which may not be long enough for SETI.

H.3.3 Cesium Atomic Fountain Standard

The recent achievement of cooling atoms to extremely low temperature by the use of lasers has been used to produce a new, very high performance, cesium standard. Atoms are laser cooled and a ball of the very cold atoms is held in a magneto-optical trap. State selection is performed by a laser pulse. The ball is then tossed upwards with a low velocity by a change in the laser conditions, and goes through a microwave cavity excited by the usual probe frequency source. The ball continues upward, slowing down due to gravity, and then, after a short distance, falls back through the cavity. Optical detection of the microwave induced transitions then takes place and is used to frequency-lock the microwave source to the hyperfine transition. The linewidth of the resonance is quite small due to the very low velocity of the atomic ball and consequent long interaction time with the microwaves.

A few of these standards are now operating in laboratories. There are no commercial units available or planned at this time.

The major characteristics of the cesium fountain are:

- resonance linewidths about 1 Hz;
- accuracy of about one part in 10^{15} – probably limited by spin exchange collisions;
- reproducibility of a few parts in 10^{16};
- good, short-term stability, of about 1 part in 10^{13} at one second – limited by shot noise;
- flicker noise of probably a few parts in 10^{16};
- drift rates probably less than 1 part in 10^{16} per day;

- requires a very good oscillator (like a hydrogen maser) for the probe signal.

There is a fair amount of uncertainty in the performance because the fountain is still under development and evaluation. It would certainly be a very good source for SETI work.

H.3.4   Trapped Mercury Ion Standard

A cloud of mercury 199 ions, trapped in a radio frequency quadrupole trap, is optically pumped with radiation from a pulsed mercury 202 discharge lamp. This does the state selection of the trapped ions. The lifetime of the trapped ions can be very long. The hyperfine resonance at 40.5 GHz is probed while the pumping light is off by a pulsed microwave signal from an oscillator/frequency-synthesizer combination. When the microwave frequency matches that of the hyperfine transition there is an increase in the fluorescence when the pumping light is turned back on. This is the detection process and is used to frequency-lock the oscillator/frequency-synthesizer combination.

The electric field from the radio frequency drive on the trap causes the ions to oscillate in position and this leads to a frequency shift due to the relativistic velocity effect, the so-called second order Doppler shift. The size of the effect depends on the size and shape of the ion cloud and can be as large as a few parts in 10^{11} for a spherical cloud. This requires that the trap operating parameters and cloud size and shape be carefully controlled. A two-dimensional trap with a linear string of ions or a single ion has almost zero shift.

Cleanliness of the vacuum system is very important. Very small traces of background gas can cause significant frequency shifts. For this reason, some groups are working with cryogenically pumped and cooled systems.

There are no commercially available units at this time.

The major characteristics of the trapped mercury ion standard are:

- resonance linewidths of less than 1 Hz;
- precision of a few parts in 10^{14} for an elongated cloud, and better than 1 part in 10^{15} for a linear chain or single ion;
- reproducibility better than accuracy;
- good, short-term stability of about 1 part in 10^{13} – limited by shot noise;
- flicker noise of a few parts in 10^{15} for elongated cloud – better for linear chain or single ion;
- drift rates of perhaps better than one part in 10^{16} per day, for a well designed system with very clean vacuum;

- requires a very good oscillator for the probe signal.

The trapped mercury ion standard would be quite good for SETI work.

Figure H.1 shows the Allan deviation for several different frequency standards.

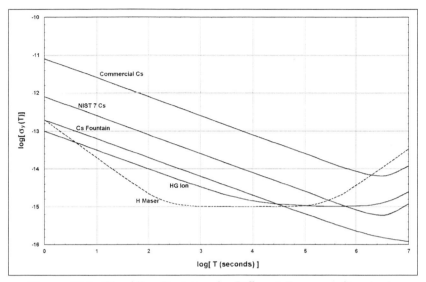

Figure H.1: The Allan Deviation for Different Frequency Standards.

H.4   Global Positioning System (GPS)

The GPS system, with its constellation of satellites, is not only an excellent navigation system but also is very useful for time and frequency. GPS and UTC time can be obtained directly to a precision much better than one microsecond. Separated sites using relatively inexpensive receivers, observing all satellites in common view at the same time, can synchronize their timescales within about 20 nanoseconds. By comparing times over several days using common view, the frequencies of the timescales can be compared and adjusted to be equal (syntonized) to better than 1 part in 10^{13}. These techniques are now in routine use by the standards laboratories all over the world. The Bureau International des Poids et Mesures (BIPM) uses the data gathered this way to realize the TAI and UTC timescales. The United States Naval Observatory (USNO) maintains the UTC timescale for the United States and is typically within about 20 nanoseconds of the BIPM timescale.

An oscillator phase-locked to the output of an inexpensive GPS receiver makes a very useful frequency/time standard. The oscillator is usually steered by comparing the timing of the once per second pulses from the GPS receiver to that of the once per second pulses derived by dividing the oscillator frequency. The oscillator frequency is controlled by a servo loop that keeps the time difference between the pulse trains to zero. The quality of the oscillator determines how well the system keeps time in the event of loss of the GPS signal or receiver failure. Oscillators used include inexpensive quartz units, high quality quartz, rubidium gas cell standards, cesium standards, and hydrogen masers. This GPS type of standard has become very prevalent in the telecommunications industry.

The clocks in the GPS satellites are purposefully phase modulated with pseudo-random low frequency noise to the extent of several hundred nanoseconds. This is a government security measure to prevent rapid, accurate position determination. It requires that the phase lock loop of the user's GPS controlled oscillator have a moderately long time constant if good short-term stability is needed directly from the receiver/oscillator combination.

GPS controlled oscillators could be quite useful for SETI, particularly if it is desired to synchronize and syntonize the timescales at separated sites.

H.5 Multiple Antenna Microwave SETI Site

It is assumed that each antenna in the array has its own down-converting receiver and that an IF signal is sent to the central, signal combining equipment. Coherent signal processing requires that the distribution of local oscillator phases delivered to the antennas be fairly narrow. Figure H.2 shows the relative power of the actual array relative to that of a perfectly fed array in dB, versus the RMS phase error for a Gaussian phase distribution. The power is down 3 dB at an RMS phase error of about 0.84 radians. If known, the phase errors could be corrected at the IF or in the signal processing at the same time as the beam-forming.

At a signal frequency of 1.5 GHz, 1 radian corresponds to about 100 ps or 3 cm distance at the speed of light. If the phase corrections are done on the local oscillator signal, line stretchers or some other form of variable phase shifters should be used. The phases must remain stable for good coherence and beam pointing.

Noise sidebands on the local oscillator that are at the same frequency as the desired signal can degrade the noise temperature of the receiver. Filtering and the use of balanced mixers reduce this effect. A phase-locked oscillator can be used as a filter.

If it is desired to know the time of arrival of signals for comparison with other sites, the time precision needed is of the order of the reciprocal, angular bandwidth of the signal, after detection and processing. Fourier Transform bin widths of the order of 1 Hz, for example, require only a few tenths of a second precision.

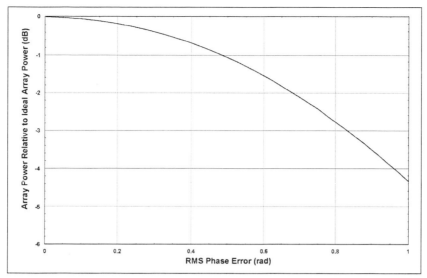

Figure H.2: Relative Power from Array with Gaussian Distribution of Phase Errors.

H.6 Spatially Separated SETI Sites

Incoherent processing, or signal comparison after detection for separated sites, was mentioned above and requires time precision between the sites of about the reciprocal angular bandwidth. The precision for moderate bandwidths is easy to obtain using GPS time comparison as described above; and optimum use of the common view technique, with receivers properly calibrated for delay, can give synchronization results as good as 20 nanoseconds.

Coherent processing between separated sites requires considerations very similar to those for the individual antennas in a multiple antenna site and those for VLBI. Each site must have its own local oscillator since it is very difficult, if not impossible, to transmit the local oscillator signal over large distances and retain good enough phase stability. An optical signal, modulated by the local oscillator, transmitted over an optical fiber, and maintained in a carefully controlled environment, could possibly work. However, this would be very expensive for any reasonable distance between sites.

Coherent processing between separated sites requires that the phase uncertainty between the local oscillators be a radian or less for the desired processing time. This requires very good local oscillators such as hydrogen masers. With these, processing times can be as long as about 40,000 seconds (11 hours).

Figure H.3 on page 366 shows the RMS random change in phase angle over time T versus T for some of the frequency standards, referred to an operating frequency of 1.5 GHz. For two identical independent standards, the coherent processing time is that time for the phase change of each to reach about 0.7 radians. Thus the hydrogen maser allows fairly long, coherent processing times.

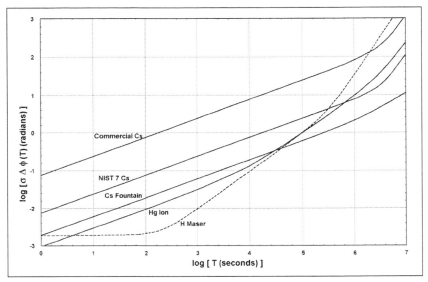

Figure H.3: RMS Phase Change at 1.5 GHz during Time T.

Appendix I

Exact Doppler Shift Removal

This appendix, by Louis K. Scheffer, describes a technique for Doppler shift removal by changing the sampling rate, which can simultaneously correct for all frequencies. This is essential for digital signal processing for SETI.

Traditional techniques for removing Doppler shifts over wide bandwidths work by dynamically changing the local oscillator frequency so that it precisely tracks the expected Doppler shift. When this oscillator is mixed with the input, the result is a constant output frequency for a constant source frequency. Unfortunately, this correction is exact for one frequency only. In digital signal processing, however, the relevant quantity is not the absolute frequency, but the ratio of the received frequency to the sampling rate. It is therefore possible to do Doppler correction by changing the sampling rate, which has the advantage that Doppler correction can be exact simultaneously for all frequencies.

I.1 Introduction

Doppler shift, on a received or transmitted frequency, is caused by the motion of the transmitter or receiver. In particular, to receive a signal that has a narrow bandwidth in the reference frame of the solar system barycenter, we must correct for the motion of the Earth. This is a fundamental problem in any narrowband SETI search [Bla92], [Wer96], [Oli71], [Hor93].

Doppler effects can be highly significant. Suppose the source is transmitting at frequency F_1, and the receiver is moving in the direction of the source with some time varying velocity $v(t)$. Then the signal will be received with frequency $F_1(1 + v(t)/c)$. This effect can be quite significant. Over the course of a 1,000 s observation, the component of the velocity in the direction of the source from a point on the equator of the Earth can change by 46 m/s. This leads to a fractional change in frequency of 1.7×10^{-7} (i.e., a 1,700 Hz shift for a 10 GHz signal). This is many times the desired bandwidth of any narrowband search, and it must be corrected.

Doppler correction is normally done by changing the frequency of the local oscillator during the observation. Consider the case of a local oscillator with a frequency F_{LO} and a source transmitting at frequency F_1. Then the resulting IF is:

$$F_1 - F_{LO} = F_{IF1} \tag{I.1}$$

If, instead of a constant frequency local oscillator, we use:

$$F'_{LO} = F_{LO} + F_1(v(t)/c) \tag{I.2}$$

then we will receive:

$$F_1(1 + v(t)/c) - F'_{LO} = F_1(1 + v(t)/c) - F_{LO} - F_1(v(t)/c) = F_1 - F_{LO} = F_{IF1} \tag{I.3}$$

The Doppler shift caused by $v(t)$ has been removed for any $v(t)$.

However, suppose the input also contains frequency F_2. The desired IF frequency for this signal is $F_{IF2} = F_2 - F_{LO}$, independent of $v(t)$. However, using the classical LO correction, we get:

$$F_2(1 + v(t)/c) - F_{LO} - F_1(v(t)/c) = F_2 - F_{LO} + (F_2 - F_1)(v(t)/c) \tag{I.4}$$

Thus, a second signal will not have the Doppler shift due to $v(t)$ removed completely. An error term will remain which is proportional to both $v(t)$ and the difference in frequency.

For very narrow bandwidth operation, this is a significant problem. For example, if we try to achieve 0.001 Hz operation, the above equation shows that only kilohertz bandwidths are possible.

I.2   Correction in the Time Domain

With a sampled data system, however, we have the option of correcting in the time domain by changing the sampling rate. To see the advantage of this approach, imagine an experiment where we look for narrowband signals by sampling an analog input, saving a sequence of samples, and performing FFT on the resulting sequence.

Suppose we have two incoming signals, 2 MHz and 3 MHz, and we sample the input at 8 MHz. The 2 MHz signal completes exactly 2 cycles per 8 input samples; the 3 MHz signal completes 3 cycles per 8 input samples. In an FFT of any length, these signals will sum perfectly coherently.

Now suppose we try this from a moving platform, and during the observation we add enough velocity to shift the frequency by 10%. (Actual shifts are much less, of course!) At the end of the observing interval, we receive 2.2 MHz and 3.3 MHz signals. Traditionally, we would shift the frequencies, using a variable IF, and keep the sampling rate constant. To do this, we must pick one frequency that will be corrected exactly, e.g., 2.0 MHz. Then, we subtract (using an IF chain) a variable frequency, starting at 0.0 MHz, at the beginning of the observation and ending at 0.2 MHz, at the end of the observation.

In this way, the original 2 MHz input remains an exactly 2 MHz output, which is still exactly 2 cycles per 8 samples. Unfortunately, the 3 MHz signal will now drift from 3.0 MHz at the beginning of the observation to 3.1 MHz at its end. It is no longer in a constant 3:8 relationship with the sampling rate. So while the energy of the 2 MHz signal still is concentrated in one FFT bin, as desired, the energy of a 3 MHz signal will be spread across many bins (about 13 bins for a 1,024 point FFT, or 13,000 bins for a 1 M point FFT).

Suppose, instead that we shift the sampling rate from 8 MHz to 8.8 MHz over the course of the observation. Now, the 2 MHz incoming signal still makes 2 cycles per 8 samples at all times. Likewise, the 3 MHz signal still makes 3 cycles per 8 samples. A simple FFT of the resulting sequence puts all the power of each signal in a single bin, as desired.

I.3   Using a Local Oscillator

The above full-band correction works only because the frequency shift is proportional to the observed frequency. However, almost every practical RF system uses a local oscillator to shift the frequencies before sampling, since the received bandwidths are much less than the received frequencies. In the IF pass band, the resulting shifts are no longer proportional to frequency and the above method will not work directly.

The solution is relatively simple. We use a local oscillator frequency of $F_{LO}(1 + v(t)/c)$, meaning we correct the local oscillator for its own frequency, not the received frequency. Then the received frequencies are:

$$F_1(1 + v(t)/c) - F_{LO}(1 + v(t)/c) = (F_1 - F_{LO})(1 + v(t)/c) = F_{IF1}(1 + v(t)/c)$$
$$(I.5)$$

$$F_2(1 + v(t)/c) - F_{LO}(1 + v(t)/c) = (F_2 - F_{LO})(1 + v(t)/c) = F_{IF2}(1 + v(t)/c)$$
$$\text{(I.6)}$$

This looks like a step backwards: Now no signal in the IF is completely Doppler corrected. However, the shift is now proportional to the IF frequency, so if we change the timescale such that:

$$t' = \frac{t}{1 + v(t)/c} \tag{I.7}$$

then we will receive F_1 and F_2 as desired. This change is easily implemented by changing the sampling rate by a factor $(1 + v(t)/c)$.

Note that the local oscillator and the sampling rate change by the same proportional amount. Therefore, they can both be derived from a common clock, and the correction only needs to be computed once.

I.4   A Numerical Example

The advantages of this approach show up most sharply with high bandwidths and narrow resolutions. Suppose we want a 200 MHz bandwidth and 0.001 Hz resolution. In this case, we need a 1,000 s observation. Suppose the band we wish to analyze is 1,400 to 1,600 MHz. We might do this by using a local oscillator of 1,375 MHz to convert the signal to a base-band of 25 to 225 MHz. Then this would be sampled at 500 MHz, digitized, and analyzed.

In the absence of Doppler shifts, the IF signal can simply be sampled then transformed (FFT) to find the presence of any constant frequency input signals.

However, the presence of a Doppler shift makes the problem harder. During a 1,000 s observation, the Doppler velocity towards a given point on the sky can change by up to 46 m/s for an observatory on the Earth's equator (40 m/s for the Earth's rotation and 6 m/s for the Earth's orbit around the Sun). This leads to a Doppler correction of 1.7×10^{-7}. During the observation, a signal at the center of the band will be Doppler shifted by:

$$1.5 \times 10^9 \text{ Hz} \times 1.7 \times 10^{-7} = 255 \text{ Hz} \tag{I.8}$$

At the upper edge of the band, the Doppler shift is:

$$1.6 \times 10^9 \text{ Hz} \times 1.7 \times 10^{-7} = 272 \text{ Hz} \tag{I.9}$$

At the lower edge of the band, it is:

$$1.4 \times 10^9 \text{ Hz} \times 1.7 \times 10^{-7} = 238 \text{ Hz} \tag{I.10}$$

I.4.1   Tuning the Local Oscillator

Conventionally, the Doppler shift is removed by dynamically tuning the local oscillator. The drift *at any specific frequency* can be completely removed by this technique. For other signals, however, some Doppler shift remains. In this example, for a signal at the band edge differing from the band center by ± 100 MHz, the remaining uncorrected drift is:

$$\pm 10^8 \text{ Hz} \times 1.7 \times 10^{-7} = \pm 17 \text{ Hz} \tag{I.11}$$

This means that a signal at the band edge will drift over 17,000 analysis bins during the course of the observation. Since, to preserve the coherence of any received signal, we need approximately 0.1 bin or less of drift, we need to break the input bandwidth into much smaller bands. These must be no more than 1,200 Hz wide.

I.4.2   Retiming

Instead, suppose we retune the local oscillator and change the sampling rate. In this case, we tune the local oscillator, based on the local oscillator frequency (not the received frequency). Then a signal at the low end of band has a remaining Doppler shift of:

$$25 \times 10^6 \text{ Hz} \times 1.7 \times 10^{-7} = 4.25 \text{ Hz} \tag{I.12}$$

At the center of the band, the Doppler shift is:

$$125 \times 10^6 \text{ Hz} \times 1.7 \times 10^{-7} = 21.25 \text{ Hz} \tag{I.13}$$

At the upper edge of the band, it is:

$$225 \times 10^6 \text{ Hz} \times 1.7 \times 10^{-7} = 38.25 \text{ Hz} \tag{I.14}$$

Note that these drifts are all proportional to the frequency. Therefore, we can also change the sampling rate by the same proportion, changing it by 85 Hz during the observation. If we do so, the ratio of the input frequency to the sampling frequency is unchanged for any signal in the input bandwidth.

Note that the correction is small in absolute terms. We have changed the sampling rate from 500,000,000 Hz to 500,000,085 Hz during the course of the observation. This small change should cause no problems to the users of the digital data.

I.5   Relativistic Doppler Shifts

The simple formula $f_0(1 + v(t)/c)$ for Doppler shifts is not correct in the full relativistic case. However, the exact solution preserves the relevant relationship. The Doppler shifts for two incoming signals are still proportional to their frequency. Hence, the desired resampling rate must be computed using the relativistic formula, but once this is done, the correction will still be exact for any frequency.

I.6   Conclusions

For systems using digital signal processing, removing the Doppler shift in the time domain has the advantage that it can exactly correct Doppler shift over any bandwidth. The correction is most easily implemented as a change in sampling rate.

Appendix J

The Fermi Paradox

This appendix, by Paul Horowitz, describes the circumstances of Enrico Fermi's 1950 observation that if, as seemed possible, the Galaxy was cluttered with intelligent beings, it was odd that none had ever made their existence known to us.

In the oral tradition of the SETI community, Fermi's legendary question, "Where are they?" has been long cherished, quoted, and debated upon. The context, of course, is alien intelligence. If technologically advanced civilizations have inhabited our Galaxy for timescales of approximately a billion years, and if some of these have engaged in interstellar travel and colonization, then why have we not seen physical evidence of their visits?

Given the uncertainty about the history of Fermi's remark, or even if it was ever made, one of the Working Group participants was startled, several years ago, by the casual remark of a colleague: "With your interest in SETI, you might enjoy hearing about that lunch at Los Alamos where Fermi asked his famous question." That colleague was Herb York. He had indeed been present and we were very interested in his recollections! For this book, we have interviewed both Herb York and Philip Morrison. The following is a synopsis of York's recollections:

The time was the summer of 1950. The Russians had exploded their first atomic bomb the previous summer. With additional troubling events in Moscow and Peking, Truman had committed the US to develop a thermonuclear weapon (the 'super', or 'H-bomb'). A nucleus of physicists (most of them veterans of the Manhattan Project) had reassembled at Los Alamos: Bethe, Fermi, Gamow, Garwin, and Teller. York was visiting from UC Berkeley, as a member of the 'measurements group', to provide key instrumentation for the thermonuclear test – the 'George shot' – scheduled for the summer of 1951.

During lunches at the Fuller Lodge, Fermi loved to pose rhetorical questions, which he then proceeded to answer. For example: If you had $100 million to spend on science, without moral restriction, what would you do? (Answer: Dig the deepest hole! We do not know much about what's under our feet.) York arrived in the midst of one of these exercises, at which Konopinski and Teller were already engaged. Out of the blue, Fermi asked: "Don't you ever wonder where everybody is?" York recalls that everyone knew that he meant ETIs. In typical fashion, Fermi then proceeded to answer his own question; in this case, by running through what we now call the Drake Equation (or Green Bank Equation). He pointed to multiple star systems as evidence of planetary systems and he talked about the probability of life arising and acquiring technology. An important part of the analysis was that, if there were others, they would be far more advanced. The odds of finding a civilization in its technological beginning phase are remote.

So, Fermi concluded, if interstellar travel is possible, it ought to be positively crowded out there. Fermi's supposition (as York remembers with a bit of uncertainty) is that interstellar travel is either impossible, or at least so difficult that nobody undertakes it because it is not worth the effort. Fermi and the group came away with the feeling that this was a strong question, one that had not been answered at that lunch. The idea that all technical civilizations inevitably destroy themselves may have been suggested but even on that eve of thermonuclear invention, there was no feeling of doom. On sum, Fermi felt, the world was a better place in 1950 than it was in 1900 – for example in terms of slavery, disease, and so on. York does not recall any further discussion of this topic.

In reminiscing about the prehistory of the Fermi Paradox, Phil Morrison points out that project scientists, in a burst of black humor, sometimes referred to supernovas as failed fission/fusion projects by ETIs. But, the wartime project predated the first reports of flying saucers, and there had been no real talk of space travel. Satellites were only a concept. Morrison also pointed out that Fermi began to complain about his memory after the war. He started keeping a notebook; it is possible that his ideas on the Paradox are recorded in them.

Appendix K

Astrometry

This appendix, by Alan P. Boss of the Carnegie Institute of Washington and George D. Gatewood of the Allegheny Observatory, discusses a wide variety of techniques. Both ground-based and space-based techniques show promise for making future discoveries of extrasolar planets through astrometric observations.

K.1   Interferometry

Interferometry has the greatest potential for astrometric detection of extrasolar planets [Sao92]. In the not very distant future, space-based interferometers, such as NASA's SIM, and ground-based interferometers such as the Palomar Testbed Interferometer (PTI) and the Keck Interferometer (KI), will replace "single-dish" astrometry of the sort one of us (GDG) pursues. SIM, for example, is planned to have an astrometric accuracy as small as a few μarc seconds in narrow angle mode. This is an accuracy that is nearly a thousand times better than what has been achieved by ground-based, single-dish telescopes. Such an accuracy would be sufficient to permit SIM to detect the presence of Earth-like planets in orbit around the nearest stars. As of the time of this writing, SIM has an estimated flight date of 2009. If past experience is any guide, space-based systems, especially those as challenging as SIM, tend to become fully operational considerably later than originally planned. SIM's launch date has already slipped by four years.

In addition, there are two ground-based interferometers that have NASA support. The PTI is mainly used at present for instrumentation and software development [Coa94]. The main limitation of the PTI is that it can only use one reference star at a time, and that one star must be quite bright and very near to the target. (The Navy's Prototype Optical Interferometer near Flagstaff is similarly restricted to very bright stars by its small [0.12 m] apertures.) Thus, the PTI does not have a significant planet survey capacity. The PTI has been successfully employed in very high precision studies of several double stars [Bod99].

The second ground-based interferometry program is based on a series of "outrigger" telescopes to be employed with the two Keck 10 m telescopes during imaging experiments, forming the KI. While the large telescopes will probably not be used often for astrometry, the outriggers will constitute a formidable astrometric search instrument. The KI is planned to have an astrometric accuracy of about 30 μarc seconds, about 30 times better than existing, single-dish, ground-based systems. Considerable progress has been made on the development of the system of four smaller (1.8 m) telescopes and site preparation, although lingering concerns exist about the permit process. First light for linking the two 10 m telescopes was achieved in March, 2001.

Europe's Very Large Telescope Interferometer (VLTI) will also be capable of astrometric planet searches. The VLTI also achieved its first fringes in March, 2001, using two small telescopes, and plans to begin astrometry with all four 8 m telescopes and three auxiliary (1.8 m) telescopes in 2003 or 2004.

K.2   CCD Astrometry

Considerable progress continues to be made in the adaptation of this fantastically capable imaging device to the field of astrometry. Most notable is its application as an astrometric instrument called Stellar Plane Survey (STEPS). It has been used at Keck and at the Palomar 5-m telescope by JPL's Stuart Shaklan, Steven Pravdo and their colleagues. STEPS is able to achieve very high precision (approximately 0.3×10^{-3} arc second) on short exposures of intrinsically faint M dwarfs [Pra96]. Advances have also been made in achieving this same level of precision for earlier spectral classes (e.g., G dwarfs similar to the Sun), the nearest members of which are too bright for STEPS.

The USNO at Flagstaff has deployed a CCD device, unofficially dubbed the "ND9 camera", which includes a nine magnitude filter that can be aligned with the image of a bright target star. The filter allows the observation of 6th through approximately 9th magnitude target stars, against a background of 16th magnitude reference stars. In rich stellar fields with a wealth of reference stars, this magnitude range can be adjusted somewhat by varying the exposure. The critical feature of the ND9 instrument is the filter, which must be completely free of light leaks so that it does not offset the photocenter of the target star. Flagstaff's Conard Dahn (1999, personal communication) has indicated that it required several manufacturing efforts and considerable time to acquire the ND9

CCD. He also indicated that the instrument has since been used to obtain parallaxes that are in agreement with the milli-arcsecond accuracy measurements of Hipparcos.

K.3   Hipparcos and FAME

Operated by the ESA from November 1989 through March 1993, the Hipparcos satellite and the resultant catalog comprise the most successful astrometric effort in history. The split-optical system utilized a Ronchi ruling metric (as in Gatewood's MAP; see Section K.5) to obtain the parallaxes of nearly 120,000 stars with a median precision of 1×10^{-3} arc second. The ability of Hipparcos to detect an astrometric perturbation was limited primarily by the mission's duration of only 3.36 years, and by its annual precision of approximately 2×10^{-3} arc second. Nevertheless, investigation of the Hipparcos data set has shown already that several radial velocity brown dwarf candidates are in fact M dwarf components of binary stars [Mar00]. This is remarkable considering the short length of the mission.

The USNO plans to fly the Full-Sky Astrometric Explorer (FAME), a low-cost astrometric successor to Hipparcos. FAME is planned to have an astrometric precision of about 0.7×10^{-3} arc second for 15th magnitude stars. FAME has been selected as a NASA Explorer mission, and it is planned to be launched in late 2004 on a five-year mission.

K.4   Hubble Space Telescope

Astrometric efforts with the Hubble Space Telescope (HST) have been almost entirely confined to use of the interferometers of the fine guidance system (FGS) instead of HST's CCD cameras. The resulting studies are among the highest precision (approximately 0.5×10^{-3} arc second) planet searches to date [Bnd98], [Bnd99]. Unfortunately these searches were of limited duration and were limited to intrinsically faint M dwarfs. The difficulty of obtaining precious HST time for lengthy astrometric surveys will likely preclude any future astrometric detection efforts with this satellite.

K.5   MAP and MAPS

One of us (GDG) is the developer and builder of a device we believe offers the best astrometric accuracy for a ground-based, single-dish search. Gatewood's

[Gat87] Multichannel Astrometric Photometer (MAP) is the only device currently being used for the astrometric study of candidate planetary systems. The 12 Channel MAP is permanently mounted on the rebuilt Thaw refractor (0.76 m (30 inch) f/18.6) of the University of Pittsburgh's Allegheny Observatory.[1]

A unique feature of the MAP is its ability to filter starlight after positional information has been transferred into frequency space. As starlight passes through the focal plane, it is modulated by an oscillating Ronchi ruling. The phase of this modulation carries the positional information. From this point on, one needs only to preserve the integrity of the modulation signal so the light may be filtered subsequently. With this feature, the MAP may be used to look at very bright stars (e.g., Sirius) with one channel and at very faint stars with the other channels. The quantum efficiency of its photometric detectors matches that of a CCD.

This feature is also the basis of MAP's ability to observe in multiple wavelength bands, and thus to overcome the astrometric effects of differential refraction. High quality astrometric data (1.0×10^{-3} arcsecond annual normal point precision) has been obtained since 1987 at Allegheny Observatory. Gatewood's Thaw MAP program is continuing and has been enlarged to include a total of 60 target stars, including several of the previously detected spectroscopic planetary companions.

In addition to MAP, Gatewood built a second MAP instrument for use with the Keck II telescope on Mauna Kea, Hawaii. The second MAP, termed MAPS, included a feature that allowed photons to be carried off by light pipes and analyzed by a spectroscope. MAP was developed by the University of Arizona's Robert McMillan, in order to combine the advantages of the radial velocity and astrometric search methods into a single device and observing program. MAPS was developed for Keck II in order to take advantage of the expected large isokinetic patch of Keck II, which is expected to scale with the aperture of a single-dish telescope [Bur92]. The Keck II MAPS effort would have obtained very high astrometric precision by comparing the seeing-related motion of the target star with reference stars near enough to share in that same motion. After the development of the related derotator, centration cameras, and guiding system, however, it was discovered that the Keck telescope's optical axis alignment capability had been significantly overestimated. The resulting large, high order, tilt-related distortions led to the discontinuation of both the MAPS and STEPS programs on the Keck II telescope.

---

[1] A new red light objective was installed in 1986.

K.6 Further References

Benedict, G. F., McArthur, B., & Nelan, E. P., et al., 1998, "Working a Space-based Optical Interferometer: HST Fine Guidance Sensor 3 Small-field Astrometry", *Astronomical Interferometry*, Proceedings of SPIE, ed., R. D., Reasenberg, 3350, 229-236.

Benedict, G. F., McArthur, B. , & Chappell, D. W., et al., 1999, "Interferometric Astrometry of Proxima Centauri and Barnard's Star Using Hubble Space Telescope Fine Guidance Sensor 3: Detection Limits for Sub-stellar Companions", *AJ*, 118, 1086-1100.

Boden, A. F., Koresko, C. D., van Belle, G. T., Colavita, M. M., Dumont, P. J., Gubler, J., Kulkarni, S. R., Lane, B. F., Mobley, D., Shao, M., Wallace, J. K., & Henry, G. W., 1999, "The Visual Orbit of Iota Pegasi", *ApJ*, 515, 356-364.

Burke, B. F. (chair), 1992, "TOPS: Toward Other Planetary Systems", NASA Solar System Exploration Division, Washington, DC.

Colavita, M. M., & Shao, M., 1994, "Indirect Planet Detection with Ground-Based Long-Baseline Interferometry", *Astrophysical Space Science*, 212, 385-390.

Gatewood, G. D., 1987, "The Multichannel Astrometric Photometer and Atmospheric Limitations in the Measurement of Relative Positions", *AJ*, 94, 213-224.

Marcy, G. W., Cochran, W. D., & Mayor, H., 2000, *Extrasolar Planets Around Main Sequence Stars, Protostars and Planets IV*, eds., V. G. Mannings, A. P. Boss, & S. S. Russell, University of Arizona Press, 1285.

Pravdo, S., & Shaklan, S. B., 1996, "Astrometric Detection of Extrasolar Planets: Results of a Feasibility Study with the Palomar 5 Meter Telescope", *ApJ*, 465, 264-277.

Shao, M., & Colavita, M. M., 1992, "Long-Baseline Optical and Infrared Stellar Interferometry", *Annual Reviews of Astronomy and Astrophysics*, 30, 457-498.

Appendix L

Archive of SETI Searches

This appendix contains an archive of SETI searches. It was originally compiled in 1974 by Jill C. Tarter, of the SETI Institute, as her first SETI research project near the end of her graduate studies. Since then, she has maintained and updated this archive.

Two special symbols († and ‡) are used for some of the entries in this archive:

† Only spot bands within the frequency range were observed.

‡ Flux Limits are based on an assumed S/N = 1.

Date:	1960
Observers:	Drake, "Ozma"
Site:	NRAO
Instrument Size (m):	26
Search Frequency (MHz):	1,420 to 1,420.4
Frequency Resolution (Hz):	100
Objects:	2 stars
Flux Limits (W/m²):	4×10^{-22}‡
Total Hours:	200
Reference:	[Dra60]
Comments:	The search used a single channel receiver.

Date:	1964 through 1965
Observers:	Kardashev & Sholomitskii
Site:	Crimea Deep Space Station
Instrument Size (m):	16 (8 antennas)
Search Frequency (MHz):	923
Frequency Resolution (Hz):	1×10^7
Objects:	2 quasars
Flux Limits (W/m^2):	2×10^{-20}
Total Hours:	80
Reference:	[Kar64], [Gin88]
Comments:	The search reported the detection of quasar CTA102 as a possible Type III civilization.

Date:	1966
Observers:	Kellermann
Site:	CSIRO
Instrument Size (m):	64
Search Frequency (MHz):	Many frequencies between 350 and 5×10^3
Frequency Resolution (Hz):	Full bandwidth for each feed
Objects:	1 Galaxy (1934-63)
Flux Limits (W/m^2):	10^{-18}
Total Hours:	Not available
Reference:	[Kel66]
Comments:	No "notch" of SETI origin was detected in galaxy 1934-63.

Date:	1968 through 1969
Observers:	Troitskii, Starodubtsev, Gershtejin & Rakhlin
Site:	Zimenkie, USSR
Instrument Size (m):	5
Search Frequency (MHz):	926 to 928, and 1,421 to 1,423
Frequency Resolution (Hz):	13
Objects:	11 stars + M31
Flux Limits (W/m^2):	2×10^{-21}‡
Total Hours:	12
Reference:	[Tro71]
Comments:	The search used 20 filters of width 100 kHz, divided into 25 channels with f=13 Hz spaced 4 kHz apart and stepped in frequency.

Date:	1968 through 1982
Observers:	Troitskii
Site:	Gorkii
Instrument Size (m):	Dipole
Search Frequency (MHz):	1×10^3, 1,874, 3,748, and 1×10^4 cm
Frequency Resolution (Hz):	NA
Objects:	All sky search
Flux Limits (W/m^2):	n/a
Total Hours:	Ongoing
Reference:	[Tro82]
Comments:	The search covered all of the sky visible by the single dipole.

Date:	1969 through 1983
Observers:	Troitskii, Bondar, & Starodubtsev
Site:	Gorkii, Crimea, Murmansk and Primorskij regions
Instrument Size (m):	Dipoles
Search Frequency (MHz):	600, 927, and 1,863
Frequency Resolution (Hz):	n/a
Objects:	All sky search for sporadic pulses
Flux Limits (W/m^2):	10^{-22} W/m^2/Hz
Total Hours:	1,200 per year (on average)
Reference:	[Tro75]
Comments:	The search used a network of isotropic detectors and cross correlation between either 2 or 4 sites over 8,000 km apart.

Date:	1970 through 1972
Observers:	Slysh, Pashchenko, Rudnitskii, & Lekht
Site:	Nançay
Instrument Size (m):	40×240
Search Frequency (MHz):	1,667 and 1,665
Frequency Resolution (Hz):	4×10^3
Objects:	5 OH Masers
Flux Limits (W/m^2):	n/a
Total Hours:	2
Reference:	[Pas71], [Pas73], [Lek75]
Comments:	The search was for a deviation from Gaussian emission statistics in 5 OH maser sources, which might indicate transmissions from another civilization.

Date:	1970 through 1972
Observers:	Slysh
Site:	Nançay
Instrument Size (m):	40×240
Search Frequency (MHz):	1,667 and 1,665
Frequency Resolution (Hz):	4×10^3
Objects:	10 nearest stars
Flux Limits (W/m^2):	n/a
Total Hours:	Not available.
Reference:	n/a
Comments:	The search looked for signals at the OH frequency, between observations of OH masers.

Date:	1971 through 1972
Observers:	Verschuur, "Ozpa"
Site:	NRAO
Instrument Size (m):	91 and 43
Search Frequency (MHz):	1,419.8 to 1,421 and 1,410 to 1,430
Frequency Resolution (Hz):	490 and 6,900
Objects:	9 stars
Flux Limits (W/m^2):	5×10^{-24} and 2×10^{-23}
Total Hours:	13
Reference:	[Ves73]
Comments:	The 384 channel correlator was online.

Date:	1972 through 1974
Observers:	Kardashev, Gindilis, Popov, Soglasnov, Spangenberg, Steinberg et al.
Site:	Caucasus, Pamir, Kamchatka, Mars 7 spacecraft
Instrument Size (m):	38, 60
Search Frequency (MHz):	371, 408, 458 and 535
Frequency Resolution (Hz):	5×10^6
Objects:	Omnidirectional
Flux Limits (W/m^2):	2×10^{-16} to 7×10^{-15}
Total Hours:	150
Reference:	[Gin88]
Comments:	This was an "Eavesdropping" search for pulses. Synchronous dispersion reception was used.

Date:	1972
Observers:	Kardashev, Popov, Soglasnov et al.
Site:	Crimea, Rt-22
Instrument Size (m):	22
Search Frequency (MHz):	8,570
Frequency Resolution (Hz):	n/a
Objects:	Galactic Center
Flux Limits (W/m^2):	n/a
Total Hours:	Not available.
Reference:	[Gin88]
Comments:	The search looked for statistical anomalies in continuum emission from the galactic center.

Date:	1972 through 1976
Observers:	Palmer & Zuckerman, "Ozma II"
Site:	NRAO
Instrument Size (m):	91
Search Frequency (MHz):	1,413 to 1,425 and 1,420.1 to 1,420.7
Frequency Resolution (Hz):	6.4×10^4 and 4×10^3
Objects:	674 Stars
Flux Limits (W/m^2):	10^{-23}‡
Total Hours:	500
Reference:	[Pal72], [She77]
Comments:	The 384 channel correlator was online.

Date:	1972 through 1976
Observers:	Bridle & Feldman "Qui Appelle?"
Site:	Algonquin Radio Observatory (ARO)
Instrument Size (m)@:	46
Search Frequency (MHz):	$22,235.08 \pm 5$
Frequency Resolution (Hz):	3×10^4
Objects:	70 stars
Flux Limits (W/m^2):	1×10^{-22}‡
Total Hours:	140
Reference:	n/a
Comments:	To date, 70 solar-type stars within 45 light years have been observed.

Date:	1973 through 1974
Observers:	Shvartsman et al. "MANIA"
Site:	Special Astrophysical Observatory
Instrument Size (m):	0.6
Search Frequency (MHz):	5.45×10^8 (5,500 Å)
Frequency Resolution (Hz):	9.9×10^{-5} (10^{-6} Å)
Objects:	21 peculiar objects
Flux Limits (W/m^2):	n/a
Total Hours:	Not available.
Reference:	[Shv77]
Comments:	This was an optical search for short pulses, of length 3×10^{-7} to 300 s, and narrow laser lines. It was a prototype for a later system using a 6 m telescope.

Date:	1973 through 1986
Observers:	Dixon, Ehman, Raub, & Kraus
Site:	OSURO
Instrument Size (m):	53
Search Frequency (MHz):	1,420.4 relative to galactic center $\pm$ 250 kHz
Frequency Resolution (Hz):	10 and 1 kHz
Objects:	All sky search
Flux Limits (W/m^2):	1.5×10^{-21} ‡
Total Hours:	100,000
Reference:	[Dix77], [Kra79]
Comments:	The receiver was tuned to the hydrogen rest frequency, relative to the galactic center (as a function of direction).

Date:	1974
Observers:	Wishnia
Site:	"Copernicus Satellite"
Instrument Size (m):	1
Search Frequency (MHz):	3×10^9
Frequency Resolution (Hz):	n/a
Objects:	3 stars
Flux Limits (W/m^2):	n/a
Total Hours:	Not available.
Reference:	[Mrp75]
Comments:	The search was for UV laser lines.

Date:	1975 through 1976
Observers:	Sagan & Drake
Site:	NAIC
Instrument Size (m):	305
Search Frequency (MHz):	1,420, 1,667 and 2,380 (B=3 MHz)
Frequency Resolution (Hz):	1,000
Objects:	Four galaxies
Flux Limits (W/m^2):	3×10^{-25}‡
Total Hours:	100
Reference:	n/a
Comments:	The search was for Type II civilizations in the Local Group of galaxies.

Date:	1975 through 1979
Observers:	Israel & DeRuiter
Site:	WSRT
Instrument Size (m):	1,500 maximum baseline
Search Frequency (MHz):	1,415
Frequency Resolution (Hz):	4×10^6
Objects:	50 star fields
Flux Limits (W/m^2):	2×10^{-23}‡
Total Hours:	400
Reference:	n/a
Comments:	Searches were conducted of 'cleaned' maps prepared for the Westerbork Synthesis Radio Telescope (WSRT) background survey. The search looked for positional coincidence between residual signals and AGK2 stars.

Date:	1976
Observers:	Clark, Black, Cuzzi, & Tarter
Site:	NRAO
Instrument Size (m):	43
Search Frequency (MHz):	8,522 to 8,523
Frequency Resolution (Hz):	5
Objects:	4 stars
Flux Limits (W/m^2):	2×10^{-24}‡
Total Hours:	7
Reference:	n/a
Comments:	VLBI high-speed tape recorder was combined with direct Fourier transformation (by software) to produce extreme frequency resolution (non-real time).

Date:	1976 through 1985
Observers:	Bowyer, et al. (U.C. Berkeley), "SERENDIP"
Site:	Hat Creek Radio Observatory (HCRO)
Instrument Size (m):	26
Search Frequency (MHz):	917 to 937, 1,410 to 1,430, 1,602 to 1,605, 1,853 to 1,873, and 5,000
Frequency Resolution (Hz):	2×500
Objects:	All sky survey
Flux Limits (W/m^2):	1.15×10^{-22}
Total Hours:	Not available.
Reference:	[Bow83]
Comments:	This was an automated survey, commensal with radio astronomical observations.

Date:	1977
Observers:	Tarter, Black, Cuzzi, & Clark
Site:	NRAO
Instrument Size (m):	91
Search Frequency (MHz):	1,665 to 1,667†
Frequency Resolution (Hz):	5
Objects:	200 stars
Flux Limits (W/m^2):	10^{-24}
Total Hours:	100
Reference:	[Tar80]
Comments:	VLBI high-speed recorder was combined with direct Fourier transformation (by software) to produce extreme frequency resolution (non-real time).

Date:	1977
Observers:	Stull & Drake
Site:	NAIC
Instrument Size (m):	305
Search Frequency (MHz):	1,664 to 1,668†
Frequency Resolution (Hz):	0.5
Objects:	6 stars
Flux Limits (W/m^2):	10^{-26}‡
Total Hours:	10
Reference:	[Tar79]
Comments:	High-speed tape was combined with an optical processor to produce extreme frequency resolution (non-real time).

Date:	1977 through 1980
Observers:	Wielebinski & Seiradakis
Site:	Max Planck Institut für Radioastronomie (MPIFR)
Instrument Size (m):	100
Search Frequency (MHz):	1,420
Frequency Resolution (Hz):	20,000,000
Objects:	3 stars
Flux Limits (W/m^2):	4×10^{-23}
Total Hours:	2
Reference:	n/a
Comments:	Candidate stars were inserted into an ongoing program, which searched for pulsed signals with periods of 0.3 to 1.5 seconds.

Date:	1978
Observers:	Horowitz
Site:	NAIC
Instrument Size (m):	305
Search Frequency (MHz):	1,420 MHz $\pm$ 500 Hz
Frequency Resolution (Hz):	0.015
Objects:	185 stars
Flux Limits (W/m^2):	8×10^{-28}‡
Total Hours:	80
Reference:	[Hor78], [Hor85]
Comments:	The search assumed that the signal frequency was corrected at the source, to arrive at rest in a heliocentric or barycentric laboratory frame.

Date:	1978
Observers:	Cohen, Malkan & Dickey
Site:	NAIC, Haystack Radio Observatory (HRO), CSIRO
Instrument Size (m):	305, 36, 64
Search Frequency (MHz):	1,665 + 1,667, 22,235, and 1,612
Frequency Resolution (Hz):	9,500, 65,000, 4,500
Objects:	25 globular clusters
Flux Limits (W/m^2):	$1.8 \times 10^{-25}, 1.1 \times 10^{-22}, 1.5 \times 10^{-24}$‡
Total Hours:	40, 20, 20
Reference:	[Coh80]
Comments:	This was a passive search for Type II and Type III civilizations, using astronomical data originally used to detect H_2O and OH masers in globular clusters.

Date:	1978
Observers:	Knowles & Sullivan
Site:	NAIC
Instrument Size (m):	305
Search Frequency (MHz):	130 to 500 (Spot)†
Frequency Resolution (Hz):	1
Objects:	2 stars
Flux Limits (W/m²):	2×10^{-24}
Total Hours:	5
Reference:	[Sul78]
Comments:	This search attempted "eavesdropping" using MKI VLBI tapes as in Black, et al., 1977.

Date:	1978
Observers:	Makovetskij, Gindilis, et al.
Site:	Zelenchukskaya, Ratan-600
Instrument Size (m):	7.4×450 (one Sector)
Search Frequency (MHz):	n/a
Frequency Resolution (Hz):	n/a
Objects:	Barnard's star
Flux Limits (W/m²):	n/a
Total Hours:	6 days
Reference:	[Gin86]
Comments:	In accordance with the "magic time" prediction by Makovetskij for B-Barnard's star and Nova Cygni 1975, a search was made for signals in September, 1978.

Date:	1978 through 1996
Observers:	Shvartsman et al. "MANIA"
Site:	Special Astrophysical Observatory
Instrument Size (m):	6
Search Frequency (MHz):	5.45×10^8 (5,500 Å)
Frequency Resolution (Hz):	9.9×10^{-5} (10^{-6} Å)
Objects:	93 Objects
Flux Limits (W/m²):	$< 3 \times 10^{-4}$ of the optical flux is variable in any object observed
Total Hours:	250
Reference:	[Shv88]
Comments:	They searched 30 radio objects with continuous optical spectra, looking for optical pulses from potential Kardashev Type II or Type III civilizations.

Date:	1979
Observers:	Cole & Ekers
Site:	CSIRO
Instrument Size (m):	64
Search Frequency (MHz):	$5,000 \pm 5$ MHz and ± 1 MHz
Frequency Resolution (Hz):	10^7 and 10^6
Objects:	Nearby F, G and K Stars
Flux Limits (W/m^2):	4×10^{-18}‡
Total Hours:	50
Reference:	[Col79]
Comments:	A search was made for simultaneous pulsed events using both 2 MHz and 10 MHz filters with detectors having a time resolution of 4μs.

Date:	1979
Observers:	Freitas & Valdes
Site:	Leuschner Observatory, UC Berkeley
Instrument Size (m):	0.76
Search Frequency (MHz):	5.45×10^8 (5,500 Å)
Frequency Resolution (Hz):	n/a
Objects:	Stable "Halo Orbits" about L4 and L5 libration points in Earth-Moon system
Flux Limits (W/m^2):	$m_v \leq 14$ (Magnitude)
Total Hours:	30
Reference:	[Fre80]
Comments:	This search attempted to discover evidence of discrete objects (such as interstellar probes) in stable orbits about L4 and L5 by studying 90 photographic plates.

Date:	1978 through 1980
Observers:	Michael J. Harris
Site:	Interplanetary Network Data
Instrument Size (m):	Pioneer Venus and Venera 11 and 12 spacecraft.
Search Frequency (MHz):	4.84×10^{12} (20KeV) to $2.42 \times^{14}$ (1 MeV)
Frequency Resolution (Hz):	n/a
Objects:	54 Gamma Ray Burst (GRB) events
Flux Limits (W/m^2):	n/a
Total Hours:	Not available.
Reference:	[Har90]
Comments:	This search attempted to find 3 GRB events in a straight line, each having the same velocity (from the e^+ e^- annihilation line) that could indicate the trajectory of an interstellar spacecraft.

Date:	1979 through 1982
Observers:	JPL, University of California at Berkeley (UCB), "SERENDIP"
Site:	DSS 14
Instrument Size (m):	64
Search Frequency (MHz):	S and X band (B=10 MHz)
Frequency Resolution (Hz):	$2 \times$ (500 Hz)
Objects:	Apparent positions of NASA spacecraft
Flux Limits (W/m^2):	8×10^{-24}‡
Total Hours:	400
Reference:	[Bow83]
Comments:	This was an automated survey, commensal to spacecraft tracking operations, using a 512 channel autocorrelator and a 100 channel correlator microprocessor control.

Date:	1979 through 1981
Observers:	Tarter, Clark, Duquet & Lesyna
Site:	NAIC
Instrument Size (m):	305
Search Frequency (MHz):	$1\,420.4 \pm 2$ and $1\,666 \pm 2$†
Frequency Resolution (Hz):	5 and 600
Objects:	200 stars
Flux Limits (W/m^2):	10^{-25}‡
Total Hours:	100
Reference:	[Tar83]
Comments:	A rapid 1-bit sampler and high-speed tape recorder were run in parallel with a 1,008 channel correlator. A direct Fourier transformation was used as in Black, et al., 1977.

Date:	1980
Observers:	Witteborn
Site:	NASA – University of Arizona, Mt. Lemmon
Instrument Size (m):	1.5
Search Frequency (MHz):	2.55×10^9 (8.5 μm) to 4.05×10^9 (13.5 μm)
Frequency Resolution (Hz):	3×10^{14} Hz (1 μm)
Objects:	20 stars
Flux Limits (W/m^2):	N magnitude excess < 1.7
Total Hours:	50
Reference:	n/a
Comments:	They searched for infrared excesses due to Dyson spheres around solar-type stars. The target stars were chosen because they were too faint for their spectral type.

Date:	1980 through 1981
Observers:	Suchkin, & Tokarev et al.
Site:	Nirfi, Gorkii, Gaish, Moscow
Instrument Size (m):	n/a
Search Frequency (MHz):	9.3 pulsed radar
Frequency Resolution (Hz):	1,500
Objects:	L4 and L5 libration points of Earth-Moon system
Flux Limits (W/m^2):	n/a
Total Hours:	20
Reference:	[Suc81]
Comments:	This search looked for radar reflections from artifacts in parking orbits around the libration points.

Date:	1981
Observers:	Lord & O'Dea
Site:	University of Massachusetts
Instrument Size (m):	14
Search Frequency (MHz):	115,000
Frequency Resolution (Hz):	2×10^4, 1.25×10^5, and 4×10^8
Objects:	North galactic rotation axis $b = 5°$ to $90°$
Flux Limits (W/m^2):	10^{-21}‡
Total Hours:	50
Reference:	[Tar85b]
Comments:	This search was for signals (near the J=1-0 CO line) from a transmitter located somewhere along the galactic rotation axis.

Date:	1981
Observers:	Tarter & Israel
Site:	WSRT
Instrument Size (m):	3,000 maximum baseline
Search Frequency (MHz):	1,420
Frequency Resolution (Hz):	4×10^6, 10×10^6
Objects:	85 star fields
Flux Limits (W/m^2):	6×10^{-22} to 6×10^{-24}
Total Hours:	600
Reference:	[Tar82]
Comments:	This was a commensal search similar to that of Israel and De Ruiter, using 'uncleaned' maps stored at Groningen and Leiden, and the AGK3 catalog.

Date:	1981
Observers:	Shostak & Tarter, "SIGNAL"
Site:	WSRT
Instrument Size (m):	3,000 maximum baseline
Search Frequency (MHz):	1,420.4 relative to the galactic center B=156 kHz
Frequency Resolution (Hz):	1,200
Objects:	Galactic center
Flux Limits (W/m^2):	10^{-24}‡
Total Hours:	4
Reference:	[Sho85]
Comments:	This search used an interferometer to search the galactic center for pulsed signals, with periods from 40 s to 2 h.

Date:	1981
Observers:	Talent
Site:	Kitt Peak National Observatory (KPNO)
Instrument Size (m):	2.1
Search Frequency (MHz):	3.54×10^8 (3,575 Å) to 5.30×10^8 (5,350 Å)
Frequency Resolution (Hz):	9.99×10^5 (10 Å)
Objects:	3 stars
Flux Limits (W/m^2):	n/a
Total Hours:	0.2
Reference:	[Gra82]
Comments:	This search was for enhanced stellar lines of praseodymium, neodymium, and zirconium, as evidence of nuclear waste being dumped into stellar atmospheres.

Date:	1981 through 1982
Observers:	Valdes & Freitas, "SETA"
Site:	KPNO
Instrument Size (m):	0.61
Search Frequency (MHz):	5.45×10^8 (5,500 Å)
Frequency Resolution (Hz):	n/a
Objects:	Earth-Moon through L5, Sun-Earth L1, L2
Flux Limits (W/m^2):	$10 \leq M_v \leq 19$ (Magnitude)
Total Hours:	70
Reference:	[Val83]
Comments:	This search attempted to see discrete artifacts (a few meters in size) in stable orbits near the Lagrange points by studying 137 III aF photographic plates.

Date:	1981 through 1988
Observers:	Biraud & Tarter
Site:	Nançay
Instrument Size (m):	40×240
Search Frequency (MHz):	1,420.4 MHz $\pm$ 320 kHz, 1,665 to 1,667
Frequency Resolution (Hz):	48.8
Objects:	343 stars
Flux Limits (W/m^2):	10^{-24}
Total Hours:	~ 600
Reference:	[Tar85]
Comments:	An eight level, 1,024 channel autocorrelator, with stepped first LO to extend the frequency coverage at 48 Hz resolution, was used.

Date:	1982
Observers:	Horowitz, Teague, Linscott, Chen & Backus, "Suitcase SETI"
Site:	NAIC
Instrument Size (m):	305
Search Frequency (MHz):	2,840.8 (B=4 kHz) and 1,420.4 (B=2 kHz)
Frequency Resolution (Hz):	0.03 (1-linear) and 0.03 (2-circular)
Objects:	250 stars and 150 stars
Flux Limits (W/m^2):	4×10^{-26} and 6×10^{-28}‡
Total Hours:	75
Reference:	[Hor85]
Comments:	A dual 64k channel, real time, micro-processor based spectrum analyzer with video archiving and swept LO frequency was used to test "magic frequencies."

Date:	1982
Observers:	Vallee & Simard-Normandin
Site:	ARO
Instrument Size (m):	46
Search Frequency (MHz):	10,522
Frequency Resolution (Hz):	1.85×10^6
Objects:	Galactic center meridian
Flux Limits (W/m^2):	10^{-19}‡
Total Hours:	72
Reference:	[Vae85]
Comments:	This search was for strongly polarized signals, by mapping a field $1/4' \times 25'$ along $l = 0°$.

Date:	1983
Observers:	Damashek
Site:	NRAO
Instrument Size (m):	92
Search Frequency (MHz):	390 ± 8
Frequency Resolution (Hz):	2×10^6
Objects:	Sky survey (pulsars)
Flux Limits (W/m^2):	2×10^{-22}
Total Hours:	700
Reference:	n/a
Comments:	16 MHz was sampled with 8 contiguous frequency channels at 60 Hz. This search was for single dispersed pulses and telemetry (bit stream) signals.

Date:	1983
Observers:	Valdes & Freitas
Site:	HCRO
Instrument Size (m):	26
Search Frequency (MHz):	$1,516 \pm 2.5$
Frequency Resolution (Hz):	4.9 and 7.6^3
Objects:	80 stars and 12 nearby stars
Flux Limits (W/m^2):	3×10^{-24}‡
Total Hours:	100
Reference:	[Val86]
Comments:	This was a search for a radioactive tritium line from nuclear fusion due to an ETI.

Date:	1983 through 1984
Observers:	Cullers
Site:	AMSETI
Instrument Size (m):	2
Search Frequency (MHz):	$\sim 1,420$ and $\leq 1,000$
Frequency Resolution (Hz):	n/a
Objects:	n/a
Flux Limits (W/m^2):	n/a
Total Hours:	n/a
Reference:	n/a
Comments:	Low noise GaAs FETs and microprocessors on satellite TV dishes were used by Silicon Valley 'hams', with NASA Ames consultation.

Date:	1983 through 1985
Observers:	Horowitz, "Sentinel"
Site:	Oak Ridge (Harvard University)
Instrument Size (m):	26
Search Frequency (MHz):	1,420.4 and 1,667.3
Frequency Resolution (Hz):	0.03 dual circular (B=2 kHz)
Objects:	Sky survey
Flux Limits (W/m^2):	5×10^{-25}
Total Hours:	Not available.
Reference:	[Hor85]
Comments:	"Suitcase SETI" was used as the backend of an automated sky survey at 2 magic frequencies over a 3 year observing period.

Date:	1983 through 1998
Observers:	Gray
Site:	Small SETI Radiotelescope
Instrument Size (m):	4
Search Frequency (MHz):	1,419.5 to 1,421.5
Frequency Resolution (Hz):	40 to 1
Objects:	Sky survey and $-27°$
Flux Limits (W/m^2):	1×10^{-22}‡
Total Hours:	Not available.
Reference:	[Gra86]
Comments:	This was a dedicated meridian transit search system that was constructed by amateurs, and operated during the evenings.

Date:	1983 through 1988
Observers:	Stephens
Site:	Interstellar Electromagnetics Institute, Hay River, NWT
Instrument Size (m):	Two 18 m × 18 m (28 m equivalent)
Search Frequency (MHz):	1,415 to 1,425
Frequency Resolution (Hz):	3×10^4
Objects:	Northern sky survey
Flux Limits (W/m^2):	$T_{sys} \sim 75$ K
Total Hours:	Not available.
Reference:	n/a
Comments:	Two surplus, 64 foot square, "troposcatter" dishes were combined for use as a dedicated amateur SETI observatory. The coverage was from 30° to 45° declination. The search was discontinued due to lack of funding.

Date:	1983
Observers:	Gulkis and Kuiper
Site:	DSS 43
Instrument Size (m):	64
Search Frequency (MHz):	8×10^3 and $2{,}380 \pm 5$
Frequency Resolution (Hz):	4×10^4
Objects:	Partial southern sky
Flux Limits (W/m^2):	2×10^{-22}‡
Total Hours:	800
Reference:	[Kui83]
Comments:	This was a sky survey of three constant declination strips from -28.9 to -34.3, by April 1983, carried out whenever the antenna was stowed.

Date:	1984
Observers:	Slysh
Site:	Satellite
Instrument Size (m):	Radiometer
Search Frequency (MHz):	3.7×10^4
Frequency Resolution (Hz):	4×10^8
Objects:	All sky 3 K BB
Flux Limits (W/m^2):	$T/T \geq 01$
Total Hours:	6,000
Reference:	[Sly85]
Comments:	The lack of fluctuations in the 3 K background radiation on angular scales of 10^{-2} sr ruled out optically thick Dyson spheres radiating more than 1 $L_\circ$ within 100 pc.

Date:	1985 through 1995
Observers:	Horowitz, "META SETI"
Site:	Oak Ridge (Harvard University)
Instrument Size (m):	26
Search Frequency (MHz):	1,420.4, 1,665.4, 1,667.3 and 2,840.8
Frequency Resolution (Hz):	.05
Objects:	Sky survey
Flux Limits (W/m^2):	1.3×10^{-24}
Total Hours:	Not available.
Reference:	[Hor86, Hor93]
Comments:	Signal processing hardware from "SENTINEL" was replicated 128 times to produce 8.4×10^6 channels. Six sequential observations of each patch of sky were made to cover 2 orthogonal circular polarizations and 3 rest frames (Sun/Earth, galactic center, and 3 K background).

Date:	1986
Observers:	Arkhipov
Site:	n/a
Instrument Size (m):	n/a
Search Frequency (MHz):	408
Frequency Resolution (Hz):	n/a
Objects:	HD21899, HD100623, HD187691 and HD187923
Flux Limits (W/m^2):	n/a
Total Hours:	Not available.
Reference:	[Ark86]
Comments:	A search was made of the 408 MHz Molonglo Survey Catalog of radio sources, and 4 solar-type stars were found within 130 arc seconds of radio source position. It was suggested that the cause of the radio emission, in these four cases, was leakage emission from an orbital industrial processing facility located about 1,000 AU from the star.

Date:	1986
Observers:	Mirabel
Site:	NRAO
Instrument Size (m):	43
Search Frequency (MHz):	4,829.620 to 4,829.776
Frequency Resolution (Hz):	76
Objects:	Galactic center and 33 nearby stars
Flux Limits (W/m^2):	6×10^{-25} to 10^{-24}‡
Total Hours:	144
Reference:	n/a
Comments:	This was a search at the H_2CO (formaldehyde) frequency. It included the star HD170493, located in front of a dark "anti-maser" cloud.

Date:	1986 through 1988
Observers:	Bowyer, Werthimer & Lampton, "SERENDIP II"
Site:	NRAO
Instrument Size (m):	92
Search Frequency (MHz):	400 to 3,500
Frequency Resolution (Hz):	1
Objects:	Sky areas observed by astronomers
Flux Limits (W/m^2):	4×10^{-24}
Total Hours:	Not available.
Reference:	[Bow88]
Comments:	This was an automated search, commensal with radioastronomical observations. It scanned the available 3.5 MHz of IF, 65 kHz at a time. It recorded events above threshold, for off-line processing.

Date:	1986 through 1989
Observers:	Dixon & Bolinger
Site:	OSURO
Instrument Size (m):	53
Search Frequency (MHz):	1,400 to 1,700
Frequency Resolution (Hz):	100, 10 and 2×10^3
Objects:	All sky search
Flux Limits (W/m^2):	1.5×10^{-22}
Total Hours:	20,000
Reference:	n/a
Comments:	n/a

Date:	1986 through 1990
Observers:	Colomb, Martin, & Lemarchand
Site:	Institute for Argentine Radioastronomy (IAR)
Instrument Size (m):	30
Search Frequency (MHz):	1,415.4, 1,425.4, 1,667
Frequency Resolution (Hz):	2.5×10^3
Objects:	78 solar-type stars.
Flux Limits (W/m^2):	5×10^{-23}‡
Total Hours:	320
Reference:	[Com92]
Comments:	78 Southern Hemisphere solar-type stars were searched at 21 cm and 18 cm.

Figure L.1: The center of the VLA with the antennas in close configuration. (Photo courtesy of VLA, NRAO.)

Date:	1987
Observers:	Tarter, Kardashev & Slysh
Site:	VLA
Instrument Size (m):	26 (9 Antennas)
Search Frequency (MHz):	1,612.231
Frequency Resolution (Hz):	6,105
Objects:	G357.3-1.3
Flux Limits (W/m^2):	n/a
Total Hours:	1
Reference:	n/a
Comments:	Remote observation was made, by the VLA staff of the Infrared Astronomical Satellite (IRAS) source near the galactic center to determine if the source could be a nearby Dyson sphere. The source was confirmed as an OH/IR star.

Date:	1987
Observers:	Gray
Site:	Oak Ridge (Harvard University)
Instrument Size (m):	26
Search Frequency (MHz):	1,420.4 MHz $\pm$ 200 kHz in heliocentric, LSR, and cosmic background rest frames
Frequency Resolution (Hz):	0.05
Objects:	Sky position corresponding to OSURO's 1977 "WOW" signal
Flux Limits (W/m^2):	1.5×10^{-24} per channel
Total Hours:	16
Reference:	[Gra94]
Comments:	This search used the "META SETI" system, [Hor86], to track the position on sky that produced the "WOW" signal at OSURO in 1977.

Date:	1988
Observers:	Bania & Rood
Site:	NRAO
Instrument Size (m):	43
Search Frequency (MHz):	8,665
Frequency Resolution (Hz):	≥ 305
Objects:	24 "Vega-like" stars with 60 μm excesses (IRAS)
Flux Limits (W/m^2):	1 Jy
Total Hours:	50
Reference:	[Ban93]
Comments:	This was a search for a narrowband, obviously artificial, signal at the frequency of $^3H_e^+$ spin-flip, being radiated by advanced ETI civilizations that have colonized their planetary systems.

Date:	1989 through 1998
Observers:	Childers & Dixon
Site:	OSURO
Instrument Size (m):	53
Search Frequency (MHz):	1,400 to 1,700
Frequency Resolution (Hz):	100, 10 and 2×10^3
Objects:	All sky search
Flux Limits (W/m^2):	1.5×10^{-22}
Total Hours:	60,000
Reference:	n/a
Comments:	Declinations $+62°$ to $-22°$ extended to $-36°$ were searched.

Date:	1990
Observers:	Blair, et al.
Site:	CSIRO
Instrument Size (m):	64
Search Frequency (MHz):	$4,462 \pm 25$ kHz
Frequency Resolution (Hz):	100
Objects:	100 solar-type stars visible only from southern hemisphere
Flux Limits (W/m^2):	2 Jy
Total Hours:	60
Reference:	[Bla91]
Comments:	This was a 'magic frequency' search at π times the hydrogen line frequency (1,420 MHz). 1,280 channels were arranged to sample three reference frames: solar barycenter, stellar barycenter and geocenter.

Date:	1990 through 1999
Observers:	Gray
Site:	Small SETI Radiotelescope
Instrument Size (m):	4
Search Frequency (MHz):	1,419.5 to 1,420.5
Frequency Resolution (Hz):	100 to 1
Objects:	n/a
Flux Limits (W/m^2):	10^{-21}
Total Hours:	Ongoing ∼6 hours/night
Reference:	[Gra82], [Gra86]
Comments:	This was a dedicated meridian transit search system, constructed by amateurs, with automated operation at night. By 1995, the system had 8,192 channels, each 1Hz wide.

Date:	1990 through 1994
Observers:	Betz
Site:	Mt. Wilson
Instrument Size (m):	1.65 m element of the Townes infrared interferometer
Search Frequency (MHz):	3×10^9 (10 μm)
Frequency Resolution (Hz):	3.5 MHz (35 m/s)
Objects:	100 nearby solar-type stars
Flux Limits (W/m^2):	1 MW transmitter out to 20 pc
Total Hours:	Ongoing
Reference:	[Bet93]
Comments:	This was a search for infrared beacons at the CO_2 laser frequency, using a narrow-band acousto-optical spectrometer.

Date:	1990
Observers:	Gray
Site:	Oak Ridge (Harvard University)
Instrument Size (m):	26
Search Frequency (MHz):	1,420.4 MHz relative to galactic standard of rest ± 200 kHz
Frequency Resolution (Hz):	0.05
Objects:	M31 and M33
Flux Limits (W/m^2):	n/a
Total Hours:	50
Reference:	Unpublished
Comments:	This study used long integrations (5 hours per HPBW) on 10^{11} stars at once, looking for low duty cycle signals.

Date:	1990 through 1998
Observers:	Lemarchand, "META II"
Site:	IAR
Instrument Size (m):	30 (one of two)
Search Frequency (MHz):	1,420.4, 1,667, 3,300
Frequency Resolution (Hz):	0.05
Objects:	Sky survey of southern skies and 90 target stars, and OH masers
Flux Limits (W/m^2):	1×10^{-23} to 7×10^{-25}‡
Total Hours:	Ongoing
Reference:	[Com95]
Comments:	This was a search for signals that have been Doppler compensated to the rest frame of the solar system barycenter, the galactic center, or cosmic ray background. The instrument was a duplicate of the META system, built by Argentinian engineers under the guidance of Professor Horowitz at Harvard, and financed by the Planetary Society. Simultaneous observations with META are carried out over the declination range $-10°$ to $-30°$. Major upgrades were made in 1996 to permit long integration times, and switching between antennas. OH masers were searched, looking for amplified signals.

Date:	1992 through 1993
Observers:	NASA "HRMS" targeted search
Site:	Arecibo
Instrument Size (m):	305
Search Frequency (MHz):	1,300 to 2,400
Frequency Resolution (Hz):	1, 7, 28 simultaneously, dual circular polarization
Objects:	25 stars
Flux Limits (W/m^2):	5×10^{-24}
Total Hours:	200
Reference:	[Tar95]
Comments:	No ETI technology was detected.

Date:	1992 through 1993
Observers:	NASA "HRMS" sky survey
Site:	Goldstone, CA
Instrument Size (m):	26, 34
Search Frequency (MHz):	1,700 and 8,300 to 8,700
Frequency Resolution (Hz):	19
Objects:	72 sky frames at X-band, and 130 sky frames at L-band, all located in three regions of the galactic plane, were repeatedly mapped in sequential circular polarization.
Flux Limits (W/m^2):	9.8×10^{-23}
Total Hours:	1,386
Reference:	[Lev95]
Comments:	No ETI technology was detected.

Figure L.2: The 1000 foot diameter Arecibo radio telescope in Puerto Rico.

Date:	1992 through 1998
Observers:	Bowyer, Werthimer & Donnelly, "SERENDIP III"
Site:	Arecibo
Instrument Size (m):	305
Search Frequency (MHz):	424 to 436
Frequency Resolution (Hz):	0.6
Objects:	Survey of 30% of the sky visible from Arecibo
Flux Limits (W/m^2):	5×10^{-25}
Total Hours:	Ongoing
Reference:	[Don95]
Comments:	4 million channels were observed. This was a commensal search occurring at twice the sidereal rate, in the backwards direction, while radio astronomers track targets using the feeds and receivers on carriage-house 1.

Date:	1993
Observers:	Steffes & DeBoer
Site:	NRAO/Tucson
Instrument Size (m):	12
Search Frequency (MHz):	203×10^3
Frequency Resolution (Hz):	32
Objects:	40 stars, and 3 locations near galactic center
Flux Limits (W/m^2):	2.3×10^{-19}
Total Hours:	25
Reference:	[Ste94]
Comments:	No artificial signals were detected near the positronium line.

Date:	1993
Observers:	Jugaku, Noguch, & Nishimura
Site:	Infrared Telescope of Institute of Space and Astronautical Science, Japan. Infrared Telescope at Xinglong Station of the Beijing Astronomical Observatory
Instrument Size (m):	1.3 and 1.26
Search Frequency (MHz):	K band 6.6×10^8 MHz (2.2 μm) and IRAS 3.6×10^9 (12 μm)
Frequency Resolution (Hz):	n/a
Objects:	180 solar-type stars
Flux Limits (W/m^2):	n/a
Total Hours:	Not available.
Reference:	[Jug95]
Comments:	This was a search for 12 μm excess radiation from IRAS catalog stars, using the K-[12] color index in an attempt to find Dyson spheres.

Date:	1993 through 1995
Observers:	Lemarchand et al., "META II Target Search"
Site:	Instituto Argentino de Radioastronomia (IAR)
Instrument Size (m):	30
Search Frequency (MHz):	1,420
Frequency Resolution (Hz):	.05
Objects:	Targeted search $< 8 - 10°$
Flux Limits (W/m^2):	10^{-24}
Total Hours:	290
Reference:	[Lem96]
Comments:	The search included all stars less than 16.3 ly from the sun, and solar-type stars at distances between 16.3 and 50 ly. A total of 80 stars were searched.

Date:	1994
Observers:	Mauersberger, Wilson, Rood, Bania, Hein & Linhart
Site:	IRAM/PICO Veleta
Instrument Size (m):	30
Search Frequency (MHz):	203,000
Frequency Resolution (Hz):	1×10^6 and 9.7×10^3
Objects:	16 stars and galactic center
Flux Limits (W/m^2):	$0.2 - 20 \times 10^{15}$ W EIRP
Total Hours:	~5
Reference:	[Mar96]
Comments:	This was a search at the positronium line, towards nearby stars and stars with infrared excess that might be Dyson spheres.

Date:	1995
Observers:	Gray
Site:	NRAO/VLA
Instrument Size (m):	27-element array of 26 m antennas
Search Frequency (MHz):	1,420
Frequency Resolution (Hz):	6,104 (B 781 kHz) and 381 (B=195 kHz)
Objects:	OSU "WOW" locale.
Flux Limits (W/m^2):	10 and 100 mJy/beam/channel
Total Hours:	4
Reference:	[Gra01]
Comments:	This was a search of the OSU "WOW" locale with 4 arc second synthesized beam.

Date:	1995 through 1998
Observers:	Kingsley
Site:	Columbus Optical SETI Observatory, Ohio
Instrument Size (m):	0.25
Search Frequency (MHz):	1.65×10^8 $(0.55\mu m)$
Frequency Resolution (Hz):	n/a
Objects:	Nearby solar-type stars
Flux Limits (W/m^2):	Transmitters with peak instantaneous power $> 10^{18}$ W
Total Hours:	Not available
Reference:	[Kin96b]
Comments:	This was a broadband optical search for short pulses ($\sim$1 ns) that instantaneously outshine the host star.

Date:	1995
Observers:	SETI Institute Project Phoenix
Site:	ATNF/PARKES and MOPRA
Instrument Size (m):	64 and 22
Search Frequency (MHz):	1,200 to 3,000
Frequency Resolution (Hz):	1
Objects:	206 stars, 1,200 – 1,750 MHz 105 stars 1,750 – 3,000 MHz
Flux Limits (W/m^2):	1.32×10^{-25} for half of the 1,750 to 3,000 MHz observations, and 1.82×10^{-25} for the other half
Total Hours:	2,600
Reference:	[Bac98], [Dre98]
Comments:	This search included an immediate two-site, pseudo-interferometric follow-up of observations of candidate signals. The targets were solar-type stars visible only from the Southern Hemisphere. No ETI technology was detected.

Figure L.3: The Parkes 64 m radio telescope in Australia, where Project Phoenix began observations in 1995.

Date:	1995
Observers:	te Lintel-Hekkert and Tarter, Phoenix Co-operative Science
Site:	ATNF/Parkes and MOPRA
Instrument Size (m):	64 and 22
Search Frequency (MHz):	1,200 to 3,000
Frequency Resolution (Hz):	1
Objects:	4 potential Dyson spheres
Flux Limits (W/m²):	1.9×10^{-25}
Total Hours:	48
Reference:	n/a
Comments:	Possible candidate Dyson spheres were selected using the following criteria: IRAS PSC sources with temperatures from 300 to 500 K, that were not identified in OH/IR or CO(1-0) surveys, and had galactic latitudes $> 5°$.

Date:	1995
Observers:	Shostak, Ekers, & Vaile, Phoenix Cooperative Science
Site:	ATNF/Parkes
Instrument Size (m):	64
Search Frequency (MHz):	1,200 to 1,750
Frequency Resolution (Hz):	1
Objects:	3 fields in the Small Magellanic Cloud (SMC)
Flux Limits (W/m^2):	1.9×10^{-25}
Total Hours:	24
Reference:	[Sho96]
Comments:	This was a search of over 10^7 stars, contained within the three fields of the SMC. The limit on detectable transmitters was 1.5×10^{18} W EIRP.

Date:	1995
Observers:	Sullivan, Wellington, Shostak, Backus, & Cordes, Phoenix Cooperative Science
Site:	ATNF/Parkes
Instrument Size (m):	64
Search Frequency (MHz):	$1,420 \pm 5$
Frequency Resolution (Hz):	1
Objects:	Galactic center and a 5° high strip $\pm15°$ longitude along the galactic plane
Flux Limits (W/m^2):	$1.5 - 10 \times 10^{-25}$
Total Hours:	48
Reference:	[Sul97]
Comments:	Multiple 30 s observations of a strip along the galactic plane and the galactic center were made in a search for repetitive signals.

Date:	1995
Observers:	Norris, Phoenix Cooperative Science
Site:	ATNF/Parkes and MOPRA
Instrument Size (m):	64 and 22
Search Frequency (MHz):	1,200 to 3,000
Frequency Resolution (Hz):	1
Objects:	Galactic center
Flux Limits (W/m^2):	1.3×10^{-25}
Total Hours:	24
Reference:	n/a
Comments:	The galactic center was searched for beacons.

Date:	1995
Observers:	Zadnik et al., Phoenix Cooperative Science
Site:	ATNF/Parkes
Instrument Size (m):	64
Search Frequency (MHz):	4,462, 4,532, 8,295, 8,393, and 8,666
Frequency Resolution (Hz):	1
Objects:	49 stars closer than 11.5 pc
Flux Limits (W/m^2):	3.5 and 5.0 Jy
Total Hours:	48
Reference:	[Zad96]
Comments:	This was a "Magic Frequency" search at $\pi \times HI$, $e \times HI$, $2\pi \times HI$, $e \times (OH + HI)$,$^3 H_e$.

Date:	1995 through 1998
Observers:	Brown, Klein & Dixon
Site:	OSURO
Instrument Size (m):	53
Search Frequency (MHz):	$1,423 \pm 1.25$
Frequency Resolution (Hz):	0.6
Objects:	All sky search
Flux Limits (W/m^2):	Unknown as yet
Total Hours:	5,000
Reference:	n/a
Comments:	This search used the SERENDIP processor to search declinations from $+8°$ to $-36°$. The program ended when OSU sold the observatory site to a golf course developer.

Date:	1995 through 1999
Observers:	Horowitz et al., BETA
Site:	Oak Ridge Observatory
Instrument Size (m):	26
Search Frequency (MHz):	1,400 to 1,720
Frequency Resolution (Hz):	0.5
3Objects:	Sky survey from $-30°$ to $+60°$ declination
Flux Limits (W/m^2):	2.2×10^{-22}
Total Hours:	Suspended in spring 1999
Reference:	[Lei00]
Comments:	This was a 'waterhole' search, using dual-beams and an omniantenna to discriminate against RFI. Project BETA is a follow-up to META. The project was interrupted when wind blew the antenna off its mount. Repairs are under way.

Date:	1996
Observers:	Biraud & Airieau
Site:	Nançay Observatory
Instrument Size (m):	40×240
Search Frequency (MHz):	$1,420 \pm 0.3$ and $1,660 \pm 2.2$
Frequency Resolution (Hz):	50
Objects:	4 stars with giant planets (51 Peg, 47 Uma, 70 Vir, Gl 229)
Flux Limits (W/m^2):	1×10^{-24}
Total Hours:	40 Hours
Reference:	n/a
Comments:	This was a search of newly discovered extrasolar planetary systems, using the observing protocol from the Biraud and Tarter 1981 to 1988 SETI program at Nançay.

Date:	1995 through present
Observers:	SETI League Project Argus
Site:	Multiple sites world-wide (about 100 sites in 2001)
Instrument Size (m):	$\sim 3 - 10$ (satellite TV dishes)
Search Frequency (MHz):	1,420 to 1,720
Frequency Resolution (Hz):	1
Objects:	All sky
Flux Limits (W/m^2):	$\sim 1 \times 10^{-21}$ (varies)
Total Hours:	Ongoing
Reference:	[Slpa]
Comments:	This implements a plan to organize up to 5,000 radio amateurs to provide continuous sky coverage for strong, transient signals, using systems that can be bought and built by individuals. The SETI League currently has 1,284 members, running 112 stations in 20 countries.

Date:	1996 through 1998
Observers:	Tilgner, Heinrichsen, Kruger, Pacher, Walker, & Wolstencroft
Site:	ISO (Infrared Space Observatory) Satellite, Isophot Photopolarimeter
Instrument Size (m):	0.6
Search Frequency (MHz):	9×10^8 (3μm) to 3×10^{10} (100μm)
Frequency Resolution (Hz):	6×10^{13} (0.2 μm, 3 μm filter band) 1.53×10^{16} (51 μm, 90 μm filter band)
Objects:	6 solar-type stars and 1 infrared excess target star
Flux Limits (W/m^2):	Approximately $30 - 90 \times 10^{-29}$ (S/N=10)
Total Hours:	1.3 hours (guaranteed schedule)
Reference:	[Til95]
Comments:	This was a search for astro-engineering products such as Dyson spheres and rings by separating their infrared spectra from that of their host star.

Date:	1996 through present
Observers:	Werthimer et al. (SERENDIP IV)
Site:	Arecibo
Instrument Size (m):	305
Search Frequency (MHz):	$1,420 \pm 50$
Frequency Resolution (Hz):	0.6
Objects:	Survey of 30% of sky visible from Arecibo
Flux Limits (W/m^2):	5×10^{-24}
Total Hours:	Ongoing
Reference:	[Wer96b]
Comments:	This is a commensal search, occurring at twice the sidereal rate in the backwards direction, while radio astronomers track targets using the Gregorian system.

Date:	1996
Observers:	Lockett, Blair & Zadnik
Site:	Perth Optical Observatory
Instrument Size (m):	1
Search Frequency (MHz):	5.24×10^8 (572 nm)
Frequency Resolution (Hz):	n/a
Objects:	24 nearby stars
Flux Limits (W/m^2):	n/a
Total Hours:	not available.
Reference:	n/a
Comments:	This was a search for microsecond laser pulses.

Date:	1996 through 1998
Observers:	SETI Institute Project Phoenix
Site:	NRAO and Woodbury, Ga
Instrument Size (m):	43 and 30
Search Frequency (MHz):	1,200 to 3,000, dual polarization
Frequency Resolution (Hz):	1
Objects:	195 stars
Flux Limits (W/m^2):	1.3×10^{-25}
Total Hours:	4,200
Reference:	[Cul00]
Comments:	Longer observations provided the same sensitivity as those achieved in Australia.

Date:	1997(A), 1999 (B)
Observers:	BAMBI (Bob And Mike's Big Investment), and SARA (Society of Amateur Radio Astronomers) members
Site:	(A) in California, (B) in Colorado
Instrument Size (m):	2.6 (A); 3 (B)
Search Frequency (MHz):	3,700 to 4,200
Frequency Resolution (Hz):	0.6
Objects:	Northern sky survey
Flux Limits (W/m^2):	No formal observing program has yet begun.
Total Hours:	Ongoing
Reference:	[Lam00]
Comments:	Amateur radio enthusiasts are using TVRO components and software FFTs to attempt a coordinated search.

Date:	1998 through 1999
Observers:	Gray & Ellingsen
Site:	Hobart (University of Tasmania)
Instrument Size (m):	26
Search Frequency (MHz):	1420.4
Frequency Resolution (Hz):	4880
Objects:	WOW locale
Flux Limits (W/m^2):	3 Jy
Total Hours:	100
Reference:	Astrophysical Journal (submitted, 2001).
Comments:	This search involved extended (14 h) observations with 30 s integrations, and looked for intermittent or periodic signals.

Date:	1998 through present
Observers:	SETI Institute Project Phoenix
Site:	Arecibo Observatory and Lovell Telescope at Jodrell Bank
Instrument Size (m):	305 and 76
Search Frequency (MHz):	1,200 to 3,000 dual polarizations
Frequency Resolution (Hz):	1
Objects:	1,000 nearby stars are Arecibo Observatory targets (500 observed to date)
Flux Limits (W/m^2):	1×10^{-26}
Total Hours:	1,300 hours to date
Reference:	[Sho99]
Comments:	This is a continuation of the NASA HRMS targeted search of 1,000 nearby stars, using real time data reduction and a pair of widely separated observatories to help discriminate against RFI.

Date:	1998 through present
Observers:	SETI Australia Southern SERENDIP
Site:	Parkes
Instrument Size (m):	64
Search Frequency (MHz):	$1,420.405 \pm 8.82$
Frequency Resolution (Hz):	.6
Objects:	Southern sky survey
Flux Limits (W/m^2):	4×10^{-24}
Total Hours:	Ongoing
Reference:	[Sto00]
Comments:	This is a comensal search that uses 2 out of 13 beams of the Parkes focal plane array to discriminate against RFI.

Date:	1998 through present
Observers:	Werthimer
Site:	Leuschner Observatory
Instrument Size (m):	0.8
Search Frequency (MHz):	300 to 650 nm
Frequency Resolution (Hz):	None
Objects:	800 solar-type stars
Flux Limits (W/m^2):	1.5×10^{-9} peak during 1 ns pulse, or 1.5×10^{-20} average per 100 second observation
Total Hours:	200 (ongoing)
Reference:	[Las94]
Comments:	This is the first optical search to use two high time-resolution photomultiplier tubes as coincidence detectors to look for nanosecond pulses.

Date:	1998 through present
Observers:	Horowitz et al. (Harvard Optical SETI)
Site:	Oak Ridge Observatory
Instrument Size (m):	1.5
Search Frequency (MHz):	350 to 700 nm
Frequency Resolution (Hz):	None
Objects:	13,000 solar-type stars of which 4,000 observed to date
Flux Limits (W/m^2):	4×10^{-9} peak in < 5 ns pulse, or 4×10^{-20} average per 500 second observation
Total Hours:	Ongoing
Reference:	[How00]
Comments:	This is a search for nanosecond laser pulses using hybrid avalanche photo diodes in coincidence. The search is commensal on nightly searches for extrasolar planets. It will soon be operated in coincidence with a cloned detector on the 0.9 m telescope at Princeton.

Date:	1998 through present
Observers:	Marcy, Reines, Butler, & Vogt
Site:	Lick, Keck
Instrument Size (m):	10
Search Frequency (MHz):	400 to 500 nm
Frequency Resolution (Hz):	Resolving power = 50,000
Objects:	600 FGK stars within 100 pc
Flux Limits (W/m^2):	1×10-13
Total Hours:	500
Reference:	[But96]
Comments:	This is a search through archival data for narrowband continuous optical laser emission lines.

Date:	1999 through present
Observers:	Werthimer & Anderson, (SETI@home)
Site:	Arecibo
Instrument Size (m):	305
Search Frequency (MHz):	$1,420.405 \pm 1.25$
Frequency Resolution (Hz):	0.6
Objects:	Data taken from SERENDIP IV through sky visible from Arecibo
Flux Limits (W/m^2):	5×10^{-25}
Total Hours:	Ongoing
Reference:	[And00]
Comments:	This is an extremely successful experiment in distributed computing. It permits more sophisticated processing of a fraction of SERENDIP IV data by harnessing the idle CPU cycles of 3 million personal and corporate computers via the Internet.

Date:	2000 through present
Observers:	Montebugnoli (SETI Italia)
Site:	Medicina
Instrument Size (m):	32
Search Frequency (MHz):	1,415 to 1,425 and 4,255 to 4,265
Frequency Resolution (Hz):	0.6
Objects:	Northern Sky
Flux Limits (W/m^2):	n/a
Total Hours:	Ongoing
Reference:	[Mon00], [Mon01]
Comments:	This is a commensal sky survey that uses the Medicina telescope and SERENDIP signal processing boards.

Date:	2000 through present
Observers:	Bhathal & Darcy
Site:	Campbelltown Rotary Observatory, Australia and Bungonia Optical SETI, Australia
Instrument Size (m):	0.4 and 0.3
Search Frequency (MHz):	550 nm
Frequency Resolution (Hz):	None
Objects:	200 solar-type stars and 25 globular clusters
Flux Limits (W/m^2):	6×10^{-9} peak during 1 ns pulse
Total Hours:	Ongoing
Reference:	[Bha00]
Comments:	This is a dedicated telescope built for SETI. It uses high time resolution photodiodes in coincidence to search for laser pulses, and uses coincidence between two telescopes separated by 20 m to reduce the false alarm rate. It will soon be teamed with a microwave search of the same objects.

Date:	2001 through present
Observers:	Drake et al., Lick Optical SETI
Site:	Lick Observatory, UC Santa Cruz
Instrument Size (m):	1 (Nickel Reflector)
Search Frequency (MHz):	550 nm
Frequency Resolution (Hz):	None
Objects:	5,039 solar-type stars (planned)
Flux Limits (W/m^2):	1×10^{-9} peak during 1 ns pulse, or 1×10^{-20} average per 100 s observation
Total Hours:	Ongoing, approximately 1 night per week
Reference:	[Dra01]
Comments:	This search uses three photodiodes in coincidence to eliminate background events.

Appendix M

Pulse Broadening in the Radio, Infrared and Optical

This appendix contains a paper prepared for the second SSTWG meeting on February 17 through 19, 1998, written by James M. Cordes of the Center for Radiophysics and Space Research, Cornell University, Ithaca, NY 14853.

M.1   Abstract

Pulse broadening from interstellar electron-density fluctuations, on scales less than 1 AU, is well known from pulsar studies. It ranges from ~ 1 μs to 1 s at 1 GHz (depending on source distance), and scales as ν^{-4}. In low-frequency surveys for pulsations, such broadening causes distant pulsars to be selected against. At infrared/optical wavelengths, plasma scattering is totally negligible. However, scattering off grains causes pulse smearing that is much larger than that seen in the radio/plasma case, because grain scattering is typically through a wide angle. For a scattering optical depth τ_s, a fraction $e^{-\tau_s}$ of the pulse arrives unscattered while a fraction $1 - e^{-\tau_s}$ is broadened in time. The implications for propagation of narrow, artificial pulses are highly wavelength and direction dependent. Roughly, optical transmission of pulses is not possible for very long path lengths through the Galaxy, i.e., more than a few kpc. At wavelengths longer than the near-infrared, i.e, for $\lambda > 2\mu$m, galactic-scale path lengths with minimal or modest loss of amplitude from pulse broadening are possible.

M.2   Summary of Radio Wave Scattering

Radio scattering is caused by refraction and diffraction from electron density irregularities in the interstellar medium (ISM). Relevant length scales are 100 km to 10 AU but there is evidence that the wavenumber spectrum for irregularities extends all the way to 1 to 10 pc with a Kolmogorov-like shape. Observable phenomena include:

- intensity scintillations;
- pulse broadening;
- angular broadening (interstellar 'seeing');
- time-of-arrival variations.

These and other effects are described further in Cordes and Lazio [Cor91a], Cordes and Lazio [Cor93], and Cordes, Lazio and Sagan [Cor97b]. Pulse broadening arises from multipath propagation and is an important selection effect in radio surveys for pulsars, which rely on detection of pulse periodicity. Pulse smearing by more than a pulse width degrades survey sensitivity in a highly frequency and distance dependent way. Figure M.1 shows the exponential broadening tail on the distant pulsar, B1849+00. This is the most heavily scattered pulsar known, with a pulse broadening time of ~ 200 ms. However, any pulsars in the galactic center, viewed through scattering regions equivalent to those that angularly broaden Sgr A* to ~ 0.7 arc second at 1.4 GHz, will show pulse broadening of *minutes* at the same frequency.

In the radio case, pulse broadening corresponds to small-angle scattering. The scattering is not accompanied by any absorption; except at low frequencies where free-free absorption may become important. Letting $\eta(s)$ be the mean-square scattering angle per unit length, the mean pulse delay is:

$$t_d = \frac{1}{2c} \int_o^D ds \; \eta(s) \left(\frac{s}{D}\right) \left(1 - \frac{s}{D}\right) \qquad (M.1)$$

It is common practice to define the wavenumber spectrum as a power law with coefficient C_n^2:

$$P_{\delta n_e}(q; s) = C_n^2(s) q^{-\alpha}, \qquad q_0 \le q \le q_1 \qquad (M.2)$$

where q_0, q_1 are wavenumber cutoffs. When C_n^2 varies on scales much longer than q_0^{-1}, and for $q_1 \gg q_0$ and $\alpha < 4$, we have:

$$\eta(s) = \left(\frac{\lambda^4 r_e^2}{4 - \alpha}\right) q_1^{4-\alpha} C_n^2(s) \qquad (M.3)$$

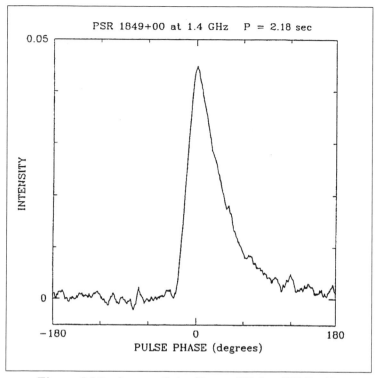

Figure M.1: Average Pulse Shape of Pulsar B1849+00.

For the local ISM, $C_n^2 \approx 10^{-3.5}$ m$^{-20/3}$, and the inner scale $l_1 = 2\pi q_1^{-1} \approx 10^2$ to 10^3 km (Spangler and Gwinn [Spa90], Moran, et al. [Mra90], Wilkinson, et al. [Wlk94]), yielding for $\alpha = 11/3$:

$$\eta \approx 3.2 \text{ mas}^2 \text{ kpc}^{-1} \left(\frac{\nu}{1 \text{ GHz}} \right)^{-4} \left(\frac{l_1}{100 \text{ km}} \right)^{-1/3} \left(\frac{C_n^2}{10^{-3.5} \text{ m}^{-20/3}} \right) \quad \text{(M.4)}$$

Figure M.2 on page 430, shows the pulse broadening time at 1 GHz for pulsars plotted against dispersion measure (DM), where $DM = \int_0^D ds\, n_e(s)$. The pulse broadening scales as $\nu^{-4\pm0.4}$ and is due to multipath propagation. The solid curve is a log-log parabolic fit, and the dashed lines denote $\pm1.5\sigma$ from this fit.

M.3 Scattering from Grains

Interstellar grains show a wide range of sizes, compositions, and diversity in dielectric properties. Interstellar extinction is a combination of scattering and

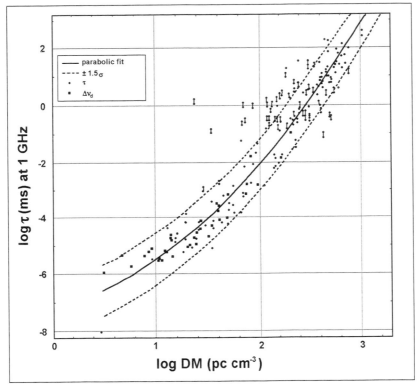

Figure M.2: Pulse Broadening Time for Pulsars.

absorption. In general, scattering can dominate absorption at a given wavelength, or vice versa. However, in our discussion we are concerned with cases where roughly equal contributions are made to the total extinction. When a photon is scattered by a grain with a size comparable to, or smaller than its wavelength, scattering is through a wide angle and has a cross section that scales with grain size (a) as a^6. Grain sizes cover a range from 0.03 μm to ~ 1 μm, and are described roughly as a power law (the MRN distribution):

$$\frac{dN}{da} \propto a^{-4\pm0.5} \qquad\qquad (\text{M.5})$$

Evidence for small grains ($a \lesssim 0.1$ μm) comes from observations of X-ray halos of size 1 to 10 arc minutes around binary X-ray sources and supernova remnants (e.g., Alcock and Hatchett [Alc78], Catura [Cat83], Molnar and Mauche [Mol86], Mauche and Gorenstein [Mau86], Mauche and Gorenstein [Mau89], and Woo et al. [Woo94]). The X-ray halos display an increasing amount of the total flux from sources of increasing distance, just as expected from grain scattering.

The corresponding pulse broadening is:

$$\Delta t \approx 4 \text{ days} \left(\frac{E}{2.3 \text{ keV}}\right)^{-2} \left(\frac{D}{10 \text{ kpc}}\right) \left(\frac{\langle a \rangle}{0.1 \ \mu\text{m}}\right)^{-2} \qquad \text{(M.6)}$$

At wavelengths between the two cutoffs of the overall grain size distribution, the general form for the angular scattering distribution from a point source consists of three components:

1. an unscattered component;

2. a component produced by forward scattering from large grains;

3. a diffuse component due to wide-angle scattering.

The general three-component form for the pulse broadening function is the same as that for the angular distribution, with characteristic timescales for the components given by equation M.1 on page 428 or its analog, taking into account large angle scattering.

For a narrow pulse (e.g., $\ll 1$ s) propagating through the Galaxy to be detectable, there must be a significant, unscattered component. *Any* scattering will be by a wide enough angle so that the corresponding pulse broadening is far too large. For reference, a scattering diameter of 1 arc second, for a source at 10 kpc distance, viewed through a medium with uniform grain density, corresponds to a pulse broadening of:

$$\Delta t = \frac{D\theta_{FWHM}^{2}}{16 ln(2) c} = 2.2 \ s \qquad \text{(M.7)}$$

Thus, the scattering of a photon will increase its travel time by many pulse widths of an infrared/optical pulsar or ETI source.

Therefore, the scattering optical depth must be small, $\tau_s \ll 1$, in order that the fractional unscattered component with amplitude $e^{-\tau_s}$ be sizable. Absorption also reduces the amplitude; so the net amplitude of the unscattered component is $e^{-\tau}$, where τ is the total optical depth.

M.4 Detectability Distances

A few useful numbers can be used to characterize the amount of scattering from grains. Locally, visual extinction is roughly 2 magnitudes, while towards the galactic center, it is about 25 magnitudes (Becklin & Neugebauer [Bck68]).

In the infrared at 2.2 μm, there are 3 magnitudes of extinction (Becklin et al. [Bck78]). This implies that narrow pulses will be reduced to 1/10 their value in the absence of extinction at a distance of 1.3 kpc (at low galactic latitudes) in the visual band. At 2.2 μm, the amplitude of pulses from objects at the galactic center are reduced to 6% of their unscattered values.

The conclusion is that optical pulses in the visual band may be detectable only a few kpc away. This requires the mean distance between civilizations to be of this order (or less), corresponding to $\lesssim$ 200 civilizations in the Galaxy. Infrared pulses, being detectable to 10 kpc, say, require only a few civilizations in the Galaxy in order that one civilization may see pulses from another.

M.5   Further References

Alcock, C., & Hatchett, S., 1978, "The Effects Of Small-Angle Scattering on a Pulse of Radiation with an Application of X-Ray Bursts and Interstellar Dust", *ApJ*, 222.

Becklin, E. E., & Neugebauer, G., 1968, "Infrared Observations of the Galactic Center", *ApJ*, 151, 145.

Becklin, E. E., Neugebauer, G., Willner, S. P., & Matthews, K, 1978, "Infrared Observations of the Galactic Center IV - The Interstellar Extinction", *ApJ*, 220, 831.

Catura, R. C., 1983, "Evidence for X-ray Scattering by Interstellar Dust", *ApJ*, 275, 645.

Cordes, J. M., & Lazio, T. J. W. L., 1991, "Interstellar Scattering Effects on the Detection of Narrow-Band Signals", *ApJ*, 376, 123.

Cordes, J. M., & Lazio, T. J. W. L., 1993, "Interstellar Scintillation and SETI", *Third Decennial US-USSR Conference on SETI*, ed. G. S. Shostak, San Francisco: ASP, 47, 143.

Cordes, J. M., & Lazio, T. J. W. L., 1997, "Finding Radio Pulsars in and Beyond the Galactic Center", *ApJ*, 475, 557.

Cordes, J. M., Lazio, T. J. W. L., & Sagan, C., 1997, "Scintillation-induced Intermittency in SETI", *ApJ*, 487, 782.

Mauche, C. W., & Gorenstein, P., 1986, "Measurements of X-ray Scattering from Interstellar Grains", *ApJ*, 302, 371.

Mauche, C. W., & Gorenstein, P., 1989, "X-ray Halos Around Supernova Remnants", *ApJ*, 336, 843.

Molnar, L. A., & Mauche, C. W., 1986, "Effects of the X-ray Scattering Halo on the Observational Properties of Cygnus X-3", *ApJ*, 310, 343.

Molnar, L. A., Mutel, R. L., Reid, M. J., & Johnston, K. J., 1995, "Interstellar Scattering Toward Cygnus X-3: Measurements of Anisotropy and of the Inner Scale", *ApJ*, 438, 708.

Moran, J. M., Greene, B., Rodriguez, L. F., & Backer, D. C., 1990, "The Large Scattering Disk of NGC 6334B", *ApJ*, 348, 147.

Spangler, S. R. S., & Gwinn, C. R., 1990, "Evidence for an Inner Scale to the Density Turbulence in the Interstellar Medium", *ApJ*, 353, L29.

Wilkinson, P. N., Narayan, R., & Spencer, R. E., 1994, "The Scatter-broadened Image of CYGNUS-X-3", *MNRAS*, 269, 67.

Woo, J. W., Clark, G.W., et al., 1994, "ASCA Measurements of the Grain-scattered X-ray Halos of Eclipsing Massive X-ray Binaries: VELA X-1 and Centaurus X-3", *ApJ Letters*, 436, L5.

Appendix N

High Power Lasers

This appendix is a paper titled "An Update on High Power Lasers", prepared by William F. Krupke for the second SSTWG meeting from February 17 through 19, 1998.

N.1 Introduction

The field of high-power, short wavelength lasers has advanced rapidly during the past half decade, and has continued to accelerate primarily because of their application as laser weapons [Blo87]; in inertial confinement fusion (ICF) [Lnd95] and strong-field physics research [Tab94]; and in the production of inertial fusion central electric power (IFE) [Ort96]. These advanced laser systems, existing now or planned for the next decade, appear to offer many of the features (requirements) suitable for SETI application. Advanced optical technologies and techniques such as nonlinear phase conjugation [Roc88], chirp-pulse-amplification [Str85], and laser diode pumping [Fan88] are aggressively being developed to simultaneously extend the energy, peak power, average powers, and radiant intensities of these types of short wavelength laser systems. On the premise that "whatever we 'terrans' can do, ETIs can certainly do", it is useful to summarize the key performance parameters of these lasers.

N.2 CW Lasers

The Strategic Defense Initiative (SDI) [Blo87] of the 1980s invested heavily in the development of megawatt class, continuous wave, HF (2.7 μm) and DF (3.8 μm) chemical lasers [Dor86]. These lasers emit output beams of high quality, with the output power appearing in a half dozen ρ-vibrational transitions.

435

Power scaling rules are by now well understood, and megawatt class power levels have been demonstrated [Blo87].

More recently, the Air Force Philips Laboratory has extensively developed the Chemical Oxygen-Iodine Laser (COIL) that emits continuous wave laser radiation in a single electronic transition of atomic iodine at 1.315 μm [McD78]. The USAF is currently developing an Airborne Laser (ABL) for boost-phase theater missile defense. The ABL comprises a megawatt-class, CW COIL with high beam quality, and a 1.5 to 2 m beam director, installed on a 747 Boeing aircraft [For97]. In support of the ABL program, a several hundred kilowatt COIL was recently reported [Ful96].

These CW lasers easily meet and exceed the power, spectral purity, and brightness levels thought necessary to be useful for at least entry level SETI.

N.3   High Peak Power, Repetitively Pulsed Lasers

Several types of high peak power lasers have been developed for use in inertial confinement fusion research: the carbon dioxide molecular laser at 10 μm; the KrF excimer gas laser at 248 nm; and the 1 μm solid-state Nd:glass laser, with nonlinear harmonic wavelength conversion to 355 nm. In recent times, only the KrF and the harmonically converted Nd:glass laser have been pursued for ICF research because of the preference of ICF targets for shorter wavelength (near UV) irradiation. An advanced, repetitively-pulsed, solid-state laser for inertial fusion central electric power (IFE) is also currently being developed. A tabulation of some key performance characteristics of these laser systems is given in Table N.1, and described below. (The Yb:S-FAP Helios and KrF Sombrero IFE driver lasers could be built in the 2010 time frame, but at present there is no commitment to do so. Each of these systems would cost about 1 billion dollars to construct).

N.3.1   KrF Lasers

A high-energy (> 1 kJ), pulsed ICF KrF excimer laser is excited using an electron beam of microsecond duration. An output pulse of a few nanoseconds is extracted from the power amplifier using a time-angle, multiplexed, optical extraction configuration [Ewe79]. The Naval Research Laboratory has recently reported achieving 4 kJ of output energy, at 248 nm, with a pulse duration of $\sim$ 4 ns, for a peak power greater than a terawatt from the KrF NIKE system [Obe96]. The spectral width of the NIKE laser emission was $\sim$ 30 cm^{-1}, (compared to the KrF gain bandwidth of $\sim$ 200 cm^{-1}) for a spectral purity of > 1,000. The NIKE laser was not designed for repetitive pulsing, but a "multi-megajoule" KrF IFE driver can be designed at > 10 Hz with good beam

Characteristic	System					
Laser Name Date	Nova 1997	NIF (2004)	Mercury 1999	Helios (2015)	Nike 1997	Sombrero (2015)
Type	SSL	SSL	SSL	SSL	Gas	Gas
Gain Medium	Nd:glass	Nd: glass	Yb:S-FAP	YB:S-FAP	KrF	KrF
Pump	Lamp	Lamp	Diode	Diode	E-beam	E-beam
Pulse Energy	$\sim$ 0.1 MJ	$\sim$ 2 MJ	0.1 kJ	$\sim$ 2 MJ	$\sim$ 2 kJ	$\sim$ 2 MJ
Pulse Duration	$\sim$1ns	$\sim$1ns	$\sim$1ns	$\sim$1ns	$\sim$1ns	$\sim$1ns
Pulse Rep. Rate	0.001 Hz	0.001 Hz	10 Hz	10 Hz	0.01 Hz	10 Hz
Wavelength	353 nm	353 nm	1,047 nm	347 nm	248 nm	248 nm
Efficiency	0.1%	0.5%	10%	10%	$\sim$ 1.5%	$\sim$ 7%
Average Power	100 W	2,000 W	1,000 W	20 MW	20 W	20 MW

NOTE: The Nova and NIF are flashlamp pumped solid-state lasers (SSLs), the Mercury laser is a diode pumped SSL.

Table N.1: Characteristics of Some Current and Projected Lasers.

quality and relatively high efficiency (7%), by convectively flowing the active laser medium [Svi92].

N.3.2   Nd:glass Lasers

The harmonic wavelength converted Nd:glass laser has been the main type of laser used worldwide for ICF research. The highest pulsed energy achieved in an ICF research laser, the Nova laser [Sim83] at the Lawrence Livermore National Laboratory, is $\sim$ 80 kJ at 1.06 μm in a $\sim$ 1 ns pulse and $\sim$ 40 kJ at 0.35 μm. These energies are obtained from 10 parallel beam lines, each with an output aperture of 46 cm. Each of the output beams is about 20 times the diffraction limit of its 46 cm exit aperture, which is sufficient for the ICF experimental domain. The spectral gain bandwidth of the Nd:glass laser medium is $\sim$ 200 cm^{-1}. The gain medium saturates in a spectrally homogeneous manner and permits the generation of an output beam with a narrow spectral width ($\ll$cm^{-1}), or with a spectral width approaching the gain bandwidth (see Section N.3.4 on page 438).

The Nova laser was designed to operate as a 'single-shot' laser facility, being capable of being fired once every few hours. Encouraged by the recent progress in ICF research using the Nova and other Nd:glass lasers, the US Department Of Energy (DOE) is funding the development and construction of the National Ignition Facility (NIF) [Van97] at LLNL. This laser will deliver 1.8 MJ of 355 nm radiation to fusion targets at a nominal peak power of 500 TW, and a typical pulse duration of 3.5 ns. It is expected that the NIF will ignite a fusion target and yield an energy gain (fusion energy generated/laser drive energy) of 10 or more. Again, the NIF is designed as single shot laser facility, consistent with its intended research use.

N.3.3   Yb:S-FAP Lasers

In anticipation of the successful demonstration of ignition and modest gain at NIF, research has been in progress to go beyond NIF. This research is to develop an ICF laser that can be repetitively pulsed, at rates suitable for IFE central electric power production ($\sim$ 10 Hz), with adequately good beam quality (less than a few times the diffraction limit), adequate efficiency (>10%), and at an affordable cost.

One approach to this goal is based on advanced solid-state laser technologies developed at LLNL [Ort96]. To achieve an order of magnitude increase in the system efficiency to $\sim$ 10%, the flashlamps used in Nd:glass lasers are replaced by efficient, high-power, reliable semiconductor laser diode pump arrays. To achieve the needed beam quality under repetitive pulsing, active gas flow cooling of thin solid-state laser disks is utilized. To mitigate the laser cost issue, a solid-state gain medium with a relatively long energy storage lifetime has been identified and developed (Yb:S-FAP crystal [Pay94] replacing Nd:glass). These component technologies have all been demonstrated individually [Mas96]. They demonstrate the functional performance of an IFE, diode-pumped, solid-state laser (DPSSL). The LLNL has initiated an effort to design, construct, and operate a kilowatt average power, repetitive-pulsed (10 Hz), 10% efficient, IFE, demonstration laser system called Mercury [Mas98]. This laser has been conceived and designed in such a manner that it should be scalable to several megajoules, with several hundred terawatt peak power, at a 10 Hz repetition rate, with a beam quality a few times the diffraction limit, and be suitable for driving an IFE power plant. The laser is designed using a large number of parallel beam lines, each with an aperture sufficient to generate the needed irradiance at the ICF target. Each beam line is expected to have a 10 cm × 15 cm output aperture and generate about 10 kJ of 355 nm radiation. This projected IFE driver (Helios [Kru96]) could be constructed within the next 20 years, with proper development of the constituent technologies.

N.3.4   Petawatt Lasers

In the past few years, considerable progress has been made in reducing the pulse duration of solid-state lasers below a picosecond, and into the femtosecond (fs) (10^{-15} s) regime, using the technique of 'Chirp Pulse Amplification' [Fan88]. To obtain femtosecond pulses simultaneously with high energy, it was necessary to find some method to avoid optical damage in the solid-state laser amplifier. This would occur as the pulse fluence increased, since the damage fluence decreases with the square root of the pulse duration. To avoid this limitation, the chirp-pulse-amplification (CPA) [Fan88] approach was conceived and developed.

In CPA, a weak, time-bandwidth-limited master oscillator pulse is generated with a linear frequency chirp. This pulse is then time stretched to several nanoseconds using a pair of gratings, and sent through the laser power amplifier as usual; boosting the pulse energy without damage, while retaining the linear frequency chirp. The amplified pulse is then recompressed using a second grating pair. Using this technique, a single beam line of the Nova laser has produced ~ 500 J pulses with a peak power $> 1,000$ TW (> 1 PW) at a pulse duration of ~ 500 fs and with high beam quality [Per96]. It is capable of being focused to an irradiance of $> 10^{21}$ W/cm^2. In the future, it should be possible to combine the CPA technique with diode-pumping of crystalline gain media (such as Yb:S-FAP) to generate repetitively pulsed, petawatt peak powers, and tens of kilowatts of average power.

N.4   Coherent Aperture Combining

The maximum dimension of the output aperture of a laser is usually limited by the phenomena of transverse parasitic oscillation. The optical damage fluence then limits the amount of pulsed energy that can be extracted from that aperture. To increase the total amount of energy obtainable, multiple beam lines are used. Because the output from each beam line is usually not phase-locked to the others, each beam line can at best diverge at the diffraction angle permitted by the aperture of each *individual* beam line, not the total emitting aperture. To overcome this limitation, the technique of aperture phase locking using non-linear phase conjugation [Roc88] has been implemented in a diffraction-limited, four-aperture, repetitively pulsed Nd:glass laser system [Per96].

N.5   Further References

Bloembergen, N. , Patel, C. K. N. , et. al, 1987, "Report to The American Physical Society of the Study Group on Science and Technology of Directed Energy Weapons", APS Study Group, *Review of Modern Physics*, 59,3.

Dane, C. B., et al., 1997, "Diffraction-limited, High-average Power, Phase-locking of Four 30 J Beams from Discrete Nd:glass Zig-zag Amplifiers", *Post Deadline Paper CPD27*, 1997 Conference on Lasers and Electro-optics (CLEO).

Dornheim, M. A., 1986, *Aviation Week & Space Technology*, August 4, 33.

Ewing, J. J., et al., 1979, "Optical Pulse Compressor Systems for Laser Fusion", *IEEE J. Quantum Electronics*, 15, 368-379.

Fan, T. Y., & Byer, R. L., 1988, "Diode Laser-Pumped Solid-State Lasers" *IEEE J. Quantum Electronics*, 24, 895-912.

Forden, G. E., 1997, "The Airborne Laser", *IEEE Spectrum*, September, 40.

Fulghum, D. A., 1996, *Aviation Week & Space Technology*, November 18, 22.

Krupke, W. F., 1996, "Diode-Pumped Solid State Lasers for IFE", *2nd Annual International Conference on Solid State Lasers for Applications to ICF*, Commissariat a l' Energie Atomique, (CEA), Paris, France.

Lind, J., 1995, *Phys. Plasmas*, 2, 3933.

Marshall, C. D., et. al, 1996, "Diode-Pumped Gas-Cooled Slab Laser Performance", OSA Conference on Advanced Solid State Lasers, (ASSL), *TOPS*, 1, 208.

Marshall, C. D., et al., 1998, "Next-Generation Laser for Inertial Confinement Fusion", OSA Conference on Advanced Solid State Lasers, (ASSL), *TOPS*, to be published.

McDermott, W. D., et al., 1978, "An Electronic Transition Chemical Laser", *Applied Physics Letter*, 32, 469.

Obenschain, S. P. , et al., 1996, "The Nike KrF Laser Facility", *Phys. Plasmas*, 3, 2098-2107.

Orth, C. D., et al, 1996, *Nuclear Fusion*, 36, 75.

Payne, S. A., et al., 1994, "Ytterbium-doped Apatite-structure Crystals: A New Class of Laser Materials", *Journal of Applied Physics*, 76.1, 497-503.

Perry, M., et al., 1996, *Paper CW14, Conference on Laser and Electro-optics, (CLEO), Technical Digest*, 307, Optical Society of America, Anaheim, California.

Rockwell, D. A., 1988, "A Review of Phase-Conjugate Solid-State Lasers", *IEEE J. Quantum Electronics*, 24, 1124-1140.

Simmons W. W., & Godwin, R. O., 1983, *J. Nucl. Tech. Fusion*, 4, 3456.

Strickland, D., & Mourou, G., 1985, *Opt. Commun.*, 56, 219.

Sviatoslavsky, I. N., et al., 1992, *Fusion Technology*, 21, 1470.

Tabak, M, et al., 1994, "Ignition and High Gain with Ultrapowerful Lasers", *Physics of Plasmas*, 1, 1626-1634.

Van Wonterghem, B. M., et. al, 1997, "Performance of a Prototype for a Large-Aperture Multipass ND:glass Laser for Inertial Confienement Fusion" *Applied Optics*, 36, 4932-4953.

Appendix O

ETI Luminosity Functions

This appendix, by Michael Lampton, Seth Shostak and John W. Dreher, discusses the ETI luminosity function. To optimize a SETI search strategy, it is necessary to estimate the relative numbers of ETIs transmitting at different luminosities.

The luminosity function of ETI transmitters is unknown. However, the detection count versus observing sensitivity for a given luminosity function and spatial distribution is obtained through the following argument:

> Use the inverse square law to establish a minimum detectable luminosity at each distance, and use an assumed spatial source density function to integrate the number of detections. For spatially, uniformly distributed objects in a Euclidean space, the expected number of detections above a given threshold flux F varies as $F^{-1.5}$, independent of the luminosity function, but dependent on the uniformity assumption. Minimizing the threshold flux maximizes the sensitivity and pays big dividends.

O.1   Sky Survey versus Targeted Search

It is worth dwelling briefly on the question of whether 'tis nobler to observe for short periods and move the telescope frequently, or to suffer long integrations to beat down the noise.

For incoherent (square law) detection, the noise spectral power density is given by the well-known radio astronomy formula:

$$S_{min} = kT_s/(A\sqrt{\beta\tau}) \tag{O.1}$$

where A = the effective collecting area of the antenna, β = the bandwidth of the receiver (assumed to be narrower than the signal bandwidth), τ = the integration time, k = Boltzmann's constant, and T_s = the system temperature.

Consider an alien transmitter at distance r, broadcasting with an EIRP (isotropic power) density P_ν. Then the received flux density:

$$S = P_\nu/(4\pi r^2) \tag{O.2}$$

and the signal-to-noise S/N:

$$S/N = P_\nu A(\beta\tau)^{1/2}/(4\pi r^2 kT_s) \tag{O.3}$$

If all transmitters have power densities P_ν, and the minimum detectable $S/N = m$, then the volume of space searched at wavelength λ is:

$$V = A^{1/2}P_\nu^{3/2}(\beta\tau)^{3/4}\lambda^2/(4\pi kT_s m)^{3/2} \tag{O.4}$$

Clearly then, if for a given total telescope time, τ, a researcher opts to observe n separate fields, the integration time per field is τ/n, and the total volume examined is:

$$V_{tot} = nV_n = A^{1/2}P_\nu^{3/2}(\beta\tau)^{3/4}n^{1/4}\lambda^2/(4\pi kT_s m)^{3/2} \tag{O.5}$$

From this, we conclude that 1) the volume of space searched increases linearly with antenna diameter, and 2) breaking the available observing time into n separate observations increases the searched volume as $n^{1/4}$.

This simple analysis slightly favors sky surveys over targeted searches. However, short observations could drop below the sensitivity required to detect any transmitters at stellar distances, in which case the only volume effectively searched would be the presumably empty spaces closest to Earth. In addition, we have assumed that all volumes of space are equally likely to house ETI broadcasters; but it is a fundamental tenet of targeted searches that there are locales worthy of special scrutiny.

The types of signals sought by targeted searches and sky surveys differ. When scanning the sky, slowly-pulsed signals cannot be efficiently found, for only CW emissions will be sure to be 'on' when the search beam passes over their source. Nonetheless, because of the inefficiency of square law detection, pulses are easier to detect than CW signals for a given power expenditure, which is why the NASA Targeted Search was attuned to pulses. They were assumed to be regularly spaced in time, invariant in pulse shape, and like the CW signals also sought by the targeted search, possibly drifting in frequency.

The choice between sky survey and targeted search is just one of the many possible strategic choices for a receiver in search of signals from an advanced technology. The correct decision (i.e., the one with the highest probability of a successful detection) must be based upon both the luminosity function of ETI transmitters and their temporal characteristics. Since neither of these functions is known, our strategy must be more complex.

If transmitters are continuous in time, then a power law luminosity function for transmitters of power P, with $N(P) \sim P^{-\alpha}$ favors a sky survey for values of $\alpha < 3/2$ [Kar73]. On the other hand, if the duty cycle of transmitters is very low, perhaps because they are being cycled among many targets, then search strategies that dwell for long times on a limited set of targets have a greater probability of success than a sky survey which rapidly passes over any direction on the sky (provided that the target set of the transmitter and receiver overlap). Of course, in either case the transmitter and receiver must be using the same frequency.

The ideal search instrument would be one that covered the whole-sky, at all times, at all-frequencies, with very good sensitivity. Astronomers in the Former Soviet Union (FSU) attempted a low sensitivity, limited frequency approximation to this ideal with clusters of dipole elements distributed across the FSU, using temporal delays to serve as both a detection criterion and as an RFI discriminant [Tro75].

Later sections demonstrate that we are now much closer to being able to implement the ideal detector, i.e., one with full spatial, temporal, frequency, and sensitivity coverage, but its enormous computational demands push out its implementation to the far horizon of this book. In the interim, hybrid receiving strategies utilizing targets of opportunity in space and time may enhance the probability of success of any receiving instrument system. The *Project Cyclops* report gave little attention to such strategies. It is time to try to codify the possibilities.

If technological civilizations have a spatial distribution that is similar to the stellar mass in the Universe, then there will be aggregates in certain directions. This could improve the probability of detection but the farther away the aggregates, the stronger the transmitter needs to be before it is detectable. Consider a volume of space V large enough that it contains several communicating ETIs.

These may be ranked in order of decreasing signal power or luminosity L, and we may write the integral luminosity function $n(L)$ as a statistical limit:

$$n(L) = number\ of\ ETIs\ brighter\ than\ L \qquad (O.6)$$

How does this function depend on L? It is necessarily a downhill function; as L increases, $n(L)$ can only decrease, since middling output ETIs will drop out of the count. A common way to parameterize luminosity functions is to use a power law:

$$n(L) = AL^{-p} \qquad (O.7)$$

Here p is some exponent greater than zero which describes the steepness of the distribution in L. The coefficient A is also positive; it is a measure of the number of ETIs per unit volume.

What else might $n(L)$ depend on? If ETIs depend on matter and energy, they will be more numerous where matter is more concentrated, and less numerous where matter is less concentrated. We might well expect, in the statistical sense of equation O.6, a proportionality with mass density ρ. This suggests separation of the constant A in O.7 into two factors; A and ρ, and write instead:

$$n(L) = A\rho L^{-p} \qquad (O.8)$$

where now ρ is a suitably averaged density of whatever part of the observable Universe we want to explore and A now characterizes the ETI count per kilogram of matter. If equation O.8 correctly separates the dependence on luminosity and on location within the Universe, then the coefficient A can be expected to be a uniform but unknown constant, in the sense of the large-scale statistical averaging inherent in the concept of luminosity function.

Figure O.1 is a plot from the astronomical literature: the log of the mass enclosed by a sphere centered on Earth having radius R meters (Figure O.1, solid line 'All-Sky'). The range of interest for log R is 17 to 26. Log $R = 17$ corresponds to the distance to the nearest stars, within which the ETI count is zero, and the assumption in equation O.6 certainly fails. The distance log $R = 26$ corresponds to redshift $= 1$ beyond which the observed matter is younger than bioevolutionary timescales.

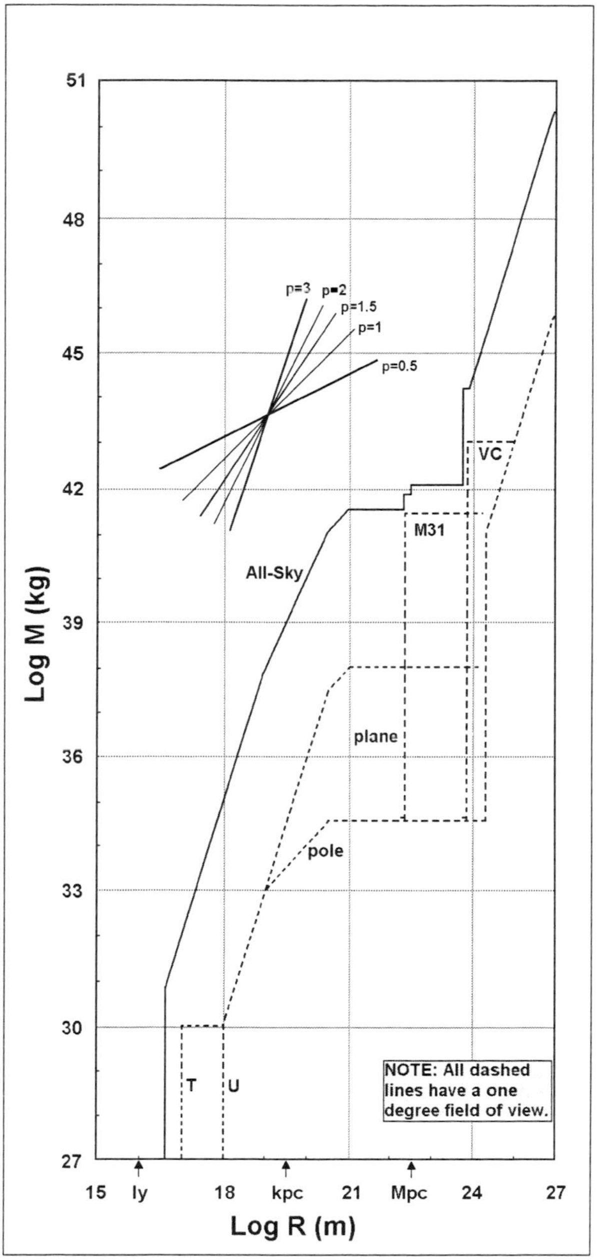

Figure O.1: Log of Mass Enclosed by Sphere.

All-Sky	the log of the mass enclosed by a sphere centered on Earth having radius R meters;
pole	average radio continuum towards the galactic poles;
1degFOV	1° field of view;
U	example of a targeted stellar search, with one nearby star plus background matter;
T	example of an untargeted search that has some field stars present but is otherwise undistinguished;
M31	local group of galaxies (Andromeda);
VC	Virgo Cluster of galaxies;
p	an exponent greater than zero, which describes the steepness of distribution in luminosity;

The lower portion of the solid curve has slope $+3$ (uniform volume density) out to the scale height of the Galaxy, followed by slope $+2$ (uniform disk density) out to beyond the galactic center, followed by slope $+1$ for the galactic halo, followed by a flattening to slope zero.

The all-sky curve continues upward with the local group of galaxies (Large Magellanic Cloud (LMC), Small Magellanic Cloud (SMC), M31, etc.) and has a major jump at the Virgo cluster of galaxies (VC) beyond which the curve joins a general $+3$ slope signifying a uniform volume density of matter.

The average mass density for matter out to distance R is:

$$\rho(R) = M(R)/(4\pi/3)R^3 \tag{O.9}$$

How many detections are implied in each distance zone? Having a mass model of the Universe in equation O.9 and a parameterized ETI model in equation O.8 on page 446 we can calculate how many ETIs are visible within any given radius. This step needs only two ingredients; the inverse square law for radiation, and some correction factor for transmitter directivity, duty cycles etc. For simplicity, pick isotropic continuous transmitters and assume that an ETI will be discovered if its flux at Earth exceeds some threshold S.

$$N_{detect} = N(flux > S) = \int_o^R 4\pi r^2 dr\, n(L > 4\pi r^2 S) \tag{O.10}$$

$$= \frac{\omega A}{(4\pi S)^p} \int \rho(r) r^{(3-2p)} \frac{dr}{r}$$

This model clearly exhibits the two parameters of the ETI luminosity function. Through the factor of S^{-p}, it shows how sensitivity governs our chances of success: If p is large, then improving S is very important, while if p is small then what we need is more patience not more sensitivity.

Figure O.1 on page 447 shows a few discovery lines: plots of how much mass it takes, at each radius, to deliver our first ETI. This calculation is straightforward. Set $N_{detect} = 1$, and solve for $M(R)$. The $M(R)$ curves are straight lines with slope $+2p$ in $[logM, logR]$ space. Examples are shown for $p = 0.5, 1, 1.5$, and 2.

Although the overall coefficient is unknown, it is clear that the smaller the value of p is, the greater the distance of first contact. For small p, strong transmitters are nearly as common as feeble ones and the inverse square law is more than made up for by the increased search volume. For $p > 1.5$, the prediction reverses, and nearby ETIs will be discovered first because the strong transmitters are so few.

Figure O.1 implies that: If first contact is from an all-sky survey, it must be either very nearby, perhaps right here in our own spiral arm, or exceedingly distant, in some moderate to high redshift object. It hinges on the power law index p exhibited by the ETI.

How about targeted searches? The sky is a lot 'lumpier' on small angular scales than the giant 'All-Sky' curve in Figure O.1. Neighboring lines-of-sight exhibit quite different amounts of matter and, therefore, have quite different probabilities of containing a detectable ETI signal. The calculations are not much different however. Let a line of sight occupy a solid angle ω. Then the number of detectable sources it contains is:

$$N_{detect} = \int_{R1}^{R2} \omega r^2 dr \; n(L > 4\pi r^2 S) \qquad (O.11)$$

With a small field of view, it is not appropriate to take a large-scale average for ρ. Instead, objects within the field break down into concentrations of matter: namely, discrete masses M_i at distances r_i. The smoothed out ρ must be evaluated based on these masses and on volume elements $\omega r^2 dr$. This leads to:

$$N_{detect} = \frac{A}{(4\pi S)^p} \sum M_i \; r_i^{-2p} \qquad (O.12)$$

Figure O.1 (dashed lines, 1degFOV one degree Field of View) shows some alternative mass versus radius curves for narrow beam work. All curves have the same angular size of one degree. The curve T is an example of a targeted stellar search: It includes one nearby star plus background matter. In contrast, U is an untargeted search that necessarily has some field stars present but is otherwise undistinguished. In directions toward the galactic pole, the mass curve levels off at a few thousand stars per square degree, until extragalactic objects begin to appear at some tens of megaparsecs. In directions, within the galactic plane, the mass curve continues to increase with a $+3$ slope until the far side of the Galaxy is reached; then it, too, flattens off.

In some directions, there are particularly large mass concentrations. The dashed curve M31 shows $M(R)$ for a line of sight towards the galaxy M31 and the dashed curve VC is the line of sight towards the Virgo cluster of galaxies. Because pointing our beam in these directions places large mass concentrations into the beam they are attractive targets for our search. Indeed, if the luminosity function slope p is of the order of 1.0, these objects contribute more detection probability to equation O.12 than does matter in our own Galaxy.

What is the expected value of the slope parameter p? Nature provides examples of both small-p and large-p phenomena. Supernovas of various kinds are steep (large p) and, indeed the historically discovered supernovas were all located in our home Galaxy. Gamma ray bursts, X-ray quasars, and other high-energy natural sources have flat p-values, often around 0.8, and, indeed, these objects are mostly found at cosmological distances.

At radio frequencies, our Galaxy is transparent (although not utterly quiet); and any observation of a low latitude target is simultaneously an extragalactic observation. Observations in the plane of our Galaxy clearly include a good deal more local matter than do high galactic latitude fields. Lines of sight that include very large amounts of matter are even more attractive. For example, a one degree field of view includes a significant fraction of the nearby spiral galaxy M31 in Andromeda. The cluster of galaxies in Virgo contains over 2,500 galaxies crowded into an angular region less than 10 degrees in extent.

At infrared/optical frequencies, observations within the plane of the Galaxy are short range due to extinction. For example in the optical, the amount of local matter being observed is limited to ~ 1 kpc; due to extinction (at low latitudes) or due to running out of Galaxy (at higher latitudes). This puts optical observations at a disadvantage with respect to radio in the sense that the amount of matter probed is not as great.

Also, optical target objects, such as edge-on spiral galaxies, introduce source extinction that has an analogous effect. For targets that are largely transparent, such as face-on spirals or ellipticals, and for lines of sight that lie more than a few degrees away from the plane of our own Galaxy, the curves of visible mass accurately follow the curves of actual mass along each line of sight. For these cases, the relative attractiveness of various search directions will resemble the conclusions for the radio regime. A search direction will be productive to the extent that the integrated amount of mass (weighted by an unknown power of distance) is large. Again, search planners need to consider the possibilities of steep-p and shallow-p cases separately. Just as in the radio, a steep-p ETI population makes nearby detections the most likely and a shallow-p (less than about 1.5) makes distant detections most likely. [Dra83]

O.2 Beacon Luminosity Function

In the study of many physical systems from neurons to quasars, it is often found that the distribution of some variable associated with a population has the form of a power law. These power laws, which are statistical in nature, are to be distinguished from the deterministic power laws such as Coulomb's law that appear to be 'hard wired' into nature. Power law distributions seem to arise from one of two processes:

- a turbulent cascade, *a la* Kalmogorov;
- a cascade of avalanches.

What really seems to be necessary is that the process be both random and scale-free (or self-similar) over a fairly wide range. The process that leads to the choice of power for an omnidirectional beacon by an alien engineer is *neither* a random *nor* a self-similar process. So, there is absolutely no reason to expect a power law distribution of outcomes.

Here are two examples contrasting distribution by random cascade with distribution by design. First is the Salpeter Initial Mass Function. This function is an idealized, empirical model of the distribution of stellar masses at the time they are born. The actual data are, of course, a lot less tidy, but the distribution is (approximately) a power law over the mass range of about 0.08 to 30 times the mass of the Sun. Much outside this range, stars like the Sun do not exist, so the distribution must drop to zero. The second example is the Initial Mass Function of passenger vehicles in the United States for the year 1990. This distribution is quite different since the mass of a vehicle is decided by its design goals, more or less. In fact it is much closer to a Gaussian than to a power law.

What kind of distribution function should we expect for the EIRP of isotropic, interstellar, radio beacons? The designer of such a device will not have nearly as precise a design goal as the designer of a passenger vehicle possesses; yet, it is not unreasonable to expect at least some constraints on the design. The first thing that would become obvious to a hypothetical designer is the gross inefficiency of using an isotropic, or to be more precise, an untargeted beacon. If we assume that the ETIs do want to build an isotropic beacon, our own experience with the receiving end suggests a range of plausible transmitter powers.

Clearly, a fairly substantial beacon is needed to do any good. Even a large search system, such as the Cyclops array, cannot do a very good job of detecting transmitters with an EIRP of approximately 10^7 W. On the other hand, even at our current, primitive level, most of our sky surveys could detect 10^{15} W at 1,000 ly (plus or minus a couple of orders of magnitude). So, it is plausible that interstellar beacons, to do their job, will have powers somewhere in the range of 10^8 to 10^{15} W. Given the number of unknown factors that enter into the design requirements of the ETI engineers; it is plausible that the resulting distribution will be centrally concentrated or approximately Gaussian. It might also be argued, by some, that the cost of a beacon is proportional to its power, and hence the distribution might well be skewed to favor more low-luminosity beacons. (That same argument could be applied to the passenger vehicles exhibited above, but not correctly it would seem.)

For use in calculating the expected number of detectable beacons as a function of sensitivity (the 'log N - log S' relationship), the range of luminosity over which a power-law approximation must be valid must be comparable to, or greater than, the range in the geometric dilution factor $(1/4\pi r^2)$. For our own Galaxy, this factor has a range of about $(10^5/4)^2 \sim 10^9$. For those wishing to consider extragalactic beacons, the range is much, much larger. Since the range of the dilution factor is much larger than the plausible range of beacon luminosities, the appropriate approximation to 'astrophysical accuracy' will be a delta function rather than a power law; *even if*, the actual distribution is a power law over a luminosity range of 10^7.

O.3   No Luminosity Function if We Are Targeted

The class of non-targeted beacons, considered previously, is extremely inefficient in the sense that virtually all of their output power never passes near a star. Even if we adopt a fairly large target zone around a star, 10 AU across, and assume that all stars are good targets, the mean free path in a random direction is about 10^{10} ly at the local stellar density, vastly greater than the actual dimensions of our Galaxy. If we consider cosmological volumes, with correspondingly lower average stellar densities, we find that a typical ray in a random direction will never pass within the target zone of any star before it intercepts the Big Bang. That is why, in previous sections, we placed such a strong emphasis on targeted beacons and, in particular, phased array transmitters.

The use of a luminosity function makes an implicit assumption; that the luminosity of an object (in this case a beacon) is a property of the object. This assumption is violated for a targeted beacon, because the luminosity depends both on the object (transmitter) and on the observer. For a given targeted beacon, if we are on the target list we see a large EIRP; but if we are not, then we see nothing. In a statistical sense, this effect might be overcome, except that it is very likely (if not almost certain) that the probability of a given star being on the target list of a beacon will be related to its distance from the beacon. As a result, it would be reasonable to expect that the 'effective' luminosity function, averaging in the probability of our being on the target list, will decline with range in some unknown way. It would be very surprising indeed, if our star were on the target list of a beacon located outside our own Galaxy. Note that targeting our entire Galaxy at once does not improve the efficiency, since virtually all the power still misses the stars.

Even worse, an ideal beacon will not even try to deliver a fixed EIRP to each target but, rather, will try to deliver a fixed flux at the target. In this case, which provides the optimal efficiency for a beacon designer, there is no $1/r^2$ effect at all, and the concept of a beacon luminosity is useless.

Appendix P

Miscellaneous Statistical Facts

This appendix, by D. Kent Cullers, contains some basic statistical facts pertaining to the analysis of astrophysical data.

The following information on the statistics of white noise and independent processes is included here for easy reference. The purpose of this appendix is just to give the reader an intuitive grasp of some basic statistical rules used to analyze astrophysical data. Many books on statistics provide detailed proofs of the following.

Independent white noise provides a good model for the digitally sampled noise from the band of a fully sampled receiver. This noise has a Gaussian distribution because contributions to it, at any given time or sample, arise from many independent, uncorrelated sources. In the limit of a large number of processes (e.g., many moving electrons, etc.), this produces a Gaussian process [Hog94].

Samples are not correlated with each other because the band is assumed *not* to be over-sampled. In other words, the noise is band limited, white, and constant in frequency up to the Nyquist frequency. For real samples, the Nyquist frequency is half the sample rate, and for complex numbers, the Nyquist frequency is the same as the sample rate.

Many readers will be familiar with the Nyquist Theorem. Basically, it states that the frequency of a wave can be unambiguously determined only if the wave frequency is limited to less than half the constant digital sampling rate.

The Nyquist Theorem can be easily illustrated. Assume that a wave is being sampled at its peaks with a frequency exactly equal to that of the sampling device. Then all samples will lie on the peak, and the wave will appear to have a frequency of zero, or not to be oscillating. Thus, the frequency is ambiguous. If the frequency of the sampler is slightly higher than that of the wave, the first sample may lie on a peak, but the second sample will be slightly beyond the peak, and so on. It will appear that the wave has very low frequency, oscillating very slowly from peak to trough to peak. A little consideration shows that one cannot differentiate between frequencies f and $s - f$, where s is the sampling rate. Thus, things clarify if frequencies lie below $s/2$. Naturally, the frequencies can lie in any band with this width, as long as the bandwidth criterion is not exceeded.

Now assume that we have time samples that are independent and Gaussian. It is apparent that the Fourier Transform of such a sequence of samples yields a set of frequencies, unambiguously defined as above, which are Gaussian and independent. Why is this? The FFT just multiplies the Gaussian samples by constants and adds them. The sum of Gaussians is always a Gaussian. Therefore, if the input to an FFT is a Gaussian variable, then the frequency output is a Gaussian variable.

Independence is slightly less obvious. Each frequency component is a set of the same Gaussian variables multiplied by a different selection from the orthogonal basis vectors of the FFT. This means that the product of two FFT coefficients is a double sum. Any two time samples which are different are uncorrelated and hence, their expectation value is zero. The only terms remaining are squares of terms with the same index. The expected value of each of these terms is the same, since the expected value of noise power does not depend on the time of measurement. Finally, summing over the orthogonal FFT vectors reduces the expected value of the product in frequency space to zero.

The following equations illustrate this, where variables enclosed within "$< \quad >$" are the expectation values of these variables. To define the FFT we have:

$$S_n = \sum_j s_j e^{i2\pi jn/N} \tag{P.1}$$

Using this definition, we desire:

$$< S_m \cdot S_n^* > = \sum_j \sum_k < s_j s_k^* > e^{i2\pi(jm-kn)/N} \tag{P.2}$$

All terms where $j \neq k$ disappear, because they are uncorrelated:

$$< S_m \cdot S_n^* > = \sum_j < s_j^2 > e^{i2\pi(m-n)j/N} = 0 \tag{P.3}$$

This sums to zero because integral numbers of sine and cosine cycles yield zero when multiplied together and summed, even for different numbers of cycles. This would not work if the noise were not white; for example, if the value of the noise was a function of time. However, the essence of the argument is sound. The input of an FFT being white Gaussian noise, insures that the output is the same.

Thus, the energy per sample in an FFT is conserved, since the number of samples into a Fourier Transform equals the number of samples out. The noise in every independent sample of an FFT is constant. In the radio spectrum, it is equal to kT_{sys}. The detectability of any signal in a given Gaussian sample, whether in frequency or time, depends on the collected energy divided by the noise temperature.

In SETI, the trick is to find algorithms that concentrate the signal in a given sample, and keep the noise dispersed uniformly throughout the band of interest. Thus, in SETI, we perform narrow transforms for detection of carriers that concentrate them in a particular Fourier sample, and samples for pulses matched to their duration.

It can be easily proved that the power spectrum is comprised of independent samples, again assuming that the noise is flat. To create a power spectrum, one adds the squares of the cosine and sine coefficients of the FFT, frequency by frequency, and thus the power is always positive. The resulting variable is exponential, with a mean equal to its standard deviation, and a variance equal to the square of this number.

Thus, a single power sample from an FFT usually has a power of unity and fluctuations at this level. It has a distribution that is:

$$f(p) = e^{-p} \tag{P.4}$$

This can be easily proved by squaring and adding two Gaussian variables and transforming the result to polar coordinates. Integrating away the theta dependence gives the required distribution.

Note that the sum of powers has a variance that grows in proportion to the sum. The uncorrelated cross terms in the sum disappear, as in the independent noise example above. This means that the standard deviation of the power grows as the square root of the sum. This yields the well-known $1/\sqrt{n}$ rule for fluctuations of the average power.

The important thing to realize is that, when performing matched filtering, we add voltage samples. The variance of the voltage sum also grows as the number of samples. However, when we take the ratio of the standard deviation of the voltage to its sum, the average voltage fluctuations vary as the inverse of the square root of the voltage. Thus, adding voltages together gives noise which decreases in power, or the square of the voltage, or as $1/n$. This is why if we can construct a matched filter to a signal, the bandwidth narrows as the observation period lengthens and the number of samples increases. However, the output of a matched filter is exponential rather than Gaussian in its power distribution, just as with an FFT. As already stated, the noise in a single Gaussian matched filter sample varies only as kT.

Photon-counting statistics are also independent. The important thing to realize is that, in this case, we are adding powers, and the improvement in S/N varies only as the square root of the time available to add photons.

In general, if more than fifty samples are added, Gaussian statistics are a good approximation. Both the power and Poisson photon distribution have an equal mean and standard deviation. Thus, $1/\sqrt{n}$ is a good rule to assess fluctuations. False alarm rates and the probability of detection can be approximated with the Gaussian curve.

Appendix Q

SSTWG Biographies

These biographical summaries were taken from information provided by the members when they joined the SETI Science & Technology Working Group. The editors realize that some members have changed positions and affiliations in the interim period, however, the information that was in place during the period of the meetings of this distinguished working group has been retained.

Dr. Peter Backus received a B.S. in Physics from Rensselaer Polytechnic Institute and a Ph.D. in Astronomy from the University of Massachusetts where he focused on radio astronomical observations of pulsars. He has been an active SETI/Project Phoenix team member since 1981, and now serves as SETI Observing Programs Manager.

Dr. John Billingham served as *Executive Secretary* of the working group. He received his medical education in his native England, graduating from Oxford University with the equivalent of an M.D. He served as a Medical Officer in the Royal Air Force for seven years. He joined NASA's Lyndon B. Johnson Space Center in Houston, Texas, as head of the Environmental Physiology Branch, and worked on the Mercury, Gemini and Apollo programs. Then he transferred to the NASA Ames Research Center in California where he headed up the Biotechnology Division, then the Extraterrestrial Research Division, and later the Life Science Division. After a brief sabbatical at Stanford, he returned to Ames where he headed up the Office for the Search for Extraterrestrial Intelligence. Upon retiring from NASA, he joined the SETI Institute as Senior Scientist, and was elected to the Board of Trustees of the SETI Institute in 1995. Billingham was SETI's recognized champion within NASA. In the early 1970's, he organized the NASA/Stanford/ASEE summer study to design a system to detect ETI life. He was successful in getting Dr. Bernard M. Oliver (then Vice

President for Research & Development at Hewlett-Packard Company) to serve as Cochair of the study. The result was the *Project Cyclops* report, which has become known as the leading SETI conceptual design study of the last century. Billingham is recognized among his peers as a visionary leader and team builder, with a wide-range of knowledge and interest in many disciplines. He made significant contributions to expanding NASA's exobiology research programs and attracting highly qualified experts into the field. He is one of a small group of scientists around the world who has spent years studying the best approaches to conduct SETI, as well as the impact of such a discovery on society, and what the international response might be after the first signal is detected.

Dr. Douglas Bock recently completed his Ph.D. in the School of Physics at the University of Sydney. He is at the Radio Astronomy Laboratory at the University of California, Berkeley, where he is observing with, and developing instrumentation for, the Berkeley-Illinois-Maryland Association Array, a millimeter interferometer in northern California.

Dr. Jaap D. Bregman received a Masters degree in Physics from the Technical University Delft, The Netherlands. He is Project Manager at Dwingeloo Radio Observatory in the Netherlands, and is involved in assessment studies and system design of the Square Kilometer radio telescope project (the SKA).

Dr. Harvey R. Butcher received a B.S. in Astrophysics from the California Institute of Technology, and a Ph.D. in Astronomy from the Australian National University, where his dissertation was on "Observational Aspects of Nucleosynthesis". He is Director of the Netherlands Foundation for Research in Astronomy and Professor of Observational Astronomy at the University of Groningen. His research interests include galaxy evolution and advanced instrumental techniques. His professional goals are to promote the Dwingeloo-Westerbork Radio Observatory as an important center for radio astronomy and to help get the Square Kilometer radio telescope project (SKA) off the ground.

Dr. John E. Carlstrom earned an A.B. in Physics from Vassar, and a Ph.D. in Physics from the University of California in Berkeley. His field of research is experimental astrophysics, and developing new astronomical techniques and instrumentation. He is a Professor of Astronomy and Astrophysics at the University of Chicago, and the Associate Director of the Center for Astrophysical Research in Antarctica (CARA).

Dr. Christopher Chyba was appointed by the SETI Institute to the Carl Sagan Chair for the Study of Life in the Universe in 1998. This seems especially fitting since he earned his doctorate under Carl Sagan at Cornell. Chyba is also affiliated with Stanford University, as Co-Director of the Center for International Security and Cooperation (CISAC), and Associate Professor (Research) in the Department of Geological and Environmental Sciences. He served on the national security staff of the White House from 1993 through 1995, and he received the Presidential Early Career Award in 1996 "for demonstrating exceptional potential for leadership at the frontiers of science and technology during the 21st century". Chyba's research centers on conditions relevant to the origin of life, and the search for life elsewhere in the Solar system. He chaired the Science Definition Team for NASA's Europa Orbiter mission to search for an ocean beneath the icy crust of Jupiter's moon Europa. At the SETI Institute, he directs the Center for the Study of Life in the Universe, where he is helping shape the SETI Institute's strategic planning for its many Life in the Universe projects. At CISAC, his research includes issues of biological terrorism and nuclear weapons material security. In 1994, while Director for International Environmental Affairs on the staff of the National Security Council, Chyba was named one of Time magazine's "Fifty for the Future".

Dr. D. Kent Cullers, Cochair of the SETI Science and Technology Working Group, is Director, SETI R&D at the SETI Institute. Blind since birth, he earned his Ph.D. from the University of California at Berkeley in 1980, becoming the first totally blind physicist in the world. He immediately joined NASA's Ames Research Center, and later, the SETI Institute, as a founding member. He designed detection algorithms for weak signals, currently used for SETI; for finding early signs of breast cancer in diagnostic mammograms; and for direct photometric detection of Earth-sized planets. A National Merit Scholar, he has been the recipient of many honors and awards, including NASA's Exceptional Engineering Achievement medal and the Federal Employee of the Year award. He's a popular guest lecturer and travels extensively. He earned his first amateur radio operator license at age eleven, in 1961, and he's still listening for that distant transmission. He's on the Board of Directors of the Sensory Access Foundation, and the Peninsula Center for the Blind and Visually Impaired.

Dr. Leonard S. Cutler is Distinguished Contributor, Technical Staff, at Hewlett-Packard Laboratories. He received his B.S., M.S., and Ph.D. in Physics from Stanford, where his thesis work was in two areas: theory of high-energy inelastic electron scattering off nuclei, and theory of noise in masers. He was Vice President of Engineering at Gertsch Products in Los Angeles, prior to joining Hewlett-Packard in 1957. His career at HP included several management positions, including Director of the Superconductivity Laboratory, then the Instruments and Photonics Laboratory, and later the Physical Research Laboratory. He has been active in areas of precision ratio transformers, frequency meters, quartz and atomic frequency standards including passive cesium, rubidium, and mercury ion standards, and active hydrogen masers. He was responsible for the design of the first all solid-state cesium beam frequency standard and has been deeply involved ever since with the theory and design of cesium atomic beam tubes and the electronics for atomic and quartz frequency standards. Among his many achievements was election to the National Academy of Engineering in 1987. In 1993, he shared the American Institute of Physics Award for Industrial Applications of Physics for the design of an atomic cesium clock that is the most precise and stable commercial time keeping device presently available in the world. In 1996 he was elected a fellow of the American Physical Society for work on atomic frequency standards and laser interferometers for precision distance measurement.

Dr. Michael M. Davis attained a BS in Physics from Yale, and a Ph.D. from Leiden University – where he defended his radio astronomy thesis in Dutch. He was Director of the Arecibo Observatory for several years, prior to accepting the current assignment of overseeing the Gregorian feed upgrade. He chaired US URSI Commission J (Radio Astronomy) from 1994 through 1996, and presently chairs the NAS/NRC Committee on Radio Frequencies.

Dr. Frank D. Drake has served as Chairman of the Board of Trustees of the SETI Institute since its chartering in 1984. He attained a B.A. in Engineering Physics from Cornell, and a Ph.D. from Harvard. He is fondly called the "Father of SETI" for several reasons: he conducted Project Ozma – the first modern organized search for ETI signals; he devised the widely-known Drake Equation – an estimate of the number of communicative ETI civilizations we might find in our Galaxy; and he constructed the "Arecibo Message of 1974" – the first interstellar message transmitted via radio waves from our planet for the benefit of any ETI civilization. In recent years he served as Director of the Arecibo Observatory, then Dean of Natural Sciences at the University of Santa Cruz. Drake is Professor of Astronomy and Astrophysics at the University of California at Santa Cruz, and splits his time between UCSC and the SETI Institute. He has chaired or served on several key boards and committees, including being a member of the three previous National Academy of Sciences/National Research Council (NAS/NRC) Astronomy Survey Committees (the "Whitford Committee", the "Greenstein Committee" and the "Field Committee"); member of the Astronomy Advisory Committee of the NSF; member of the National Advisory Board, Center for National Policy; Chairman, U.S. National Committee for the International Astronomical Union; Chairman of the Board on Physics and Astronomy, National Research Council (1988 through 1992); and President, Astronomical Society of the Pacific (1988 through 1990). Among his many awards and achievements was election to the National Academy of Sciences in 1972. He co-authored a popular SETI book with Dava Sobel, he has published many articles in journals and magazines, is a popular lecturer around the world, and is frequently called upon by the media for scientific comment, verification or critique when new discoveries are announced.

Dr. John W. Dreher earned a BA in Physics and a Ph.D. in Astronomy from the University of California at Berkeley. He was involved in construction projects at both Hat Creek Observatory in California and the Very Large Array in New Mexico prior to joining the physics faculty at the Massachusetts Institute of Technology (MIT). In 1989 he joined the SETI program, first at NASA Ames Research Center and then the SETI Institute, where he is currently the Systems Engineer on Project Phoenix.

Professor Ronald D. (Ron) Ekers *Chair* of the SETI Science & Technology Working Group sessions. A native of Australia, he graduated from the University of Adelaide then gained his Ph.D. in astronomy at the Australian National University. In 1988 he was appointed to his present position of Foundation Director of CSIRO's ATNF. His professional career has taken him to the California Institute of Technology; the Institute of Theoretical Astronomy in Cambridge, U.K.; the Kapteyn Laboratory in Groningen, The Netherlands; and the National Radio Astronomy Observatory, New Mexico, USA. He was director of the VLA, the major national radio telescope in the USA, from 1980 until 1987. He has been a member of various advisory and visiting committees in the United States of America (USA), United Kingdom (UK), Canada, France, Germany and The Netherlands. He is a member of the national committee for Astronomy and an Adjunct Professor at the Australian National University. He was elected a Fellow of the Australian Academy of Science and Foreign Member of the Royal Dutch Academy of Science in 1993. His research interests include extragalactic astronomy, especially cosmology, and galactic nuclei, radio astronomical techniques and image formation theory.

Dr. Sandra Faber has been a University Professor (since 1996) and Professor of Astronomy and Astrophysics (since 1972) at the University of California at Santa Cruz (UCSC). She earned a B.A. in Physics from Swarthmore College, and a Ph.D. in Astronomy from Harvard University. Faber is renowned for her research in extragalactic astronomy and cosmology. She has worked closely on two of the major optical astronomy ventures of recent years: the Hubble Space Telescope and the W. M. Keck Observatory. She was Cochair of the Keck Observatory Science Steering Committee (1985 through 1993). Among her many achievements was election to the National Academy of Sciences in 1985 and the American Academy of Arts and Sciences in 1987. She has been on the Board of Directors of Annual Reviews since 1988. She is on the Board of Trustees of the Carnegie Institute of Washington, and of the SETI Institute.

Dr. J. Richard (Rick) Fisher received a BS in physics from Penn State and a Ph.D. in astronomy from the University of Maryland. His thesis was done at the Clark Lake Radio Observatory under the mentorship of Bill Erickson. Rick joined the NRAO scientific staff at Green Bank, West Virginia, immediately following graduate school and has held various positions such as Head of the Electronics Division, Site Director, and Project Manager of several instrumentation projects. His current title at NRAO is Scientist. In 1978 he took leave from the NRAO to spend time at the Division of Radio Physics, CSIRO in Australia and at the Raman Research Institute in Bangalore, India. His research interests include galaxies, pulsars, antenna design, RFI excision techniques, and signal processing. He has chaired or served on several key boards and committees, including The International Union of Radio Science, URSI (Chairman, Commission J of U.S. National Committee, 1987 through 1990), National Science Foundation Advisory Committee for Astronomical Sciences (1986 through 1989), National Research Council's Committee on Radio

Frequencies (1982 through 1989), International Astronomical Union, American Astronomical Society, Institute of Electrical and Electronics Engineers (Antenna Society), American Association for the Advancement of Science, American Association of Physics Teachers, and the Arecibo Scientific Advisory Committee (1981 through 1984).

Dr. W. Daniel (Danny) Hillis is a Disney Fellow and Vice President, Research and Development, at Walt Disney Imagineering. Hillis is well known as an inventor, scientist and computer designer. He designed some of the fastest computers in the world, and pioneered the concept of massively parallel computers that is now the basis of most new super computer designs. Holder of over 40 patents, he was co-founder and Chief Scientist of Thinking Machines Corporation. He is also an adjunct professor at MIT. His awards are many, including being named a Fellow of the American Academy of Arts and Sciences.

Dr. Paul Horowitz received both an A.B. and a Ph.D. from Harvard, where he has been Professor of Physics and of Electrical Engineering since 1974. His research interests span a wide range, including experimental astrophysics, X-ray and particle microscopy, studies of the E. Coli rotary engine, optical interferometry, the search for ultra-heavy matter, radon measurement techniques, and new technologies for humanitarian demining. Currently his group at Harvard is running the "BETA" 250 million channel waterhole SETI, and also a search for short optical pulses from nearby solar-type stars.

Dr. Kenneth I. Kellermann received a B.S. in Physics from MIT and a Ph.D. in Physics and Astronomy from Caltech. He is the Chief Scientist at the National Radio Astronomy Observatory in Charlottesville, Virginia. He is also Research Professor at the University of Virginia and an Outside Scientific Member of the Max Planck Society in Germany. Kellermann's research interests are radio galaxies, quasars, cosmology, radio telescopes, and the history of radio astronomy. He is also interested in amateur radio. His claim to fame in SETI was having the first published SETI observations in a refereed scientific journal.

Greg Klerkx is Director of Development at the SETI Institute, and leads the SETI Institute's fund-raising efforts to ensure the continuation of SETI and Life in the Universe research activities. He holds a B.S. in journalism from Bowling Green State University, and held several positions in journalism prior to moving to the development arena.

Dr. Michael Lampton attained a B.S. from Caltech and a Ph.D. in Physics from the University of California at Berkeley. His fields of research have included upper atmosphere and auroral zone physics, X-ray astronomy, extreme ultraviolet astronomy, and the development of photon-counting detectors for use in these fields. He spent over a decade in NASA's payload specialist astronaut program, and as a space science newscaster for NASA's Public Affairs Office. Lampton now holds the position of Senior Research Fellow in the Space Sciences Laboratory at the University of California at Berkeley, and he is involved in experiments to examine relationships between the solar wind, the Earth's magnetosphere, and the upper atmosphere.

Dr. Larry Lesyna joined the newly formed NASA SETI team (then called the Interstellar Communication Study Group) at Ames Research Center as a part-time student intern in the mid 1970's during his undergraduate years at Caltech. He then worked with the SETI group at Ames for a year, and then returned to school to begin graduate studies at Stanford. He received a Ph.D. in Applied Physics, and successfully used magnetic refrigeration techniques to cool an astronomical detector system to the lowest temperature ever, a fraction of a degree above absolute zero. He then joined Grumman Aerospace's Corporate Research Center and is currently a Staff Scientist at Lockheed Martin's Advanced Technology Center in Palo Alto, CA. His research interests include extragalactic astronomy and its associated instrumentation, low temperature solid-state physics, and of course SETI.

Dr. David E. Liddle is co-founder, president and CEO of Interval Research Corporation, a high-technology research lab studying technologies designed to be meaningful to people in the future. During and after his education (B.S., E.E. University of Michigan; Ph.D. Computer Science, University of Toledo, Ohio), he spent his professional career developing technologies for interaction and communication between people and computers, in activities spanning research, development, management and entrepreneurship. He spent a decade at the Xerox Palo Alto Research Center, then founded Metaphor Computer Systems whose technology was first adopted by IBM and the company was ultimately acquired by IBM. His affiliations are many: Chairman of the Board of Trustees of the Santa Fe Institute, Director at Sybase, Broderbund Software, Starwave and Ticketmaster, and a consulting professor of Computer Science and member of the advisory committee of the School of Engineering at Stanford University. Among his many honors is election as a Senior Fellow of the Royal College of Art.

Dr. Nathan Myhrvold is chief technology officer, reporting to Microsoft CEO Bill Gates as a member of the Executive Committee. Myhrvold holds a doctorate in theoretical and mathematical physics and a master's degree in mathematical economics from Princeton University. He also has a master's degree in geophysics and space physics and a bachelor's degree in mathematics, both from the University of California. His education has also included certificates in mountain climbing, formula car racing, photography, and French cooking. Before joining Microsoft in the mid 1980's, he worked with Professor Stephen Hawking on research in cosmology, quantum field theory in curved space time, and quantum theories of gravitation. In May 1993, Myhrvold joined the Board of Trustees of the Institute for Advanced Study in Princeton, New Jersey, one of the world's foremost institutions for pure scientific research. He serves on the advisory board of Princeton University's Department of Physics. When he takes time off, Myhrvold works as an assistant chef at one of Seattle's leading French restaurants. He has competed twice in the world championship of barbecue in Memphis, Tennessee, winning first and second place titles.

Dr. Greg Papadopoulos is Vice-president and Chief Technology Officer for Sun Microsystems. Before joining Sun, he was Senior Architect and Director of Product Strategy for Thinking Machines Corporation (TMC) in Cambridge, Massachusetts, and an Associate Professor at MIT, where he taught electrical engineering and computer science. At MIT, he was the recipient of the National Science Foundation's Presidential Young Investigator Award. He also co-founded three companies: PictureTel (video conferencing), Ergo Computing (high-end PCs) and Exa Corporation (fluid dynamics). Papadopoulos received a B.A. in systems science from the University of California at San Diego, and an M.S. and Ph.D. in electrical engineering and computer science from MIT.

Thomas Pierson is Chief Executive Officer of the SETI Institute. He pursued undergraduate studies in aerospace engineering at the University of Oklahoma, before switching to business administration and earning a Bachelor's degree with dual majors in management and accounting. Later, he earned an MBA from San Francisco State University, writing a distinguished thesis on the effect of differing leadership styles in higher education management. Pierson began his business career as a finance officer at the University of Oklahoma, then moved to California where he established a university foundation for faculty research for Sonoma State University. He later served for eight and one-half years as Associate Director of the much larger Fredric Burk Foundation at San Francisco State University. After meeting several SETI scientists and getting to know the objectives and underpinning arguments for the search for life beyond Earth, he conceived the idea of the SETI Institute and founded the corporation on November 20, 1984. He has served as Chief Administrator and Corporate Secretary since the SETI Institute's incorporation, and has received numerous awards, including NASA's Public Service Medal. As CEO of the SETI Institute, Pierson most prizes the team accomplishment represented by the NASA Public Service Group Achievement Award given by NASA Administrator Dan Goldin to the SETI Institute for excellence in furthering research and education in the fields of SETI and Life in the Universe.

Dr. Zoya Popovic is an Associate Professor in the Department of Electrical and Computer Engineering at the University of Colorado in Boulder. She received her Dipl. Ing. degree from the University of Belgrade, Serbia, Yugoslavia, and the M.S. and Ph.D. degrees from Caltech. Her doctoral thesis was on large-scale quasi-optical microwave power-combining. She joined the faculty of the University of Colorado in Boulder in 1990, where she has developed undergraduate and graduate electromagnetics and microwave laboratory courses. Her research interests include microwave and millimeter-wave quasi-optical techniques and design and analysis of active microwave circuits and antennas for high frequency communication applications.

Dr. Alan Roy received a Ph.D. in astrophysics from the University of

Sydney in 1995, and is currently serving in a postdoc position at the Very Large Baseline Array (VLBA) at the National Radio Astronomy Observatory in Socorro, New Mexico. His research interests are: active galaxies, especially Seyfert galaxies; looking for obscuring tori in nearby Seyferts using the VLBA; starburst activity in galaxies; using radio recombination lines to trace gas dynamics and plasma conditions; ultraluminous infrared galaxies; and radio astronomy instrumentation and imaging techniques.

Dr. Louis (Lou) K. Scheffer earned a B.S. in Engineering and Math and an M.S. in Electrical Engineering from the California Institute of Technology. He then earned a Ph.D. in Electrical Engineering from Stanford. Since 1981, he has been a Fellow at Cadence Design Systems, where he is responsible for the software architecture of digital IC design tools. Previously, he was with Hewlett-Packard, where he designed and coded VLSI analysis programs and was an editor. He helped design and layout digital filter chips for a DSP-based spectrum analyzer. He has also taught courses in Computer Aided Design of Integrated Circuits at UC Berkeley and Stanford University.

Dr. Seth Shostak is the Public Programs Scientist at the SETI Institute. He holds a degree in physics from Princeton University, and a doctorate in astronomy from the California Institute of Technology. For much of his career, he has conducted radio astronomy research on galaxies. For a dozen years, he worked at the Kapteyn Astronomical Institute, in Groningen, The Netherlands, using the Westerbork Radio Synthesis Telescope. He has written a book about SETI, plus numerous articles in journals and magazines. He is a popular voice-of-science on radio and TV nationwide, and he presents lectures around the globe. Among many other titles, Shostak is the Managing Editor of the SETI Institute's newly formed SETI Press.

Dr. Richard (Rick) Smegal joined the SETI Institute in the final phase of an Industrial Postdoctoral research program in Victoria, Canada. A native of Winnipeg, he obtained both a B.S. and an M.S. in Electrical Engineering from the University of Manitoba. He earned a Ph.D. in Electrical Engineering from the University of Alberta, specializing in instrumentation for aperture synthesis radio astronomy. His doctoral research was carried out at the Dominion Radio Astrophysical Observatory in Canada and involved upgrading the DRAO synthesis telescope to an imaging polarimeter. His current research interests include instrumentation for radio astronomy and the application of computational intelligence to such systems.

Dr. Anthony (Tony) A. Stark received B.S. degrees in physics and astronomy from the California Institute of Technology, and M.A. and Ph.D. degrees in astrophysical sciences from Princeton University. His Ph.D. thesis in millimeter-wave radio astronomy, "Galactic Kinematics of Molecular Clouds", was under the direction of Dr. Arno A. Penzias. He then joined the Radio Physics Research Department at Bell Laboratories headed by Dr. Robert W. Wilson, where he developed high frequency radio receivers and conducted several extensive radio astronomy survey projects. His interest in millimeter- and submillimeter-wave astronomical observations led to pioneering efforts in Antarctic astronomy. Since 1991, he has been an astronomer at the Smithsonian Astrophysical Observatory, where he serves as Principal Investigator of the Antarctic Submillimeter Telescope and Remote Observatory (AST/RO), a user-facility astronomical observatory located at the geographic south pole. His interests include radio telescope design and instrumentation, radio astronomical observations of the interstellar medium, and observational cosmology.

Dr. Richard (Rick) Stauduhar received a B.S. in electrical engineering and a Ph.D. in mathematics from the University of California at Berkeley with the thesis, "The Automatic Determination of Galois Groups". He was a post doctoral fellow at the University of Alberta; a staff member at the Space Sciences Laboratory at the University of California at Berkeley, and co-developer of a widely used method of speech synthesis. He was a lecturer in Computer Science at the University of California at Berkeley, and worked as a consultant on a wide variety of computer projects. In 1989 he joined the technical staff at the SETI Institute, where he has worked on SETI algorithm development and implementation in hardware and software.

Dr. Jill Cornell Tarter was appointed by the SETI Institute to the Bernard M. Oliver Chair for the Search for Extraterrestrial Intelligence in 1997. She attended Cornell University on a full scholarship from Proctor and Gamble, Inc., and completed the five-year professional engineering course in four years, earning a Bachelor of Engineering Physics Degree with Distinction. Her fifth-year project involved spatial calibration of a cosmic ray spark chamber using the 2 GEV synchrotron accelerator. She earned a Master's Degree and a Ph.D. in astronomy from the University of California at Berkeley, where her major field of study was theoretical high-energy astrophysics. Her thesis discussed the observable properties (or lack thereof) of small brown dwarf stars that never successfully fuse hydrogen and the observability of interstellar gases stripped from the interiors of galaxies as they interact with other galaxies and their surroundings within rich clusters of galaxies. As a graduate student at the University of California at Berkeley, Tarter became involved in SERENDIP, a small commensal search for ET radio signals using the Hat Creek Observatory. That led to her joining the SETI team at NASA Ames Research Center, then later the SETI Institute, where she now heads the SETI research effort. Tarter is an international SETI ambassador, and is held in high regard among her peers in the science community. Among her many awards is the Lifetime Achievement Award from Women in Aerospace for her contribution to the field of exobiology.

Douglas D. Thornton is a specialist at the Radio Astronomy Laboratory at the University of California, Berkeley. He was employed previously at Tektronix, Sylvania WDL, and Philco WDL. He received his B.S. degree in electrical engineering from Oregon State University in 1957, and an M.S. in electrical engineering from the University of California at Berkeley in 1961.

His primary interests have been in centimeter and millimeter radio astronomy and the associated instrumentation problems. These have included low-noise microwave receiver design, frequency control problems, and computer control systems.

Dr. Charles H. Townes, who received the Nobel Prize for his role in the invention of the maser and the laser, is a Professor in the Graduate School at the University of California, Berkeley, and engaged in research in astrophysics. He is known for a variety of researches involving the interaction of electromagnetic waves and matter, and also as a teacher and government advisor. He graduated with highest honors from Furman University, earning a bachelor of science degree in physics and a bachelor of arts degree in modern languages. He completed a master's degree at Duke University, and received a Ph.D. degree at the California Institute of Technology. He was a staff member of the Bell Telephone Laboratories, Associate Professor and Professor at Columbia University, Vice President and Director of Research at the Institute for Defense Analysis, and Provost and Professor of Physics at MIT, prior to being named University Professor of Physics at the University of California at Berkeley in 1967. Dr. Townes' principal scientific work is in microwave spectroscopy, nuclear and molecular structure, quantum electronics, radio astronomy, and infrared astronomy; he is presently most active in the latter two fields. He has the fundamental patent on masers, and with A. L. Schawlow, the basic patent on lasers. In addition to the Nobel Prize, he has received a number of awards and honors, including the National Medal of Science and membership in the National Academy of Sciences, National Academy of Engineering, The Royal Society of London, The Russian Academy of Sciences, The Pontifical Academy of Science, National Inventors Hall of Fame, and Engineering and Science Hall of Fame, as well as honorary degrees from twenty-five colleges and universities.

Dr. Arnold van Ardenne is head of the Technical Laboratory at the Netherlands Foundation for Research in Astronomy (NFRA) in Dwingeloo, The Netherlands. NFRA is heavily involved in rejuvenating the Westerbork Synthesis Radio Telescopes into a VLBI station and in setting up the R&D program for the next generation of radio telescopes; i.e., the Square Kilometer Array Interferometer. He studied at the University of Twente in the Netherlands and received a physics degree on research on superconducting planar infrared detectors. He subsequently worked on several electromagnetic modelling problems in microwave components in the field of phased array antennas at the National Defense Physics Laboratories until he joined NFRA.

Dr. Sander (Sandy) Weinreb received both a B.S. and a Ph.D. in Electrical Engineering from MIT. He is currently the Research Professor in the Five College Radio Observatory, and in the Department of Physics and Astronomy of the University of Massachusetts. Weinreb spent the early portion of his career at the National Radio Astronomy Observatory, where he headed the Electronics Division and later served as Assistant Director. He was responsible for design, construction, operation and maintenance of radio astronomy receivers at the Green Bank, West Virginia and Kitt Peak, Arizona observatories. He led the design of the electronics system of the VLA at Socorro, New Mexico, and he led the addition of 8.4 GHz cooled HEMT receivers to the VLA to augment the DSN and receive images from the NASA Voyager Neptune encounter mission. Weinreb taught courses in microwave theory and techniques as a visiting professor at the University of California and the University of Virginia. His major technical accomplishments include: (1) developed coarse-quantization, digital correlation techniques that have been widely used in radio astronomy systems for the past 30 years; (2) co-discoverer of the first radio molecular line, OH, observed in radio astronomy (1963); (3) introduced cooled field-effect transistor and HEMT amplifiers to radio astronomy (1980); and (4) led the electronics design of the VLA. He has been involved in many national and international activities.

Dr. William J. Welch was named to the Watson and Marilyn Alberts Chair in the Search for Extraterrestrial Intelligence at the University of California at Berkeley in 1998. For over a dozen years, he served as the Vice Chairman of the SETI Institute's Board, and he remains a Board Trustee. Welch received a B.A. from Stanford University and a Ph.D. from the University of California at Berkeley, where his thesis was on electromagnetic scattering. He has been at UC Berkeley since 1960, as Professor of Electrical Engineering since 1969, and Professor of Astronomy since 1971. He was Director of the Radio Astronomy Laboratory from 1971 through 1996. He has chaired and been a member of a number of key scientific Committees and Advisory Boards, such as the Advisory Panel on Astronomy for the NSF, the Advisory Board for the Arecibo Observatory, and the Visiting Committees for the Owens Valley Radio Observatory of the California Institute of Technology, the MIT Haystack Observatory, and the NRAO. He was Chairman of Commission J (US) of URSI, and a representative of the American Astronomical Society to the U.S. National Committee. He served as a trustee of Associated Universities, Inc., and as a board member for BIMA, the Berkeley-Illinois-Maryland-Association Millimeter Array. Among his honors, Welch was the recipient of the Docteur Honoris Causa from the Universite de Bordeaux I in 1979, and was elected to the National Academy of Sciences in 1999.

Daniel Werthimer is project scientist of the SERENDIP SETI program at the Space Sciences Laboratory, University of California, Berkeley. He also oversees the newly formed SETI@home project, and is president of Techne Instruments, a small corporation which designs scientific instrumentation. He was Associate Professor in the Engineering and Physics Departments of San Francisco State University from 1977 through 1982, and has been a Visiting Professor at Beijing Normal University, the University of St. Charles in Marseille, and Eotvos University in Budapest. Werthimer also conducts annual science education workshops with UNESCO and the International Center for Theoretical Physics which has allowed him to teach at universities in Peru, Egypt, Ghana,

Ethiopia, Zimbabwe, Uganda, and Kenya. He has published numerous scientific papers in the fields of SETI, radio astronomy, instrumentation and science education, and is co-editor of the recent book *Astronomical and Biochemical Origins and the Search for Life in the Universe.*

Figure Q.1: Vera Buescher, of the SETI Institute, is the workshop coordinator par excellence.

Figure Q.2: The SETI Institute's indefatigable Chris Neller, who put it all together as workshop co-coordinator.

Acronyms

$\sim$	Approximately or "on the order of".
$\approx$	Approximately equal to.
α	1. Often used in astronomy for Right Ascension (RA); 2. Often used in SETI for postdetection signal-to-noise ratio.
$\alpha - \delta$	Right Ascension-Declination. This gives the position of an object in the sky in the celestial equatorial coordinate system. Not a complete specification without epoch designation, because of precession, nutation, etc.
μ	Micro or 10^{-6}; as in micron (μm), or microsecond (μs), or micro-arc second (μarc second.
ω	Angular frequency ($2\pi\upsilon$).
1hT	One Hectare (Radio) Telescope, which will consist of many small antennas providing a collecting area totalling 10,000 m^2, equivalent to the area of a single dish 113 m in diameter. meter dish. The R&D installation, consisting of seven dishes, began operating in April 2000.
A/D	Analog to Digital (converter).
ABL	Airborne Laser.
AIRES	Airborne Infrared Echelle Spectrometer (for SOFIA).
AO	Arecibo Observatory.
APD	Avalanche Photodiode.
ARO	Algonquin Radio Observatory in Ontario, Canada.
AST	Antarctic Submillimeter Telescope.

ATA   Allen Telescope Array. This telescope array was formerly called the 1hT, and 1hT is used throughout this book. 1hT was the name used during the SSTWG meetings and for some time thereafter.

ATNF   Australia Telescope National Facility.

AU   Astronomical Unit; 1 AU $\approx 1.5 \times 10^{11}$ meters, or ≈ 93 million miles.

az   Azimuth; direction angle in the horizontal plane. Measured eastward from North.

az-el   Azimuth-elevation; type of telescope mounting.

bel   Unit measure of the power ratio equal to $log_{10} (P_1/P_2)$. (See dB.)

BETA   Billion-channel Extraterrestrial Assay.

BIPM   Bureau International des Poids et Mesures (International Bureau of Weights and Measures).

BW   Bandwidth.

CARA   Center for Astrophysical Research in Antarctica.

CATV   Community Antenna Television (Cable TV).

CCD   Charge-Coupled Device.

CDMA   Code Division Multiplex Access.

CDR   Critical Design Review.

CFA   1. Cooled FET Amplifier. (See FET.) 2. Center for Astrophysics, Cambridge,MA

CGS4   Cooled Grating Spectrometer. A 1 to 1.5 μm multipurpose, two-dimensional, grating spectrometer containing a 256 X 256 InSb array, installed in a cryostat cooled by liquid nitrogen and closed cycle coolers. It is used with the UK Infrared Telescope on Mauna Kea, Hawaii.

CISAC   Center for International Security and Cooperation.

CMB   Cosmic Microwave Background.

CMOS   Complementary Metal Oxide Semiconductor.

CO   Carbon Monoxide.

COIL   Chemical Oxygen-Iodine Laser.

COSETI   Columbus Optical SETI; the optical SETI project operated by Dr. Stuart Kingsley in Columbus, Ohio.

CPA	Chirp-Pulse-Amplification.
cps	Counts per second.
CPU	Central Processing Unit.
CSHELL	Cryogenic Echelle Spectrometer (1-5 μm). A spectrometer at the Infrared Telescope Facility.
CSIRO	Commonwealth Scientific Industrial Research Organization (Australia).
CW	1. Continuous Wave; an essentially single-frequency EM wave; 2. Carrier Wave.
dB	Decibel; a measure of the power ratio equal to 10 $log_{10}(P_1/P_2)$ or 20 $log_{10}(V_1/V_2)$ or 0.1 bel. (See bel.)
dBi	Decibel antenna gain (referred to gain of an isotropic antenna).
DC	Direct Current.
DFT	Direct Fourier Transform.
DM	Dispersion Measure.
DOE	Department of Energy.
DPSSL	Diode-Pumped Solid-State Laser.
DRAM	Dynamic Random Access Memory.
DSN	Deep Space Network. This is part of NASA; it is used to run most USA deep space missions.
DSP	Digital Signal Processing.
EIRE	Effective Isotropic Radiated Energy (joules).
EIRP	Effective Isotropic Radiated Power (watts).
el	Elevation; angle of a direction above the horizon.
EME	Earth-Moon-Earth.
EMI	Electromagnetic Interference.
ESA	European Space Agency.
ESTEC	European Space research and Technology Centre for the European Space Agency, with headquarters in Noordwijk, Holland.

ET	Extraterrestrial. In this book, "ET" is used only as an abbreviation or acronym for "extraterrestrial". When referring to an intelligent extraterrestrial life form, who might be able to communicate with us, the acronym "ETI" is used.
ETI	Extraterrestrial Intelligence. In this book, ETI refers to an intelligent extraterrestrial life form who might be able to communicate with us.
EUV	Extreme Ultraviolet.
EUVE	Extreme Ultraviolet Emission.
EVLA	Expanded Very Large Array.
EXES	Echelon Cross Echelle Spectrograph, being built in Texas for the SOFIA project.
f	1. Frequency; 2. femto-(10^{-15}) 3. function.
F/D	The focal length of a lens, mirror or antenna divided by its diameter, which is also known as the Focal Ratio or 'f' number.
FAME	Full-Sky Astrometric Explorer.
FET	Field Effect Transistor. FET is a solid-state device which, when properly used in an amplifying system cooled to cryogenic temperatures (Cooled-FET-Amplifier, CFA), can provide amplification throughout the microwave window with exceptionally low intrinsic noise.
FFT	Fast Fourier Transform.
FGS	Fine Guidance System.
FLOPS	Floating point Operations Per Second.
FM	Frequency Modulatation.
FOM	Figure-of-Merit.
FPGA	Field Programmable Gate Array.
FSK	Frequency Shift Keying.
FSU	Former Soviet Union.
FTE	Full Time Equivalents.
FWHM	Full Width at Half Maximum.
FY	Fiscal Year.

GBT	Green Bank Telescope. A 100 meter radio telescope of novel design with an off-axis feed, designed to operate at frequencies up to 90 GHz.
GHz	Gigahertz; 10^9 Hz.
GLONASS	Global Navigation Satellite System, built by the Former Soviet Union (FSU). It is similar in purpose to the GPS.
GMRT	Giant Metrewave Radio Telescope; an array of thirty 45 meter antennas located in India.
GNP	Gross National Product.
Gop	10^9 operations, or one billion operations. The operations are assumed to be complex operations, unless otherwise stated.
Gops	Gigaop, one billion (10^9) operations per second, usually complex operations.
GPS	Global Positioning System, built by the USA, and originally known as NAVSTAR. It is similar in purpose to GLONASS.
GRB	Gamma Ray Burst.
GSPS	Geosynchronous Satellite Positioning System.
GW	Gigawatt = 10^9 W.
h	1. Plank's constant; 6.6252×10^{-34} J s; 2. hour of time (60 minutes or 3,600 seconds); 3. hour of arc or angle ($15°$).
HA	Hour Angle.
HA-DEC	Hour Angle-declination; coordinates for the celestial coordinate system based on Earth's equatorial plane and the observer's prime meridian (north-zenith-south). It is also used to designate an equatorial (EQ) telescope mounting.
HCRAO	Hat Creek Radio Astronomy Observatory, operated by the Radio Astronomy Laboratory of the University of California at Berkeley.
HCRO	Hat Creek Radio Observatory in Castel, California.
H I	The 21 cm hyperfine line in the radio spectrum, which is from H I or neutral hydrogen.
HEMT	High Electron Mobility Transistor.

HESS   High Energy Stereoscopic System; a project for a next-generation system of Imaging Atmospheric Cerenkov Telescopes for the investigation of cosmic gamma rays in the 100 GeV energy range.

HIRES   High Resolution Spectrometer for the visible band (on the Keck telescope).

HRMS   High Resolution Microwave Survey.

HRMS IWG   High Resolution Microwave Survey Investigators Working Group.

HRO   Haystack Radio Observatory in Westford, Massachusetts.

HST   Hubble Space Telescope.

Hz   Hertz; cycles-per-second; formerly referred to as cps or c/s.

IAA   International Academy of Astronautics.

IAR   Instituto Argentino de Radioastronomia (Argentinian Institute of Radio Astronomy) in Villa Elisa, Argentina.

IAU   International Astronomical Union.

IC   Integrated Circuit.

ICF   Inertial Confinement Fusion; an approach to initiating nuclear fusion using banks of lasers.

IF   Intermediate Frequency.

IFE   Inertial Fusion central Electric power.

IPS   Interplanetary Scintillation, caused by plasma irregularities in the solar wind.

IR   Infrared.

IRAS   Infrared Astronomical Satellite.

IRTF   Infrared Telescope Facility, a 3.0 meter NASA telescope at the summit of Mauna Kea, Hawaii.

ISM   Interstellar Medium.

ISO   Infrared Space Observatory.

ISS   Interstellar Scintillation.

J	Joule. The SI unit of work and energy equal to the work done when the point of application of a force of one newton (N) moves, in the direction of the force, a distance one meter (m).
JPL	Jet Propulsion Laboratory of the California Institute of Technology. Part of NASA. It runs most USA deep space missions.
Jy	Jansky. A unit of spectral flux density widely used in radio astronomy and sometimes elsewhere: (10^{-26} Wm^{-2} Hz^{-1}) – named for Karl Guthe Jansky. He essentially began radio astronomy with his paper "Electrical Disturbances Apparently of Extraterrestrial Origin" [Jan33].
k	Boltzmann's constant; 1.3806×10^{-23} J/K.
K	The temperature scale with 0 K (zero kelvin) corresponding to absolute zero (-273.15 °C or -459.67 °F). The usual degree sign (°) is not used. The SI unit of thermodynamic temperature.
K-band	Frequency band between 18 and 26.5 GHz.
K_a-band	Frequency band between 26.5 and 40 GHz.
K_u-band	Frequency band between 12.4 and 18 GHz.
kHz	Kilohertz; 10^3 Hz.
KI	Keck Interferometer.
KL	Karhunen Loeve transform – a mathematical operation.
KPNO	Kitt Peak National Observatory in Tucson, Arizona.
kW	Kilowatt; 10^3 W.
L-band	Frequency band between 1.12 and 1.73 GHz.
LEO	Low Earth Orbit.
LLNL	Lawrence Livermore National Laboratory.
LMC	Large Magellanic Cloud
LNA	Low-Noise Amplifier, or preamplifier. In SETI, it usually implies an FET amplifier cooled to liquid nitrogen temperatures or below.
LO	Local Oscillator.
LP	Linear Polarization.

ly	Light year; 9.4605×10^{12} m, which is the distance light travels in free space in one year; 3.26 ly = one parsec (pc).
m	1. Mass, or mass density. 2. Apparent magnitude, in astronomy, namely the apparent brightness of a star. 3. Meter.
M	Absolute magnitude, in astronomy, is the apparent magnitude that an object would have if it were located 10 pc from the Sun.
MACHO	Massive Compact Halo Object.
MANIA	Multichannel Analyzer of Nanosecond Intensity Alterations. An instrument developed by Victor F. Shvartzman at the Special Astrophysical Observatory in Russia (1973 through 1974), and used to search for optical pulses. It will be used with the 2.15 m telescope at Complejo Astronomico El Leoncito (Astronomical Complex at El Leoncito), San Juan Province, Argentina.
MAP	Multichannel Astrometric Photometer.
mas	μarc second or milli-arcsecond.
MCP	Microchannel Plate.
META	Mega-channel Extraterrestrial Assay capable of monitoring more than eight million channels.
META/BETA	Mega-channel Extraterrestrial Assay / Billion-channel Extraterrestrial Assay
META1	Mega-channel Extraterrestrial Assay, Site 1, at Oak Ridge, MA.
META2	Mega-channel Extraterrestrial Assay, Site 2, in Argentina.
MHz	Megahertz; 10^6 Hz.
MIPS	Million Instructions Per Second; a measure of computational speed.
MIT	Massachusetts Institute of Technology.
MMIC	Monolithic Microwave Integrated Circuit.
MOPS	Million Operations Per Second.
MOS	1. Multi-Object Spectrograph. 2. Metal Oxide Semiconductor.
Mpc	Megaparsec = 10^6 pc.

MPIfR	Max Planck Institut für Radioastronomie (Max Plank Institute of Radioastronomy) in Bonn, West Germany.
mux	Multiplexer, or multiplex.
mW	Milliwatt; 10^{-3} Watts.
MW	Megawatt; 10^6 Watts.
NAIC	National Astronomy and Ionosphere Center.
Nançay	Observatoire de Nançay (Observatory of Nancay) in Nançay, France.
NAS	National Academy of Sciences.
NASA	National Aeronautics and Space Administration, the space agency of the United States.
NCRA	National Center for Radio Astrophysics, operated by the Tata Institute of Fundamental Research in India.
NEA	Near Earth Asteroid.
NEAR	1. Near Earth Asteroid Rendezvous. 2. A NASA space probe that orbited and then landed on asteroid EROS. It has been renamed to honor E.M. Shoemaker.
NEXRAD	Next Generation Weather Radar.
NFRA	Netherlands Foundation for Research in Astronomy.
NGC	New General Catalog.
NGST	Next Generation Space Telescope. A 6 m aperture class telescope, planned for launch about 2010.
NIF	National Ignition Facility.
NIRSPEC	Near-Infrared Spectrometer (on the Keck telescope).
NRAO	National Radio Astronomy Observatory.
NRC	National Research Council (USA).
NSF	National Science Foundation.
NSS	New Search System, a new modular signal processing system for Project Phoenix.
NTRS	National Technology Roadmap for Semiconductors.
NTSC	National Television Standards Committee.
OGLE	Optical Gravitational Lensing Experiment.
OH	Hydroxyl radical. It produces a set of four lines at 18 centimeters in the electromagnetic spectrum.

OMT	Ortho Mode Transducer.
OSS	Omnidirectional SETI System. (OSS can also mean Omnidirectional Sky Survey.)
OSU	Ohio State University.
OSURO	Ohio State University Radio Observatory.
pc	Parsec; $\approx 3.085 \times 10^{16}$ meters, or 3.262 ly.
PC	1.Printed Circuit. 2. Personal Computer.
PDR	Preliminary Design Review.
PMT	Photomultiplier Tube; a type of optical detector.
PSF	Point Spread Function; the relative distribution of energy in the focal plane of a telescope system.
PTI	Palomar Testbed Interferometer.
PW	Petawatt = 10^{15} W.
Q	Proportional to energy stored in circuit/energy dissipated in circuit during one cycle. It originated early in the development of electrical engineering as a measure of the sharpness of tuning of a resonant circuit.
QE	Quantum Efficiency.
RA	Right Ascension.
RA-DEC	Right Ascension-Declination. This is a celestial coordinate system (see RA and Declination).
RAID	Redundant Array of Inexpensive Disks.
RAL	UC Berkeley Radio Astronomy Laboratory.
R&D	Research & Development.
RF	Radio Frequency.
RFI	Radio Frequency Interference.
RFP	Request For Proposal.
RMS	Root Mean Square.
RO	Remote Observatory.
s	Second. SI base unit of time.
S/N	Signal-to-Noise ratio.
SARA	Society for Amateur Radio Astronomers.
SDI	Strategic Defense Initiative.

SEMATECH	Semiconductor Manufacturing Technology consortium.
SERENDIP	Search for Extraterrestrial Radio Emissions from Nearby Developed Intelligent Populations (one of the first SETI projects).
SETI	Search for Extraterrestrial Intelligence.
SI	Systeme Internationale d'Unites (the international system of units.
SIA	Semiconductor Industry Association.
SIM	Space Interferometer Mission.
SKA	Square Kilometer Array. A planned array with a total area of a million square meters.
SMC	Small Magellanic Cloud.
SNR	Signal-to-Noise Ratio.
SOFIA	Stratospheric Observatory for Infrared Astronomy.
SPIE	Society of Photo-Optical Instrumentation Engineers. It is also known as "SPIE, the International Society for Optical Engineering".
SSL	Solid State Lasers.
SSPM	Solid-State Photomultiplier.
SSTWG	SETI Science and Technology Working Group.
sr	Steradian; solid angle.
STEPS	Stellar Plane Survey.
STJ	Superconducting Tunnel Junction.
SZM	Shielded Zone of the Moon. The far side of the Moon, which is shielded from Earth's radio leakage.
TAI	International Atomic Time, (French).
TMC	Thinking Machines Corporation.
TVRO	Television Receiving Only.
TW	Terawatt = 10^{12} W.
UCB	University of California at Berkeley.
UCSC	University of California at Santa Cruz.
UHF	Ultra High Frequency.
UKIRT	United Kingdom Infrared Telescope.

URSI   Union Radio-Scientifique Internationale (International Union of Radio Science).

USNO   US Naval Observatory.

UTC   Coordinated Universal Time.

UV   Ultraviolet.

VC   Virgo Cluster.

VCSEL   Vertical Cavity Surface Emitting Laser.

VERITAS   Very Energetic Radiation Imaging Telescope Array System.

VHF   Very High Frequency.

VLA   Very Large Array. An array of 27 parabolic dishes in New Mexico with each dish having a diameter of 25 m. It was built in the 1970s, and has an upper limit (reception) of about 50 GHz.

VLBI   Very Long Baseline Interferometry.

VLTI   Very Large Telescope Interferometer.

VOR   VHF Omnidirectional Receiver.

WSRT   Westerbork Synthesis Radio Telescope, operated by the Netherlands Foundation for Research in Astronomy.

ZAMS   Zero Age Main Sequence.

ZD   Zodiacal Dust. The dust in our solar system which causes Zodiacal Light.

Glossary

Absolute Magnitude (M)	The apparent magnitude an object would have if it were located at a distance of ten parsecs.
Aerostat	A static lighter-than-air device, such as a tethered balloon or dirigible.
Albedo	A measure of reflectivity; the degree to which a planet, asteroid, etc. reflects light.
Ames	NASA Ames Research Center, Mountain View, California.
Antenna Feed	In a reflecting (mirror) antenna system, the device which exchanges energy between the mirror system and a transmission line system leading to a transmitter or a receiver.
Apparent Magnitude (m)	The observed brightness of a distant source of radiation ($-2.5\ log_{10}\ (\frac{P_1}{P_2})$). One magnitude is a power ratio equal to $-(100^{-5})$ or about -2.512, relative to the mean brightnesses of several groups of standard stars as they would be observed above the Earth's atmosphere. The relevant spectral band must always be given along with the magnitude. Astronomical magnitude varies inversely with brightness.
Arecibo	Short for Arecibo Observatory, the principal NAIC facility. A 305 meter radio telescope in Puerto Rico. It was built in the 1960s in a sinkhole, and is not fully steerable. It has been upgraded several times, operates up to 10 GHz, and can look at sources within 20 degrees of the zenith.
Argus Project	An amateur radio SETI effort of the SETI League, directed by Dr. Paul Shuch.
Astronomical Unit (AU)	The mean Earth to Sun distance. (1 AU $\approx 1.5 \times 10^{11}$ meters, or ≈ 93 million miles.)

Azimuth (az)	Direction angle in the horizontal plane. Measured eastward from North .
Azimuth Elevation (az-el)	Type of telescope mounting.
Band	A segment of the spectrum with a specified width (frequency or wavelength).
Bandwidth	Frequency width of a signal passband between 3 dB (half-power) points unless otherwise noted (see Power Bandwidth).
Beacon	An omnidirectional, continuous, or regularly intermittent signaling device to get the attention of anyone listening.
Beamwidth	1. Angular 3 dB width of main lobe of the directivity pattern of an antenna (unless otherwise noted); 2. Usually the angular distance between half-power points as measured in the E- and H- (principal) planes containing the pointing (Boresight) axis. Often designated by the more awkward "full HPBW", or "full width HPBW", or just HPBW, rather than BW or $\theta_{1/2}$.
Bel (B)	Unit measure of power ratio equal to the common logarithm of the ratio log_{10} $(\frac{P_1}{P_2})$.
Billion	10^9; one thousand million; or 1,000,000,000.
Bin	A segment of a band and spectrum with a defined width (e.g., 1 Hz) from which a sample is collected for a defined time period (assumed to be 1 s, unless otherwise specified). A bin represents one sample collected at one time from a specified band.
Binwidth	Half-power resolution bandwidth of an analyzer output channel.
Blackbody	A totally absorbing body that does not reflect radiation. In thermal equilibrium a blackbody absorbs and radiates at the same rate; the radiation will just equal absorption when thermal equilibrium is maintained.
Blueshift	The shift to shorter wavelengths of light radiated from an object caused by its moving towards the observer.
Boltzmann's Constant	The number that relates the average energy of a molecule to its absolute temperature. It is approximately 1.3806×10^{-23} J/K.

C-band	Frequency band between 3.95 and 5.85 GHz. Contains adjacent portions of the S-band and the X-band.
Carrier	A signal on which another, usually more complex signal can be impressed. A carrier is often a very good approximation to a pure tone, and so stable that its 3 dB bandwidth over ten seconds is less than 0.1 Hz. Typical stability is better than 10^{-11} of the frequency, when controlled by an atomic frequency standard.
Cassegrain Antenna	A directional microwave antenna with a feed or subreflector that radiates forward toward a semi-spherical reflector, located in front of the focus of a parabolic main reflector.
Cassegrain Telescope	1. A telescope in which light rays are reflected from two mirrors, a large concave mirror and a small convex secondary mirror; 2. A reflecting telescope that has a paraboloidal primary mirror and hyperboloidal secondary mirror and a principle focus just in front, in, or just behind the primary mirror. Invented by a French physician in 1672, it is by far the most common optical, infrared, and radio telescope architecture in use today.
Celestial Equator	Great circle on the Celestial Sphere formed by intersection with Earth's equatorial plane.
Celestial Sphere	An imaginary sphere of infinite radius on which all celestial bodies seem to be projected.
Cepheid	A pulsating variable star. The period of pulsation is directly related to a Cepheid's intrinsic brightness, making observations of these stars one of the most powerful tools for determining distance.
Cerenkov Radiation	Radiation emitted when a charged particle moves through a medium faster than the speed of light in that medium. When particles enter the atmosphere at speeds exceeding that of light in air, flashes can be observed.
Chirality	At the molecular level, the tendency of all known life forms to display a 'handedness'.
Coherence	A systematic or methodical connectedness or interrelatedness.
Columbus Optical SETI	The optical SETI project operated by Dr. Stuart Kingsley in Columbus, Ohio (COSETI).

Commensal Search   A serendipitous SETI observing mode in which part or all of a SETI receiving system operates in parallel with an astrophysical (or other) observing program. Where the radio telescope is pointed, and for how long, is determined by the astrophysical (or other) requirements. This is a useful way to do SETI when higher priority SETI observations are not possible. Such searches are sometimes called *piggyback* or *parasitic* searches.

Continuous Wave   An essentially single-frequency electromagnetic wave.

Coulomb's Law   States that the force between two electrically charged particles is proportioned to the product of their magnitudes, and inversely proportional to the square of the distance between them. The force between like charges is a repulsion, and the force between unlike charges is an attraction.

CSIRO   Commonwealth Scientific and Industrial Research Organization, Australia.

Cyclops Design Study   The "1971 Summer Stanford University/NASA Ames Faculty Fellowship Program in Engineering Systems Design" that resulted in the publication of the *Project Cyclops* report. It was conducted at Ames Research Center by B. M. Oliver and J. Billingham. Also see *Project Cyclops* report and Cyclops system.

Cyclops System   A proposed array of up to twenty-five hundred radio telescopes, dedicated to SETI activities, that was studied under NASA auspices in 1971 but never built. (See NASA CR 114445, 2nd ed. 1973.) Also see *Project Cyclops* report and Cyclops Design Study.

Dark Matter   A form of matter which has not been directly observed but has been detected by its gravitational effects.

Decibel Antenna Gain   (dBi)Refers to an isotropic antenna.

Declination   1. Angular distance of an object above (North or positive) or below (South or negative) the celestial equator. It is the celestial equivalent of latitude for establishing positions of planets and stars; 2. One of the co-ordinates, with right ascension, that defines the position of a heavenly body. It has a value between $0°$ and $90°$ and is labeled North (+) or South (-), with $0°$ being on the celestial equator.

Detection	Detection is the conversion of a complex (vector) electromagnetic wave, for example, $A[\cos(2\pi\nu\tau + \phi) + j\sin(2\pi\nu\tau + \phi)]$, into a scalar electromagnetic wave which is some (usually simple) function of one or more of the parameters of the complex wave.
	In the case of a digital analog of an electromagnetic wave, where the detection is not further qualified, the digitized output of the detection process is usually a digital analog of either the absolute value of the amplitude, $|A|$, or of its square, A^2, which is a measure of the total power in the electromagnetic wave.
	In SETI, unless stated otherwise, the simple term "detection" implies the generation of this measure of the total power of the electromagnetic wave.
Doppler Compensation	Allowance made for the Doppler shift.
Doppler Effect	The apparent frequency change of radiant energy, varying with the relative velocity of the source and the observer.
Doppler Shift	The shift in the frequency of light or sound caused by the relative motion of the source and observer.
Drake Equation	An equation used to estimate the average number of civilizations in the Universe, formulated in 1961 by Frank Drake. (See page xxxv.)
Dyson Sphere	Freeman Dyson suggested that a sufficiently advanced civilization would inhabit the interior surface of a sphere built to enclose its sun, thereby utilizing all of its energy output.
Effective Isotropic Radiated Energy (EIRE)	It is the product of the antenna gain in a given direction and the real transmitted energy, and is measured in joules. If the direction is not specified, then the direction of maximum gain is assumed.
Effective Isotropic Radiated Power (EIRP)	It is the product of the antenna gain in a given direction and the real transmitted power, and is measured in watts. If the direction is not specified, then the direction of maximum gain is assumed. EIRP was formerly referred to as ERP.
Elevation	Angle of a direction above the horizon.

Euclidean Geometry	The geometry developed by Euclid about 300 BC. Euclidean geometry, like all geometries, deduces certain results from a set of starting assumptions. One of the critical assumptions of Euclidean geometry is that for any given straight line and a point not on that line, there is exactly one line that can be drawn through that point parallel to the first line. One of the results of Euclidean geometry is that the interior angles of any triangle sum to 180 degrees. Euclidean geometry is the geometry of conventional three-dimensional space.
Exoplanet	A planet orbiting a star other than our Sun. Sometimes they are referred to as "extra-solar planets".
Fast Fourier Transform (FFT)	Name for the Cooley-Tukey algorithm used in digital computing.
Field Effect Transistor (FET)	A solid-state device which, when properly used in an amplifying system cooled to cryogenic temperatures (Cooled-FET-Amplifier, CFA), can provide amplification for frequencies throughout the microwave window with exceptionally low intrinsic noise.
Figure-of-Merit	A numerical quantity, based on one or more parameters of a system that represents a measure of quality, efficiency or effectiveness. It is a value that can be assigned to an option for comparison with other options.
Footprint	The area on the ground covered by, for example, a beam or a piece of equipment.
Fourier Transform (Fourier Principle)	It shows that all repeating waveforms can be resolved into sine or cosine wave components consisting of a fundamental and a series of harmonics at multiples of this frequency. It can be extended to prove that non-repeating waveforms occupy a continuous frequency spectrum.
Fraunhofer Lines	Specifically, any of the 754 dark lines on the solar spectrum observed by Joseph von Fraunhofer in 1817, and later shown to be absorption lines caused by the presence of different elements in the Sun's atmosphere. The nine most prominent he labeled with capital letters (from the red end) A, B, C, D, E, F, G, H, and K. The A and B bands (at 7,600 and 7,100 Å) are now known to be groups of telluric lines due to O_2 absorption in Earth's atmosphere, and C and F are respectively known as H_α and H_β. Generally, any absorption line detected in a stellar atmosphere.

491

Frequency Shift Keying (FSK)	Frequency modulation in which the modulating signal shifts the output frequency between predetermined values. Usually, the instantaneous frequency is shifted between two discrete values termed the "mark" and "space" frequencies. This is a noncoherent form of FSK. Coherent forms of FSK exist in which there is no phase discontinuity in the output signal.
GAIA	GAIA is an ambitious space observatory for astronomy, adopted within the scientific program of the European Space Agency in October 2000. It plans to measure the positions of more than 1 billion stars with unprecedented accuracy in a global stellar census of our Galaxy and its nearest neighbors. In addition, it will obtain multi-colour photometry as crucial diagnostic data for all stars observed, along with radial velocities for the brighter objects to complete the kinematic data. As a result, the distances and motions of the stars in our Galaxy will be determined with extraordinary precision, allowing astronomers to determine the three-dimensional structure of our Galaxy, and space velocities of its constituent stars. It will further our understanding of our Galaxy's origin and evolution. GAIA should be launched around 2010 through 2012, and will be operated for 5 years.
Gain	A power ratio, P_{out}/P_{in}, though on occasion it is useful to talk of amplifier voltage gain. When the gain of a circuit element is less than unity, the term attenuation is more common than gain. With antennas, directional gain is calculated with respect to the gain of a purely theoretical, loss free, isotropic antenna. In practice, two terms are used to express the directional property of an antenna, gain and directivity. Gain is equal to the (usually theoretically calculated) directivity minus the effective (or equivalent) I^2R loss between the incident (or emitted) free-space electromagnetic wave and the antenna terminals which meet the attached receiver (or transmitter). This loss, due to finite conductivity effects and the unaccounted for scattering and depolarization phenomena, can be significant in some cases.
Galactic Bulge	Galaxies tend to be approximately lens-shaped; the center thicker than the edge. The bulge refers to the thicker central portion of the galaxy.

Galactic Longitude	Established circa 1960 when the position of the NGP and the zero longitude point were redefined.
Gaussian Noise	The usually undesirable electromagnetic disturbances whose density function follows a "normal" bell-shaped or Gaussian distribution.
Goldstone	One of the major NASA tracking stations, in California.
Gravitational Lens	A distant object that deflects light by gravitation as described by the general theory of relativity. It is analogous to a lens in optics. The prediction of a gravitational lensing effect in general relativity theory has been confirmed in observations on quasars.
Ground Scattering	That part of the signal received by a radio telescope that has been reflected from the nearby ground.
Habitable Zone	The region around a star within which the temperature is presumed to be suitable for life to exist; if a body is too far from a star, then it is too cold for life; if a body is too close to a star, then it is too hot for life. The interpretation has changed over time.
Hamming Window	A weighting function used with Fourier Transform spectral analysis, which is useful for resolving closely spaced frequencies.
Hanning Window	A weighting function used with Fourier Transform spectral analysis, which is particularly useful where a large dynamic range between spectral lines occurs.
Hectare	10,000 square meters (2.471 acres).
Helios Laser	Lasers used in the Helios Laser Fusion Facility at Los Alamos National Laboratory.
Hertz (Hz)	A unit of frequency. Cycles-per-second (cps or c/s) is the classic English usage, now outmoded by international agreement.
Hertzsprung-Russell Diagram	Plot of stellar absolute magnitudes versus stellar photospheric temperature.
Heterodyne	A detection method, used extensively in radio astronomy, in which the wave nature of light is used. The method usually involves combining the measured wave with a local oscillator or reference wave and looking for the signal at the difference frequency.
High Frequency	Decametric wave range of 3 to 30 MHz.

Hipparcos	High-Precision Parallax Collecting Satellite. A European space probe astrometric mission from November 1989 through March 1993. It accurately determined positions for nearly 120,000 stars in three dimensions.
Hot Jupiters	Extrasolar planets that are similar to Jupiter, but hotter.
Hour Angle	The angle (measured westward) between the meridian and an hour circle. The hour angle of a star depends both on time and on the observer's location. It can be determined by subtracting the star's right ascension from the local sidereal time.
Hour Angle-Declination	HA-DEC. Coordinates for the celestial coordinate system based on Earth's equatorial plane and the observer's prime meridian (north-zenith-south). It is also used to designate an equatorial telescope mounting .
Hubble Telescope	The Earth-orbital optical telescope launched in April 1990.
Integrated Circuit (IC)	A small package that contains the functions of many electrical components.
Interferometer	1. A device on an infrared or optical telescope that measures the angular diameter of a star by the spacing of its interference fringes. 2. A radiotelescope, consisting of two or more dishes that measure minute angular distances, as small as 1 milli-arcsecond, from interference between received radio waves.
Isokinetic Patch	The portion of the sky where the stars appear to move together as a result of atmospheric turbulence.
Isotropic	Having one or more physical properties that are the same regardless of the direction of measurement.
K-band	Frequency band between 18 and 26.5 GHz.
K_a-band	Frequency band between 26.5 and 40 GHz.
K_u-band	Frequency band between 12.4 and 18 GHz.
Kardashev Type	Kardashev defined possible civilizations according to

Civilizations	their ability to control energy. Type I could control or harness all radiation falling on their planet from their sun (for Earth, about 10^{17} W), Type II could control or harness the entire energy output of their sun (for our Sun, about 10^{27} W), Type III could control the energy of their entire galaxy (for the Milky Way, about 10^{38} W).
KARST	The Chinese version of the SKA, so named for its location in a limestone karst formation.
Keck Telescope	A pair of 10 meter diameter optical telescopes on Mauna Kea in Hawaii.
Kelvin	The SI unit of thermodynamic temperature equal to the fraction 1/273.16 of the thermodynamic temperature of the triple point of water. The temperature scale with 0 K (zero Kelvin) corresponding to absolute zero (-273.15 °C or -459.67 °F). The usual degree sign (°) is not used.
KT Boundary	The geologic boundary between the Cretaceous period (K) and the Tertiary period (T), which occurred about 64-66 million years ago.
KT Boundary Event	The Cretaceous-Tertiary boundary event that coincided with the extinction of the dinosaurs.
KT Extinction	Refers to the extinction of dinosaurs at the K-T boundary.
L-band	Nominally, frequency band between 1.12 GHz and 1.73 GHz.

Lagrangian Points	Five points in the orbital plane of two massive particles in circular orbits around a common center of gravity, where a third particle of negligible mass can remain in equilibrium. Three of the points are on the line passing through the centers of mass of the two bodies – L_2 beyond the most massive body, L_1 the point at which control transfer occurs between the two bodies, and L_3 beyond the less massive body. All three of these points are in unstable equilibrium. The other two (L_4 and L_5) are stable, and are located at the two points in the orbit of the less massive component leading and trailing it by $60°$, and are equidistant from the two main components. These are points in space where the pull of gravity, by two or more bodies such as the Earth and Moon, is equalized and where another object would remain in relatively stable position. Jupiter's Trojan asteroids orbit around the Jupiter-Sun L_4 and L_5 points.
Laser	A device used for light amplification by stimulated emission of radiation (LASER). Lasing can occur throughout most of the electromagnetic spectrum: at frequencies in the radio, microwave, infrared, visual, UV, X-ray, and perhaps even γ-ray regions. Analogous to Maser.
Leakage	Radiated energy, e.g., radio and TV signals, that could be detected at a distance from the planet of origin.
Lensing	See Gravitational lensing and Nanolensing.
Light Bucket	An inexpensive mirror, usually with spherical rather than parabolic optics, used when precise focus is not required.
Light Year (ly)	The distance light travels in free space in one year. It is equal to 9.4605×10^{15} m. Also, 3.262 ly = one parsec.
Low-Noise Amplifier	In SETI, it usually implies an FET amplifier cooled to liquid nitrogen temperatures or below (LNA).
Magic Frequencies	Frequencies (wavelengths) which are likely to be interesting to intelligent life. See *Water Hole*.
Maser	A device for microwave amplification by stimulated emission of radiation; analogous to Laser.
Milky Way	A bright band of stars in the night sky; another name for our Galaxy.

Modulation	The process of impressing one signal on another. Usually one signal bears information of interest while the other is a carrier or subcarrier which defines where and how the combined signal appears in the spectrum.
Moore's Law	The observation made in 1965 by Gordon Moore, co-founder of Intel, that the number of transistors per square inch on integrated circuits had doubled every year since the integrated circuit was invented. Moore predicted that this trend would continue for the foreseeable future. In subsequent years, the pace slowed down, but data density has doubled approximately every 18 months, and this is the current definition of Moore's Law, which Moore himself has blessed. Most experts, including Moore himself, expect Moore's Law to hold for at least another two decades.
Muons	A muon is a lepton that is essentially a more massive electron. They are short-lived elementary particles with negative electrical charge. They resemble electrons, but are 207 times more massive. A muon was formerly called a mu-meson, but now is classified as a lepton. Muons are produced when cosmic rays enter the upper atmosphere, and in high-energy accelerators.
Multiplexing	When two or more effectively continuous signals pass apparently simultaneously through the same signal circuit elements, they are said to be multiplexed. Signals may be multiplexed in time (TDM), in frequency (FDM), by the use of orthogonal polarizations (PMX), by spread spectrum techniques, or any combination of these approaches.
n/a	Not applicable.
Nanolensing	Possible variation in gravitational microlensing of a distant star caused by its planets. (See Gravitational Lensing.)

Nebula	Originally a fixed, extended and somewhat fuzzy white haze observed in the sky with a telescope. Many of these objects can now be resolved into clouds of individual stars and have been identified as galaxies. The gaseous nebulae, however, cannot be resolved into individual stars and consist, for the most part, of interstellar dust and gas. In some gaseous nebulae, the gas atoms have been ionized by ultraviolet radiation from nearby stars, and light is emitted as these ions interact with the free electrons in the gas. These are called emission nebulae. In the dark nebulae, there are no nearby stars and these objects are consequently dark; they can only be detected by what they obscure. Some nebulae, such as the one in Orion's sword, contain stars being born. Others, called planetary nebulae, are gas clouds ejected from dying stars.
Nyquist Theorem	States that the digital sampling rate should be twice the highest frequency contained in the signal being digitized. Any lower rate will lose information.
One Hectare Telescope	One Hectare (Radio) Telescope, which will consist of many small antennas, providing a collecting area totalling 10,000 m^2, equivalent to the area of a single dish 113 m in diameter. The R&D installation, consisting of seven dishes, began operating in April 2000.
Octave	The span over which the frequency doubles or halves; e.g., middle C is 262 Hz; the C one octave above it is 524 Hz. The observed electromagnetic spectrum covers a range of 22 decades or about 73 octaves – from about 10 to about 10^{23} Hz.
Optical Link	An optical transmission channel, including any repeaters or regenerative repeaters, designed to connect two electronic or optoelectronic communications terminals. An optical link is sometimes considered to include the terminal optical transmitters and receivers.
Order of Magnitude	A value expressed to the nearest power of ten.
Order of the Dolphin	An informal association of the attendees of the first conference on SETI in 1961 at the Green Bank Observatory in West Virginia, during which the Drake Equation was first elucidated.

Origins	A NASA program to pursue fundamental questions about the evolution of the Universe, including the development of life.
Parameter Space	In SETI, a combination of target directions, frequency range, modulation, sensitivity and time.
Parkes	The Parkes 64 m radio telescope of the Australia Telescope National Facility (ATNF).
Parsec (pc)	Short for parallax second. Distance corresponding to an astronomical parallax of one arc second; distance at which one AU subtends one arc second of angle. It is equal to 3.26 ly, or 3.09×10^{16} m.
Phoenix	Targeted microwave search currently being conducted at Arecibo, so named because it represented the rebirth of SETI searches after Congress terminated the NASA survey.
Pink Noise	Similar to White Noise, but with an excess of power at lower frequencies. See *White Noise*
Planck's Constant (h)	The fundamental constant equal to the ratio of the energy (E) of a quantum to its frequency (ν) or $E = h\nu$. It is equal to 6.6252×10^{-34} Js.
Poisson Distribution	The distribution of count n of the number of randomly occurring events in a fixed time interval. It has many applications as a distribution function for the occurrence of rare events. The probability distribution can be expressed as $f(n) = \frac{k^n e^{-k}}{n!}$, where $n = 0, 1, 2, 3, ...$ and k is the average rate per unit time, valid when $k \ll 1$ The Poisson distribution is an approximation to the binomial distribution used when the probability of success in a single trial is very small and the number of trials is very large.

Polarization	A plane electromagnetic wave which exhibits different properties in different directions at right angles to the line of propagation is said to be polarized. In radio physics, radio astronomy, and SETI, but not in classical optics, these properties are described as though looking along the direction of propagation and in relation to the directional behavior of the electric vector. All physical radiators are intrinsically polarized and, in the general case, elliptically polarized. If the axial ratio of the emitted wave is nearly infinite or nearly zero, the radiation is said to be linearly polarized. If the axial ratio is unity, the polarization is said to be circular, and the electric vector at a fixed point along the direction of propagation rotates through 360 degrees once every cycle of the electromagnetic wave. If this rotation is clockwise (as seen from the transmitter) the angular momentum vector is in the direction of propagation, the helicity is said to be positive and the electromagnetic wave to be right circularly polarized (RCP); and vice versa for LCP.
Prime Meridian	A great circle on the celestial sphere passing through the celestial poles and the observer's zenith.
Project Cyclops Report	The 1972/1973 published report resulting from the Cyclops Design Study. (Also see Cyclops Design Study and Cyclops system.)
Project Orion	A design study of a system, for detecting extrasolar planets. An account of the 1976 summer study directed by D. C. Black, from Ames, and R. Piziali, from Stanford University. A study spun-off by the 1975 through 1976 Science Workshops on Interstellar Communication, chaired by Philip Morrison. See NASA SP-436 (1980).
Project Ozma	The first systematic, high-sensitivity search for manifestations of ETI life, conducted by Frank Drake in 1960.
Project SERENDIP	A commensal SETI arrangement developed at UC Berkeley and carried out at the Hat Creek Radio Astronomy Observatory (HCRAO) and at JPL-Goldstone. SERENDIP taps the IF output of a receiver making radio astronomical observations and performs part of a multichannel spectral analysis. The resulting data are tape recorded and the power spectrum calculation is completed off-line.
Proper Motion	The transverse motion of one star relative to other stars.

Proterozoic   The eon of geologic time or the corresponding system of rocks that includes the interval between the Archean and Phanerozoic eons. It probably exceeds in length all of the subsequent geologic time periods. It is marked by rocks that contain fossils indicating the first appearance of eukaryotic organisms (e.g., algae).

Quasar   A class of astronomical objects that appear on optical photographs as starlike but have large redshifts quite unlike those of stars. They were first observed in 1961 when it was found that strong radio emission was emanating from many of these starlike bodies. Over 600 such objects are now known and their redshifts can be as high as 4. The redshifts are characteristic of the expansion of the Universe. If the redshifts are cosmological, as favoured by most astronomers, quasars are the most distant objects in the Universe, some being up to 10^{10} light years away. The exact nature of quasars is unknown but it is believed that they are the nuclei of galaxies in which there is violent activity. The luminosity is so much greater than the rest of the galaxy that the source appears point-like. It has been proposed that the power source in a quasar is a supermassive black hole accreting material from the stars and gas in the surrounding galaxy.

Racemic   A mixture of usually equal quantities of the (+), d- or dextrorotatory and (-), l- or levorotatory entantiomers of a molecule that has chirality, and which is optically inactive.

Radial Velocity   The "line of sight" motion of a body or galaxy relative to the Sun.

Radian   A unit of angular measure, approximately equal to 57.295 degrees, which is equal to the angle formed at the center of a circle by two radii cutting off an arc whose length is equal to the radius of the circle; $2\pi\ radians = 360°$.

Redshift   The reddening of light radiated from an object caused by its moving away from the observer.

Right Ascension   The longitude-like component of the astronomer's equatorial coordinate system. See Declination.

Salpeter Initial A simple functional interpolation for the distribution

Mass Function	by mass of newly formed stars. The Salpeter function (the number of stars formed per unit mass range) is proportional to $m^{-2.35}$, where m is the mass of a star.
Seeing	Describes the blurring of a stellar (point-like) image due to turbulence in the Earth's atmosphere, both at high altitudes and within the telescope dome. Seeing estimates are often given in terms of the full-width in arc seconds of the image at the points where the intensity has fallen to half its peak value. The typical value at a good site is a little better than 1 arc second.
SEMATECH	Semiconductor Manufacturing Technology consortium. SEMATECH and International SEMATECH are unique endeavors of 14 semiconductor manufacturing companies from seven countries. It recently entered into globalization with the formation of International SEMATECH from the original consortium. SEMATECH companies cooperate, before competing, in key areas of semiconductor technology, sharing expenses and risk.
Sensitivity	Minimum detectable signal flux with acceptable false alarm probability.
SETI	Search for Extraterrestrial Intelligence. A passive exploration of the microwave, infrared /optical regions of the spectrum for artificial signals which might be present among the natural extraterrestrial signals.
SETI@home	The first SETI project that uses personal computers as a distributed network, conceived by David Anderson and Dan Werthimer at the University of California at Berkeley. The URL is `http:setiathome.ssl.berkeley.edu`.
SETI League	A membership-supported nonprofit educational and scientific organization searching for ETI.
Shannon's Theorem	Defines the capacity of a network in terms of its bandwidth and signal-to-noise ratio (C. E. Shannon, 1948).
Signal	In general, any electromagnetic wave; but may refer selectively to, e.g., radiation from a pulsar, terrestrial RFI (QRM or QRN), radar transmission, or band-limited thermal noise.

Signal-to-Noise Ratio(SNR)(S/N)

Commonly, the ratio of the power of a particular signal to the combined power of all other signals simultaneously present at the input to the detection processor. SNR is maximized by use of a matched filter. After detection, the symbol α is commonly used for the corresponding ratio.

Simple Measure of Search Merit

The product of the number of detectable stars (as a function of the assumed transmitter power in Watts EIRP) and the number of decades of frequency (the number of bands) observed.

Sinusoid

Shaped like, or varying like a sine curve or wave.

Square Law Detector

A device producing an output current or voltage which is proportional to the square of the input signal voltage.

Steradian (sr)

The solid angle which encloses a surface area on a sphere equal to the square of the radius of the sphere. There are 4π steradians in a sphere.

Water Hole

The part of the radio spectrum between 1.4 GHz (hydrogen) and 1.7 GHz (hydroxyl radical). Together, they imply the presence of water. The band has comparatively little background noise within it.

White Noise

Completely random and uncorrelated noise with the power spread equally across all frequencies.

Windowing

A weighting function applied to a time series of data values. (e.g., see Hamming and

Function

Hanning Windows.)

Working Group

The SSTWG.

Zenith

Local vertical direction.

Zodiacal Light

A faint glow that extends away from the Sun in the ecliptic plane of the sky, visible to the naked eye in the western sky shortly after sunset or in the eastern sky shortly before sunrise. Its spectrum indicates it to be sunlight scattered by interplanetary dust. (Pioneer 10 has determined that its brightness varies inversely as the square of the distance out to 2.25 AU and then decreases more rapidly.) The zodiacal light contributes about a third of the total light in the sky on a moonless night.

Bibliography

[Alc78]   Alcock, C., & Hatchett, S., 1978, "The Effects Of Small-Angle Scattering on a Pulse of Radiation with an Application of X-Ray Bursts and Interstellar Dust", *ApJ*, 222.

[All63]   Allen, C. W., 1963, *Astrophysical Quantities*, University of London, The Athlone Press.

[Alm95]   Almar, I., 1995, "The Consequences of Discovery: Different Scenarios", *Progress in the Search for Extraterrestrial Life*, ASP Conference Series 74, 499.

[Alv80]   Alvarez, L. W., Kasting, J. F., & Alvarez, W., et al., 1980, "Extraterrestrial Cause for the Cretaceous-Tertiary Extinction", *Science*, 108, 4448.

[And00]   Anderson, D., 2000, "Internet Computing for SETI", *Bioastronomy '99 - A New Era in Bioastronomy*, Proceedings of a Conference held on the Kohala Coast, Hawaii, ASP Conference Series 213, 511. See http://setiathome.ssl.berkeley.edu.

[Ark86]   Arkhipov, A. V., 1986, "Academy of Sciences Ukranian SSR Institute of Radiophysics and Electronics", *preprint* No. 303.

[Bac98]   Backus, P., 1998, "The Phoenix Search Results at Parkes", *Acta Astronautica*, 42, 10-12, 651-654.

[Bai98]   Bailey, J., Chrysostomou, A., Hough, J. H., Gledhill, T. M., McCall, A., Clark, S., Menard, F., & Tamura, M., 1998, "Circular Polarization in Star-Formation Regions: Implications for Biomolecular Homochirality", *Science*, 281, 672.

[Ban93]   Bania, T. M, & Rood, R. T., 1993, "Search for Interstellar Beacons at the 3He+ Hyperfine Transition Frequency", *Third Decennial US-USSR Conference on SETI*, Santa Cruz, California, August 1991, ASP Conference Series 47, ed. S. Shostak, Astronomical Society of the Pacific, San Francisco, 357-365.

[Bck68]   Becklin, E. E., & Neugebauer, G., 1968, "Infrared Observations of the Galactic Center", *ApJ*, 151, 145.

[Bck78] Becklin, E. E., Neugebauer, G., Willner, S. P., & Matthews, K, 1978, "Infrared Observations of the Galactic Center IV - The Interstellar Extinction", *ApJ*, 220, 831.

[Bck97] Becklin, E. E., 1997, "The Far Infrared and Submillimetre Universe", *ESA Symposium*, SP-401, Grenoble.

[Bec96] Beckwith, S. V. W., & Sargent, A. I., 1996, "Circumstellar Disks and the Search for Neighbouring Planetary Systems", *Nature*, 383, 139.

[Ben96] Bennet, D., & Rhie, S. H., 1996, "Detecting Earth-Mass Planets with Gravitational Microlensing", *ApJ*, 472, 660.

[Ben99] Bennet, D., & Rhie, S. H., 1999, "Astronomers Observe Distant Solar System with Possible Neptune-like Planet", *Press Release*, January 8, University of Notre Dame.

[Bet86] Betz, A. L., 1986, "A Direct Search for Extraterrestrial Laser Signals", *Acta Astronautica*, 13.10, 623-629.

[Bet93] Betz, A. L., 1993, "A Search for Infrared Laser Signals", *Third Decennial US-USSR Conference on SETI*, ASP Conference Series 47, ed. G. S. Shostak, 373.

[Bha00] Bhatal, R. See http://www.coseti.org/ragbir00.htm.

[Bil99] Billingham, J., et al., 1999, *Societal Implications of the Detection of an Extraterrestrial Civilization*, SETI Press, Mountain View, California.

[Bir83] Biraud, F., 1983, "SETI at the Nançay Telescope", *Acta Astronautica*, 10, 759.

[Bla91] Blair, D. G., Norris, R., Wellington, K. J., Williams, A., & Wright, 1991, "A Test for the Interstellar Contact Channel Hypothesis in SETI", *Bioastronomy: The Search for Extraterrestrial Life*, eds. J. Heidmann, & M. Klein, Springer-Verlag, Berlin, LNP 390, 271-279.

[Bla92] Blair, D. G., et al., "A Narrow-Band Search for Extraterrestrial Intelligence (SETI) Using the Interstellar Contact Channel Hypothesis", *MNRAS*, Monthly Notices, 257.1, 105.

[Bla93] Blair, D. G., & Zadnik, M. G., 1993, "A List of Possible Interstellar Communication Channel Frequencies for SETI", *A&A*, 278, 669.

[Blo87] Bloembergen, N., Patel, C. K. N., et. al, 1987, "Report to The American Physical Society of the Study Group on Science and Technology of Directed Energy Weapons", APS Study Group, *Review of Modern Physics*, 59.3, 1-200.

[Bnd98]   Benedict, G. F., McArthur, B., & Nelan, E. P., et al., 1998, "Working a Space-based Optical Interferometer: HST Fine Guidance Sensor 3 Small-field Astrometry", *Astronomical Interferometry*, Proceedings of SPIE, ed., R. D., Reasenberg, 3350, 229-236.

[Bnd99]   Benedict, G. F., McArthur, B. , & Chappell, D. W., et al., 1999, "Interferometric Astrometry of Proxima Centauri and Barnard's Star Using Hubble Space Telescope Fine Guidance Sensor 3: Detection Limits for Sub-stellar Companions", *AJ*, 118, 1086-1100.

[Bod99]   Boden, A. F., Koresko, C. D., et al., 1999, "The Visual Orbit of IOTA Pegasi", *ApJ*, 515, 356.

[Bon00]   Bond, I. H., et al., 2000, "The Very Energetic Radiation Imaging Telescope Array System (VERITAS), Proposal to DOE/NSF".

[Bos96]   Boss, A. P., 1996, "Forming a Jupiter-like Companion for 51 Pegasi, Lunar and Planetary", *Science*, 27, 139.

[Bow83]   Bowyer, S, Zeitland, G. M., Tarter, J., Lampton, M., & Welch, W. J., 1983 , "The Berkeley Parasitic SETI Program", *Icarus*, 53, 147-155.

[Bow88]   Bowyer, S., Wertheimer, D., & Lindsay, V., 1988, "The Berkeley Piggyback SETI Program: SERENDIP II", *Bioastronomy: The Next Steps*, ed. G. Marx, Kluwer Academic Publishers, Dordrecht, The Netherlands & New York, NY.

[Bow00]   Bowyer, S., Korpela, E., Werthimer, D., Lampton, M., Cobb, J., & Lebofsky, M., 2000, "Final Results of the SERENDIP III Search", *Bioastronomy 99: A New Era in Bioastronomy*, ASP Conference Series 213, eds. G. Lemarchand & K. Meech, 473.

[Bra74]   Bracewell, R. N., 1974, *The Galactic Club: Intelligent Life in Outer Space*, W. H. Freeman and Company, San Francisco.

[Brn99]   Brand, S., 1999, *The Clock of the Long Now*, Basic Books, New York.

[Bta98]   For BETA Project information
           See http://seti.harvard.edu/seti/beta.html.

[Bulge]   Derue, F., & Alfonso, C., et al., 1999, "Observation of Microlensing Towards the Galactic Spiral Arms. Eros II 2 year Survey", *Astronomy and Astrophysics*.
           See http://xxx.lanl.gov/abs/astro-ph/9903209.

[Bur92]   Burke, B. F. (chair), 1992, "TOPS: Toward Other Planetary Systems", NASA Solar System Exploration Division, Washington, DC.

[But96]   Butler, R. P., Marcy, G. W., Williams, E., McCarthy, C., Dosanjh, P., & Vogt, S. S, 1996, "Attaining Doppler Precision of 3 meters/sec", *PASP*, 108, 500.
See http://seti.ssl.berkeley.edu/opticalseti/.

[But97]   Butler, R. P., & Marcy, G. W., 1997, "The Lick Observatory Planetary Search", *Astronomical and Biochemical Origins and the Search for Life in the Universe*, Proceedings at IAU Colloquium 161, eds. C. Cosmovici, S. Bowyer, & D. Werthimer, Editrice Compositari, Bologna, 331.

[Cam63]   A. G. W., Cameron, ed., 1963, *Interstellar Communication*, W. A. Benjamin Inc., New York, NY.

[Car83]   Carter, B., 1983, "The Anthropic Principle and its Implications for Biological Evaluation", *Philosophical Transactions of the Royal Society of London*, Series A, 310, 352.

[Cat83]   Catura, R. C., 1983, "Evidence for X-ray Scattering by Interstellar Dust", *ApJ*, 275, 645.

[Chy92]   Chyba, C. F., & Sagan, C., 1992, "Endogenous Production, Exogenous Delivery, and Impact-shock Synthesis of Organic Molecules: an Inventory for the Origins of Life", *Nature*, 355, 125.

[Chy98]   Chyba, C. F., & Bradley, C., 1998, "Radar Detectability of a Subsurface Ocean on Europa", *Icarus*, 134, 292.

[Coa94]   Colavita, M. M., & Shao, M., 1994, "Indirect Planet Detection with Ground-Based Long-Baseline Interferometry", *Astrophysical Space Science*, 212, 385-390.

[Coc59]   Cocconi, G., & Morrison, P., 1959, "Searching for Interstellar Communications", *Nature*, 183, 844.

[Coh80]   Cohen, N., Malkan, M., & Dickey, J., 1980, "A Passive SETI in Globular Clusters at the Hydroxyl and Water Lines", *Icarus*, 41, 198-204.

[Coh95]   Cohen, N., & Charlton, D., 1995, "Polychromatic SETI", *Progress in the Search for Extraterrestrial Life*, ASP Conference Series 74, ed. G. S. Shostak, 313.

[Col79]   Cole, T. N., & Ekers, R. D., 1979, "A Survey for Sharply Pulsed Emissions", Proceedings of the Astronomical Society of Australia, 3, 328.

[Com92]   Colomb, F. R., Maartin, M. C., & Lemarchand, G. A., 1992, "SETI Observational Program in Argentina", *Acta Astronautica*, 26, 3/4, 211-212.

[Com95]   Colomb, F. R., Hurrel, E. E., Lemarchand, G. A., & , 1995, "Re-
           sults of Two Years of SETI Observations with META II", *Progress
           in the Search for Extraterrestrial Life*, ASP Conference Series 74,
           ed. S. Shostak, Astronomical Society of the Pacific, San Francisco,
           345-352.
           See `http://www.planetary.org/html/UPDATES/seti/META2`
           `/default.html`.

[Cor91a]   Cordes, J. M., & Lazio, T. J. W. L., 1991, "Interstellar Scattering
           Effects on the Detection of Narrow-Band Signals", *ApJ*, 376, 123.

[Cor91b]   Cordes, J. M., & Lazio, T. J. W. L., 1991, "Interstellar Scintillation
           and SETI", *NAIC Report*, No. 289.

[Cor91c]   Cordes, J. M., 1991, "Statistics of Scintillation Modulated SETI
           Signals, High Resolution Microwave Survey Investigators Working
           Group (HRMS IWG)", internal memo.
           See `http://astro.cornell.edu/SPIGOT/papers`
           `/seti/iwg_jmc2.ps`.

[Cor93]   Cordes, J. M., & Lazio, T. J. W. L., 1993, "Interstellar Scintillation
           and SETI", *Third Decennial US-USSR Conference on SETI*, ed. G.
           S. Shostak, San Francisco: ASP, 47, 143.

[Cor97a]   Cordes, J. M., & Lazio, T. J. W. L., 1997, "Finding Radio Pulsars
           in and Beyond the Galactic Center", *ApJ*, 475, 557.

[Cor97b]   Cordes, J. M., Lazio, T. J. W. L., & Sagan, C., 1997, "Scintillation-
           induced Intermittency in SETI", *ApJ*, 487, 782.

[Cos97]   C. Cosmovici, S. Bowyer, & D. Werthimer, eds., 1997, *Astronomical
           and Biochemical Origins and the Search for Life in the Universe*,
           Proceedings at IAU Colloquium 161, Bologna.

[Cro97]   Cronin, J. R., & Pizzarello, S., 1997, "Exantiomeric Excesses in
           Meteoritic Amino Acids", *Science*, 275, 951.

[Cul85a]   Cullers, D. K., Linscott, I. R., & Oliver, B. M., 1985, "Signal
           Processing in SETI", *ACM/IEEE-CS* Joint Issue, 28.11, 1151.

[Cul85b]   Cullers, D. K., Oliver, B. M., Wolfe, J. H., 1985, "Sensitive De-
           tection of Narrowband Pulses", *J. Br. Interplanet. Soc.*, 138, 6,
           278-279.

[Cul86]   Cullers, D. K., 1986, "Sensitive Detection of Narrow-band Pulses",
           Acta Astronautica, 13.1, 31.

[Cul00]   Cullers, D. K., 2000, "Project Phoenix and Beyond", *Bioastronomy
           '99 - A New Era in Bioastronomy*, Proceedings of a Conference held
           on the Kohala Coast, Hawaii, 2-6 August 1999, ASP Conference

Series 213, 451.
See http://www.seti.org.

[Dah99]   Dahn, C. C., 1999, "USN OCCD Parallaxes for White Dwarfs:
            What's Available now ... and Future Prospects", 11th European
            Workshop on White Dwarfs, ASP Conference Series 169, eds. J. E.
            Solheim & E. G. Meistas, 24.

[Dai96]   Daishido, T., Tanaka, N., Sudo, S., Suzuki, M., Saito, Y.,
            Agehama, R., Adachi, M., & Watanabe, N., 1996, "Design of
            (2+1+2)D FFT for Interferometric Pulsar Survey", Pulsars: Prob-
            lems and Progress, ASP Conference Series 105, eds. S. Johnson, M.
            A. Walker, & M. Bailes, 19.

[Dan97]   Dane, C. B., et al., 1997, "Diffraction-limited, High-average Power,
            Phase-locking of Four 30 J Beams from Discrete Nd:glass Zig-
            zag Amplifiers", Post Deadline Paper CPD27, 1997 Conference on
            Lasers and Electro-optics (CLEO).

[Daw88]   Dawkins, R., 1988, The Blind Watchmaker, Penguin Science, Lon-
            don.

[Deh99]   DeHorta, A. Y., Werthimer, D., Stootman, F. H., & Welling-
            ton, K. J., 1999, "New Developments in the Southern SERENDIP
            Project", Bioastronomy 99: A New Era in Bioastronomy, ASP
            Conference Series 213, eds. G. A. Lemarchand & K. J. Meech, 497.

[Des94]   Des Marais, D. J., 1994, "Tectonic Control of the Crustal Or-
            ganic Carbon Reservoir During the Precambrian", Chemical Ge-
            ology, 114, 303.

[Des98]   Des Marais, D. J., 1998, "Earth's Early Biosphere and its Evolu-
            tion", Origins, ASP Conference Series 148, eds. C. E. Woodward,
            J. M. Shull, & H. A. Thronson, Jr., 415.

[Dic96]   Dick, S., 1996, The Biological Universe: The Twentieth Century
            Extraterrestrial Life Debate and the Limits of Science, Cambridge
            University Press.

[Dix77]   Dixon, R. S., & Cole, D. M., 1977, "A Modest All-sky Search
            for Narrowband Radio Radiation Near the 21-cm Hydrogen Line",
            Icarus, 30, 267.

[Dix95]   Dixon, R. S., 1995, "Argus: A Future SETI Telescope", Progress
            in the Search for Extraterrestrial Life, ASP Conference Series 74,
            ed. G. S. Shostak, 355.

[Don93]   Donnely, C., Bowyer, S., Werthimer, D., Ng, D., & Cobb, J., 1993,
            "The Berkeley SETI Program: SERENDIP III", 182nd AAS Meet-
            ing 51.1, 25, 886.

[Don95]   Donnelly, C., Bowyer, S., Werthimer, D., & Malina, R. F., 1995, "Forty Trillion Signals from SERENDIP: The Berkeley SETI Program", *Progress in the Search for Extraterrestrial Life*, ASP Conference Series 74, ed. S. Shostak, Astronomical Society of the Pacific, San Francisco, 285.

[Dor86]   Dornheim, M. A., 1986, *Aviation Week & Space Technology*, August 4, 33.

[Doy93]   Doyle, L., McKay, C., Whitmire, D., Matise, J., Reynolds, R., & Davis, W., 1993, "Astrophysical Constraints on Exobiological Habitats", *Third-Decennial US-USSR Conference on SETI*, ASP Conference Series 47, ed. G. S. Shostak, 199.

[Doy00]   Doyle, L. et al. 2000, Ap.J. 535, "Observational Limits on Terrestrial-sized Inner Planets around the CM Draconis System Using the Photometric Transit Method with a Matched-Filter Algorithm", pp 338-349.

[Dra60]   Drake, F. D., 1960, "How Can We Detect Radio Transmissions from Distant Planetary Systems?", *Sky & Telescope*, 39, 140.

[Dra61a]   Drake, F. D., 1961, "Project Ozma", *Phys. Today*, 14, 40.

[Dra61b]   Drake, F. D., 1961, "Project Ozma", *Interstellar Communication: A Collection of Reprints and Original Contributions*, ed., A. G. W. Cameron, 1963, W. A. Benjamin, NY, 176.

[Dra65]   Drake, F. D., 1965, *Current Aspects of Exobiology*, eds. G. Mamikunian, & M. H. Briggs, Pergamon Press, 324.

[Dra78]   Drake, F. D., & Helou, G., 1978, "The Optimum Frequencies for Interstellar Communication as Influenced by Minimum Bandwidth", *NAIC Report*, 76.

[Dra81]   Drake, F. D., 1981, "Quantitative Estimates of the Probability of Success of SETI Programs", *NAIC Report*, 155.

[Dra83]   Drake, F. D., 1983, "Estimates of the Relative Probability of Success of the SETI Search Program", *SETI Science Working Group Report*, eds. F. D. Drake, J. H. Wolfe, & C. L. Seeger, *NASA Technical Paper*, 2244.

[Dra01]   Drake, F. D., et al., *Lick Optical SETI* See http://seti.ucolick.org/optical.

[Dre97]   Dreher, J. W., & Cullers, D. K., "SETI Figure of Merit", *Astronomical and Biochemical Origins and the Search for Life in the Universe*, Proceedings at IAU Colloquium 161, eds. C. B. Cosmovivi, S. Bowyer, & D. Werthimer, Editrice Compositari, Bologna, 711.

[Dre98]   Dreher, J., 1998, "The Phoenix Signal Detection System", *Acta Astronautica*, 42, 10-12, 635-640.

[Dut98]   Dutie, Y., & Dumas, S., 1998, "Active SETI: Target Selection and Message Conception", *Poster Paper*, presented at AAS Meeting No. 193.

[Dys93]   Dyson, F., 1993, "Concluding Remarks", *Third Decennial US-USSR Conference on SETI*, ASP Conference Series 47, ed. G. S. Shostak, 433.

[Eik95]   Eikenberry, S. S., Fazio, G. G., & Ransom, S. M., 1995, "An SSPM-based High-speed Infrared Photometer for Astronomy" *Infrared Detectors and Instrumentation for Astronomy*, Proceedings of SPIE, ed., A. M., Fowler, 2475, 210.

[Eik96a]   Eikenberry, S. S., Fazio, G. G., & Ransom, S. M., 1996, "An SSPM-Based High-Speed Near-Infrared Photometer for Astronomy", *PASP*, 108, 939.

[Eik96c]   Eikenberry, S. S., Fazio, G. G., Ransom, S. M., Middleditch, J., Kristian, J., & Pennypacker, C. R., 1996, "Infrared-to-ultraviolet Wavelenght-dependent Variations within the Pulse Profile Peaks of the Crab Nebula Pulsar", *ApJ* 467, L85.

[Eik97a]   Eikenberry, S. S., Fazio, G. G., & Ransom, S. M., 1997, "The Infrared to Gamma-Ray Pulse Shape of the Crab Nebula Pulsar", *ApJ*, 476, 281.

[Eik97b]   Eikenberry, S. S., Fazio, G. G., Ransom, S. M., Middleditch, J., Kristian, J., & Pennypacker, C. R., 1997, "High Time Resolution Infrared Observations of the Crab Nebula Pulsar and the Pulsar Emission Mechanism." *ApJ*, 477, 465.

[Enc99]   Encounter 2001. See http://seti.harvard.edu/seti/beta.html.

[Eri98]   Erickson, et al., 1998, "Science Instrument Interfaces on SOFIA", *Infrared Astronomical Instrumentation*, Proceedings of SPIE, eds. Y. P. Bely, & J. B. Breckinridge, 3354, 930.

[Esh97]   Eshleman, V., 1997, "Non-repeatable SETI Signals due to Gravity - Plasma Lenses of Stars", *Astronomical and Biochemical Origins and the Search for Life in the Universe*, Proceedings at IAU Colloquium 161, eds. C. B. Cosmovici, S. Bowyer & D. Werthimer, Editrice Compositari, Bologna, 741.

[Ewe79]   Ewing, J. J., et al., 1979, "Optical Pulse Compressor Systems for Laser Fusion", *IEEE J. Quantum Electronics*, 15, 368-379.

[Ext00]   Extrasolar Planet Search. See http//:exoplanets.org.

[Fan88]   Fan, T. Y., & Byer, R. L., 1988, "Diode Laser-Pumped Solid-State Lasers" *IEEE J. Quantum Electronics*, 24, 895-912.

[Fis00]   Fisher, R. J., & Bradley, R. F., 2000, "Full-Sampling Focal Plane Arrays", *Imaging at Radio Through Submillimeter Wavelengths*, ASP Conference Series 217, eds. J. G. Mangum & S. J. E. Radford, 11.

[For97]   Forden, G. E., 1997, "The Airborne Laser", *IEEE Spectrum*, September, 40.

[Fre80]   Freitas, R. A., & Valdes, F., 1980, "A Search for Natural or Artificial Objects Located at the Earth-Moon Libration Points" *Icarus*, 42, 442-447.

[Fre85]   Freitas, R. A., 1985, "Observable Characteristics of Extraterrestrial Technological Civilizations", *JBIS*, 38, 106.

[Ful96]   Fulghum, D. A., 1996, *Aviation Week & Space Technology*, November 18, 22.

[Gat87]   Gatewood, G. D., 1987, "The Multichannel Astrometric Photometer and Atmospheric Limitations in the Measurement of Relative Positions", *AJ*, 94, 213-224.

[Gek96]   Gekman, C. S., & Lauglin, G., 1996, "The Prospects for Earthlike Planets Within Known Extrasolar Planetary Systems", *PASP*, 108, 1018.

[Gin86]   Gindilis, L. M., 1986, "Radio Astronomy and Search for Extraterrestrial Civilizations-Development of Investigations in the USSR", *Trudy Gaish*, 58, 87-118.

[Gin88]   Gindilis, L. M., Dubinskij, B. A., & Rudinitskij, G. M., 1988, "SETI Investigations in the USSR", *Paper #IAA-88-544*, presented at IAF Congress, Bangalore, India.

[Gol76]   Gold, T., 1976, "Through a Looking Glass", URSI Meeting, Amhurst, MA.

[Gon80]   Gondhalekar, et al., 1980, "Observations of the Interstellar Ultraviolet Radiation Field from the S2/68 Sky-Survey Telescope", *A&A*, 85, 272-280.

[Got93]   Gott, J. R., 1993, "Implications of the Copernican Principle for our Future Prospects", *Nature*, 363, 315.

[Gou89]   Gould, S. J., 1989, *Wonderful Life: The Burgess Shale and the Nature of History*, W. W. Norton, New York.

[Gra82]   Gray, R. H., Dixon, et al., 1982, preprint "21 cm Radio Emissions with Geometric Fine Structure", cited by Freitas in *JBIS*, 38, 106, (1985).

[Gra86]   Gray, R. H., 1986, "Small SETI Systems", *The Search for Extraterrestrial Intelligence*, Proceedings of NRAO Workshop at NRAO, Green Bank, West Virginia, May 1985, eds, K. Kellermann, & G. Seielstad, NRAO/AUI Publishers, 205.

[Gra94]   Gray, R. H., 1994, "A Search of the 'Wow' Locale for Intermittent Radio Signals", *Icarus* 112, 485-489.

[Gra01]   Gray, R. H., 2001, & Marvel, K.B., "A VLA search for the Ohio State 'Wow'", *ApJ*, 546, 1171 - 1177.

[Gul85]   Gulkis, S., 1985, "Optimum Search Strategy for Randomly Distributed CW Transmitters", *The Search for Extraterrestrial Life: Recent Developments*, IAU Symposium 112, ed. M. D. Papagiannis, 411-417.

[Haa98]   Haas, M. R., et al., 1998, "Preliminary Optical Design of AIRES: an Airborne Infrared Echelle Spectrometer for SOFIA", *Infrared Astronomical Instrumentation*, Proceedings of SPIE, ed., A. M., Fowler, 3354, 940.

[Hal97]   Hale, A. S., Gatewood, G. D., Hale, D. D. S., Persinger, W. T., & McMillan, R. S., 1997, "The Multichannel Astrometric Photometer with Spectrograph: An Instrument for Characterizing Planetary Systems", Proceedings at 20th Annual Lunar and Planetary Science Conference, 1359.

[Ham98]   1998, *Photomultiplier Tube and Hybrid Detector Data Sheets*, Hamamatsu Corp.

[Han96]   Hankins, T. H., Ekers, R. D., & O'Sullivan, J. D., 1996 "A Search for Lunar Radio & Caron Cerenkov Emission from High-Energy Neutrinos", *MNRAS*, 283, 3, 1027-1030.

[Har90]   Harris, M. J., 1990, "A Search for Linear Alignments of Gamma Ray Burst Sources", *JBIS*, 43, 551.

[Has97]   Harrison, A., 1997, *After Contact: The Human Response to Extraterrestrial Life*, Plenum Press, New York.

[Hei00]   Heidmann, J., 2000, "Sharing the Moon by Thirds: An Extended SAHA Crater Proposal", *Advances in Space Research*, 26, 371.

[Hen96]   Henry, T. J., Soderblom, D. R., Donahue, R. A., & Baliunas, S. L., 1996, "A Survey of CA II H and K Chromospheric Emission in Southern Solar-type Stars", *AJ*, 111, 439.

[Hes99]   The HESS project. See
           `http://www.mpi-hd.mpg.de/hfm/HESS/HESS.html`.

[Hof00]   Hofmann, W., 2000, "The High Energy Stereoscopic System
           (HESS) Project", *GeV-TeV Gamma Ray Astrophysics Workshop:
           Towards a Major Atmospheric Cerenkov Detector VI*, AIP Confer-
           ence Proceedings 515, eds. B. L. Dingus, M. H. Dalamon & D. B.
           Keida, 500.

[Hog94]   Hogg, R. V., & Craig, A. T., 1994, *Introduction to Mathematical
           Statistics*, Prentice Hall, New Jersey, NJ.

[Hor78]   Horowitz, P., 1978, "A Search for Ultra-Narrowband Signals of
           Extraterrestrial Origin" *Science*, 201, 733-735.

[Hor85]   Horowitz, P., & Forster, J., 1985, "Project Sentinel: Ultra-
           Narrowband SETI at Harvard/Smithsonian", *The Search for Ex-
           traterrestrial Life: Recent Developments*, ed. M. D. Papagiannis,
           D. Reidel Publishing Co., Dordrecht, 291- 303.

[Hor86]   Horowitz, P., 1986, *The Search for Extraterrestrial Intelligence*,
           Proceedings of NRAO Workshop at NRAO, Green Bank, West Vir-
           ginia, May 1985, eds. K. Kellermann, & G. Seielstad, NRAO/AUI
           Publishers, 99.

[Hor93]   Horowitz, P., & Sagan, C., 1993, "Five Years of Project META: An
           All-Sky Narrow-Band Radio Search for Extraterrestrial Signals",
           ApJ, 415, 218-235.

[How00]   Howard, A., Horowitz, P., et al., 2000, "Optical SETI at Harvard-
           Smithsonian", *Bioastronomy 99: A New Era in Bioastronomy*, ASP
           Conference Series 213, eds. G. A. Lemarchand & K. Meech, 545.
           See `http://seti.harvard.edu/oseti/index.html`.

[Hrw81]   Harwit, M., 1981, *Cosmic Discovery – The Search, Scope & Her-
           itage of Astronomy*, Basic Books, Inc., New York.

[Isb60]   Isbell, D. E., 1960, *IRE Trans.*, AP-8, 260.

[Jan33]   Jansky, K. G., 1933, "Electrical Disturbances Apparently of Ex-
           traterrestrial Origin", presented at the IRE meeting, in Washing-
           ton, D.C.

[Jug95]   Jugaku, J., Noguchi, K., Nishimura, S., 1995, "A Search for Dyson
           Spheres Around Late-Type Stars in the Solar Neighborhood",
           Progress in the Search for Extraterrestrial Life, ASP Conference
           Series 74, ed. S. Shostak, Astronomical Society of the Pacific, San
           Francisco, 381-385.

[Kar64]   Kardashev, N., 1964, "Transmission of Information by Extraterrestrial Civilizations", *Soviet A. J.*, 217.

[Kar73]   Kardashev, N., 1973, "Communication with Extraterrestrial Intelligence", *Paper from the CETI Byurakan Conference*, ed. C. Sagan, MIT Press, Cambridge, MA, 240.

[Kas93]   Kasting, J. F., 1993, "Earth's Early Atmosphere", *Science*, 259, 920.

[Kas97]   Kasting, J. F., 1997, "Habitable Zones Around Low Mass Stars and the Search for Extraterrestrial Life", *Planetary and Interstellar Processes Relevant to the Origin of Life*, ed. D. C. B. Whittet, Kluwer Academic Publishers, Dordrecht, The Netherlands & New York, NY.

[Kel66]   Kellermann, K. I., 1966, *Australian Journal of Physics*, 19, 195.

[Kel85]   K. I. Kellermann, & G. A. Seielstad, eds., 1985, "The Search for Extraterrestrial Intelligence, Proceedings of an NRAO Workshop", NRAO, Green Bank, West Virginia, May 20-22, 1985, NRAO/AUI.

[Ker99]   Kerr, R., 1999, "Early Life Thrived Despite Earthly Travails", *Science*, 284, 2111.

[Kil89]   Kildal, P. S., Skyttemyr, S. A., 1989, "Diffraction Analysis of a Proposed Dual Reflector Feed for the Spherical Reflector Antenna of the Arecibo Observatory", *Radio Science*, 24, 601.

[Kin96]   S. A. Kingsley & G. A. Lemarchand, eds., 1996, *The Search for Extraterrestrial Intelligence, (SETI) in the Optical Spectrum II*, Proceedings of SPIE, 2704.

[Kin96b]   Kingsley, S. A., 1996, "Prototype Optical SETI Observatory", *The Search for Extraterrestrial Intelligence. (SETI) in the Optical Spectrum II*, Proceedings of SPIE, eds., S. A., Kingsley & G. A., Lemarchand, 2704, 102.
            See http://www.coseti.org

[Kir00]   Kirschvink, J. L., Gaidos, E. J., et al., 2000, "Paleoproterozoic Snowball Earth: Extreme Climate and Geological Change and its Biological Consequences", Proceedings at the National Academy Sciences, 97, 1400.

[Koc98]   Koch, D. G., Bowchi, W., Dunhan, E., Jenkins, J., Marriott, J., & Reitsma, H. J., 1998, "Kepler: A Space Mission to Detect Earth-Class Exoplanets", *Space Telescopes and Instruments V*, Proceedings of SPIE, eds., P. Y., Bely & J. B., Breckinridge, 3356, 599.

[Kra79]   Kraus, J. D., 1979, "We Wait and Wonder", *Cosmic Search*, 1, 3, 32.

[Kru96]   Krupke, W. F., 1996, "Diode-Pumped Solid State Lasers for IFE", 2nd *Annual International Conference on Solid State Lasers for Applications to ICF*, Commissariat a l' Energie Atomique, (CEA), Paris, France.

[Kui83]   Kuiper, T., & Gulkis, S., 1983, *The Planetary Report*, 3, 17.

[Lac94]   Lacy, J. H., Achtermann J. M., 1994, "Astronomy With A Mid-Infrared Echelle Spectrograph", *Infrared Astronomy with Arrays, The Next Generation*, ed. I. S. McLean, ASSL, 190, 85.

[Lam91]   Lampton M., 1991, *Scientific American*, November, 67.

[Lam00]   Lampton, M., 2000, "Optical SETI: The Next Search Frontier", *Bioastronomy '99 - A New Era in Bioastronomy*, ASP Conference Series 213, 565.
            See http://seti.ssl.berkeley.edu/opticalseti.

[Lan94]   Lang, K., 1994, "Radio Evidence for Nonthermal Particle Acceleration on Stars of Late Spectral Type", *ApJ Suppl.*, 90, 753.

[Las94]   Lash, R., & Fremont, M., 1994, "Up and running at 4 GHz: the SETI-capable Radio Telescope", *Radio Astronomy, the journal of the Society of Amateur Radio Astronomers*, June/July 1994, 1-6.
            See http://www.bambi.net/sara/bambi.htm.

[Lat93]   Latham, D. W., Soderblom, D. R., 1993, "Strategies for SETI Target Selection", *The Search for Extraterrestrial Intelligence (SETI) in the Optical Spectrum*, Proceedings of SPIE, ed., S. A., Kingsley, 1867, 20.

[Lau99]   Laufer, D., Notesco, G., Bar-Nun, A., & Owen, T., 1999, "From the Interstellar Medium to Earth's Oceans via Comets - An Isotopic Study of HDO/H_2O", *Icarus*, 140, 446.

[Leg00]   Leger, A., 2000, "Strategies for Remote Detection of Life - DARWIN-IRSI and TPF Missions", *Advances in Space Research*, 25, 2209.

[Lei00]   Leigh, D., & Horowitz, P., 2000, "Strategies, Implementation and Results of BETA", *Bioastronomy 99: A New Era in Bioastronomy*, ASP Conference Series 213, eds. G. Lemarchand & K. Meech, 459.
            See http://seti.harvard.edu/seti/beta.html.

[Lek75]   Lekht, et al., 1975, "Investigations of Statistical Properties of OH Maser Sources", *Pis'ma V Astronomicheskii Zhurnal*, 1, 29-32.

[Lem94]   Lemarchand, G., 1994, "Passive and Active SETI Strategies Using the Synchronization of SN1987A", *Astrophysics and Space Science*, 214, 209.

[Lem96]   Lemarchand, G. A., 1996, "SETI From the Southern Hemisphere", *SETI Quest*, 2, 2, 13-18.

[Lev95]   Levin, S., Olsen, E. T., Backus, C., & Gulkis, S., 1995, "The NASA HRMS Sky Survey X-B and Observations: A Progress Report", *Progress in the Search for Extraterrestrial Life*, ASP Conference Series 74, ed. S. Shostak, Astronomical Society of the Pacific, San Francisco, 470-477.

[Lnd95]   Lind, J., 1995, *Phys. Plasmas*, 2, 3933.

[Lin96]   Lin, D. N. C., Bodenheimer, P., & Richardson, D. C., 1996, "Orbital Migration of the Planetary Companion of 51 Pegasi to its Present Location", *Nature*, 380, 660.

[Liv62]   Livingstone, M. S., & Blewett, J. P., 1962, *Particle Accelerators*, McGraw Hill Book Company.

[Lun94]   Lungren, S. C., 1994, Ph.D. Thesis, Cornell University.

[Mac94]   Maccone, C., 1994, "The Karhunen-Loève Transform: A Better Tool Than the Fourier Transform for SETI and Relativity", *JBIS*, 47.1.

[Mai93]   Maihara, et al., 1993, "Observations of the OH Airglow Emission", *PASP*, 105, 940-944.

[Mak80]   Makovetskii, R. V., 1980, "Mutual Strategy for CETI Call Signals", *Icarus*, 41, 178.

[Mar96]   Mauersberger, R., Wilson, T. L., Rood, R. T., Bania, T. M., Hein, H., & Linkart, A., 1996, "SETI at the Spin - Flip Line Frequency of Positronium", *Astron. Astrophys.*, 306, 141-144.

[Mar00]   Marcy, G. W., Cochran, W. D., & Mayor, H., 2000, *Extrasolar Planets Around Main Sequence Stars. Protostars and Planets IV*, eds., V. G. Mannings, A. P. Boss, & S. S. Russell, University of Arizona Press, 1285.

[Mas96]   Marshall, C. D., et. al, 1996, "Diode-Pumped Gas-Cooled Slab Laser Performance", OSA Conference on Advanced Solid State Lasers, (ASSL), *TOPS*, 1, 208.

[Mas98]   Marshall, C. D., et al., 1998, "Next-Generation Laser for Inertial Confinement Fusion", OSA Conference on Advanced Solid State Lasers, (ASSL), *TOPS*, to be published.

[Mau86] Mauche, C. W., & Gorenstein, P., 1986, "Measurements of X-ray Scattering from Interstellar Grains", *ApJ*, 302, 371.

[Mau89] Mauche, C. W., & Gorenstein, P., 1989, "X-ray Halos Around Supernova Remnants", *ApJ*, 336, 843.

[May97] Mayor, M., Queloz, D., Udry, S., & Halbwachs, J. L., 1997, "From Brown Dwarfs to Planets", *Astronomical and Biochemical Origins and the Search for Life in the Universe*, Proceedings at IAU Colloquium 161, eds., C. B. Cosmovici, S. Bowyer, & D. Werthimer, Editrice Compositari, Bologna, 313.

[McD78] McDermott, W. D., et al., 1978, "An Electronic Transition Chemical Laser", *Applied Physics Letter*, 32, 469.

[McL95] McLean, I. S., Becklin, E. E., Figer, D., Larson, S., & Graham, J., 1995, "NIRSPEC: A Near-Infrared Cross-Dispersed Echelle Spectrograph for the Keck II Telescope", *Infrared Detectors and Instrumentation for Astronomy*, Proceedings of SPIE, ed., A. M., Fowler, 2475, 350-358.

[Mel85] Melia, F., & Frisch, D. H., 1985, "Mutual Help on SETIs", *QJRAS*, 26, 147.

[Mic90] M. Michaud & J. Tarter, guest eds., 1990, *Acta Astronautica Special Issue: SETI Post Detection Protocol*, 21, 2.

[Mol86] Molnar, L. A., & Mauche, C. W., 1986, "Effects of the X-ray Scattering Halo on the Observational Properties of Cygnus X-3", *ApJ*, 310, 343.

[Mol95] Molnar, L. A., Mutel, R. L., Reid, M. J., & Johnston, K. J., 1995, "Interstellar Scattering Toward Cygnus X-3: Measurements of Anisotropy and of the Inner Scale", *ApJ*, 438, 708.

[Mon00] Montebugnoli, S., et al., 2000, "SETItalia", *Bioastronomy 99: A New Era in the Search fo Life*.

[Mon01] Montebugnoli, S. See http://www.ira.bo.cnr.it/setiweb.

[Mor73] Morimoto, H., Hirbayashi, H., & Jauaku, J., 1973, "Preferred Frequency for Interstellar Communication", *Nature*, 276, 694.

[Mra90] Moran, J. M., Greene, B., Rodriguez, L. F., & Backer, D. C., 1990, "The Large Scattering Disk of NGC 6334B", *ApJ*, 348, 147.

[Mrd99] D. Morrison, & G. K. Schmidt, eds., 1999, "Astrobiology RoadMap", See http://astrobiology.arc.nasa.gov/roadmap/roadmap.pdf.

[Mrp75]   Morrison, P., 1975, *Letter to directors of Radio Observatories dated 08/29/75*, NASA SP-419, 204.

[Mrp77]   P. Morrison, J. Billingham & J. Wolfe, eds., 1977, *The Search for Extraterrestrial Intelligence: SETI*, NASA SP-419; 1979, Reprinted with trivial deletions by Dover Publications, New York, NY.

[Mou98]   Mourou, G. A., Barty, C. P. J., & Perrt, M. D., 1998, "Ultrahigh-Intensity Lasers: Physics of the Extreme on a Tabletop", *Physics Today*, 51.1, 22-28.

[Myr94]   Mayr, E., 1994, "Does It Pay to Acquire High Intelligence?", *Perspectives in Biology and Medicine*, 150.

[Nrc72]   *Astronomy and Astrophysics for the 1970s*, Report of the NRC Astronomy Survey Committee, Jesse L. Greenstein, Chairman, 1972.

[Nrc82]   *Astronomy and Astrophysics for the 1980's*, Report of the Astronomy Survey Committee, George B. Field, Chairman, 1982.

[Nrc91]   *The Decade of Discovery in Astronomy and Astrophysics*, Report of the Astronomy Survey Committee, John H. Bahcall, Chairman, 1991.

[Obe96]   Obenschain, S. P. , et al., 1996, "The Nike KrF Laser Facility", *Phys. Plasmas*, 3, 2098-2107.

[Ohm74]   Ohm, E. A., 1974, "A Proposed Multiple Beam Microwave Antenna for Earth Stations and Satellites", *Bell Systems Technical Journal*, 53, 1657.

[Oli71]   B. M. Oliver & J. Billingham, eds., *Project Cyclops. A Design Study of a System for Detecting Extraterrestrial Life*, NASA CR-114445, 1972, revised 1973, reprinted with additions 1996.

[Oli90]   Oliver, B., 1990, "Letter to the Editor", *S&T*, March.

[Oli90b]   Oliver, B. M., 1990, "A Review of Interstellar Rocketry Fundamentals", *JBIS*, 43, 259.

[Ort96]   Orth, C. D., et al., 1996, "A Diode Pumped Solid State Laser Driver for Inertial Fusion Energy", *Nuclear Fusion*, 36, 75.

[Pal72]   Palmer, P., & Zuckerman, B., 1972, *The NRAO Observer*, 13, 6, 26.

[Pan81]   Pankonin, V., & Price, R. M., 1981, "Radio Astronomy and Spectrum Management – the Impact of WARC-79", *IEEE*, EMC-23, 308.

[Pap85]   M. D., Papagiannis, ed., 1985, *The Search for Extraterrestrial Life: Recent Developments*, Proceedings at IAU Symposium 112, D. Reidel, Dordrecht, Holland.

[Pap95]   Papagiannis, M. D., 1995, "The Search for Extraterrestrial Technologies in Our Solar System", *Progress in the Search for Extraterrestrial Life*, ASP Conference Series 74, ed. G. S. Shostak, 425.

[Par80]   Paresce & Jakobsen, 1980, "The Diffuse UV Background", *Nature*, 288, 119-126.

[Par80b]   Parkonin, V., 1980, "Protecting Radio Windows for Astronomy", *S&T*, 61, 308.

[Pas71]   Paschenko, et al., 1971, "Measurement of One-Dimensional Function of Distribution for Signals from Galactic Sources" *Astronomicheskii Tsirkulyar*, No. 626, 1-3.

[Pas73]   Paschenko, et al., 1973, "Investigation of the Density of Probability for Interstellar Hydroxyl Radio Lines", *Uchebynkh Zavedenii-Radio Fizika*, 16, 1344-1349. F

[Pay94]   Payne, S. A., et al., 1994, "Ytterbium-doped Apatite-structure Crystals: A New Class of Laser Materials", *Journal of Applied Physics*, 76.1, 497-503.

[Pen94]   Pennypacker, C. R., 1994, "High Time Resolution Infrared Observations of the Crab Nebula Pulsar", *ApJ*, 431, L43-L46.

[Per96]   Perry, M., et al., 1996, *Paper CW14, Conference on Laser and Electro-optics, (CLEO), Technical Digest*, 307, Optical Society of America, Anaheim, California.

[Per96b]   Perry, M., 1996, "Crossing the Petawatt Threshold", *Science & Technology Review*, 4-11.

[Pet87]   Petroff, M. D., Stapelbroek, M. G., & Kleinhans, W. W., 1987, "Detection of Individual 0.4-28 Micron Wavelength Photons Via Impurity-impact Ionization in a Solid-state Photomultiplier", Letter, *Applied Physics Letters*, 51, 406-408.

[Pil95]   Pilachowski, C., Dekker, H., Hinkle, K., Tull, R., Vogt, S., Walker, D. D., Diego, F., & Angel, R., 1995, "High Resolution Spectrographs for Large Telescopes", *PASP*, 107, 983.

[Pos89]   Pospieszalski, M., 1989, "Modeling of Noise Parameters of MOSFETs and MODFETs and Their Frequency and Temperature Dependence", *IEEE MTT Trans.*, 37, 1340.

[Pos93]   Pospieszalski, M., 1993, "Millimeter-Wave Cryogenically-Coolable
            Amplifiers Using AlInAs/GaInAs/InP HEMTs", *IEEE MTT-S Digest*, 515.

[Pra96]   Pravdo, S., & Shaklan, S. B., 1996, "Astrometric Detection of Extrasolar Planets: Results of a Feasibility Study with the Palomar 5 Meter Telescope", *ApJ*, 465, 264-277.

[Pri63]   de Solla Price, D. J., 1963, *Little Science, Big Science*, Columbia University Press.

[Pri86]   de Solla Price, D. J., 1986, *Little Science, Big Science – and Beyond*, Columbia University Press.

[Rch98]   Richter, M. J., Lacy, J. H., et al., 1998, "EXES: An Echelon Cross Echelle Spectrograph for SOFIA", *Infrared Astronomical Instrumentation*, Proceedings of SPIE, eds. Y. P. Bely, & J. B. Breckinridge, 3354, 962-972.

[Ric98]   Rickett, B. J., 1998, "Interstellar Scintillation and Intraday Variablility", *Radio Emission from Galactic and extragalactic Compact Sources*, ASP Conference Series 144, eds. J. A. Zensus, G. B. Taylor, & J. M. Wrobel, 269.

[Roc88]   Rockwell, D. A., 1988, "A Review of Phase-Conjugate Solid-State Lasers", *IEEE J. Quantum Electronics*, 24, 1124-1140.

[Rum66]   Rumsey, V. H., 1966, *Frequency Independent Antennas*, Academic Press, New York, NY.

[Sag73]   Sagan, C., 1973, *Communication with Extraterrestrial Intelligence*, MIT Press, Cambridge, MA.

[Sag75]   Sagan, C., & Drake, F. D., 1975, "In Search for Extraterrestrial Intelligence", *Scientific American*, 232, 80.

[Sag93]   Sagan, C., & Horowitz, P., 1993, "Five Years of Project META: An All-Sky Narrow-band Radio Search for Extraterrestrial Signals", *ApJ*, 415, 218.
            See http://seti.harvard.edu/seti/harvard_seti.html.

[Sao92]   Shao, M., & Colavita, M. M., 1992, "Long-Baseline Optical and Infrared Stellar Interferometry", *Annual Reviews of Astronomy and Astrophysics*, 30, 457-498.

[Sao98]   Shao, M., 1998, "SIM: the Space Interferometry Mission", *Astronomical Interferometry*, Proceedings of SPIE, ed., R. D., Reasenberg, 3350, 536.

[Sca58]   Schawlow, A., & Townes, C., 1958, "Infrared and Optical Masers", *Physical Review*, 112.6, 1940-1949.

[Sce94]   Scheffer, L., 1994, "Machine Intelligence, the Cost of Interstellar Travel, and Fermi's Paradox", *QJRAS*.

[Sch61]   Schwartz, R. N., & Townes, C. H., 1961, "Interstellar and Interplanetary Communication by Optical Masers", *Interstellar Communication*, ed. A. G. W. Cameron, W. A. Benjamin, New York, NY, 190, 4772, 205.

[Sema]   Sematech.
            See http://www.sematech.org/public/index.htm.

[Ses88]   Sesslcr, A. M., 1988, "New Particle Acceleration Techniques", *Phys. Today*, 41, 26-37.

[SetXX]   *SETI Algorithm Book*, unpublished, available from the SETI Institute.

[Sha95]   Shaklan, S., Sharman, M. C., & Pravdo, S. H., 1995, "High-precision Measurement of Pixel Positions in a Charge Coupled Device", *Appl. Astrophysics*, 6672.

[She77]   Sheaffer, R., 1977, "1977 SETI Progress Report", *Spaceflight*, 19, 9, 307-310.

[Shk66]   Shklovskii, I. S., & Sagan, C., 1966, *Intelligent Life in the Universe*, Holden-Day, Inc.

[Shl95]   Sholomitskii, G. B., 1965, "IAU Information Bulletin on Variable Stars", *New York Times editorial, February 27, 1965*, 36.

[Sho85]   Shostak, G. S., & Tarter, J. C., 1985, "SIGNAL Search for Intelligence in the Galactic Nucleus with the Array of the Lowlands", *Acta Astronautica*, 12.5, 369.

[Sho95]   G. S., Shostak, ed., 1995, *Progress in the Search for Extraterrestrial Life*, ASP Conference Series 74, San Francisco.

[Sho96]   Shostak, G. S., Ekers, R., & Vaile, R., 1996, "A Search for Artificial Signals from the Small Magellanic Cloud", *AJ*, 112, 164 - 166.

[Sho97]   Shostak, G. S., 1997, "A New Class of SETI Targets", *Astronomical and Biochemical Origins and the Search for Life in the Universe*, Proceedings at IAU Colloquium 161, eds. C. B. Cosmovici, S. Bowyer & D. Werthimer, Editrice Compositari, Bologna, 719.

[Sho99]   Shostak, G. S., & Tarter, J. C., 1999, "Project Phoenix Enters Adulthood", *Paper No. IAA-99-IAA.9.1.01*, presented at 50th International Astronautical Congress, Amsterdam, The Netherlands, to be published in a special issue of *Acta Astronautica*.

[Sho00]   Shostak, G. S., 2000, "SETI Merit and the Galactic Plane", *Paper No. IAA-98-IAA.9.1.08*, presented at the 49th International Astronautical Congress in 1998, Melbourne, Australia, published in a special issue of *Acta Astronautica*, 46, 10-12, 649-654.

[Shu97]   Shuch, H. P., 1997, "Project Argus and the Challenge of Real Time All-Sky SETI", *Astronomical and Biochemical Origins and the Search for Life in the Universe*, Proceedings at IAU Colloquium 161, eds., C. B., Cosmovici, S. Bowyer & D. Werthimer, Editrice Compositari, Bologna.

[Shv77]   Shvartsman, V. F., et al., 1977, "The MANIA Experiment", *Communications of the Special Astrophysical Observatory*, 19, 5.

[Shv88]   Shvartsman, V. F., 1988, "SETI in Optical Range with the 6 m Telescope (MANIA)", *Bioastronomy: The Next Steps*, ed., G., Marx, Kluwer Academic Publishers, Dordrecht, The Netherlands & New York, NY, 389-390.

[Shv93]   Shvartsman, V. F., et al., 1993, "Results of the MANIA Experiment: An Optical Search for Extraterrestrial Intelligence", *Third Decennial US-USSR Conference on SETI*, ASP Conference Series 47, ed. G. S. Shostak, 381.

[Sim83]   Simmons W. W., & Godwin, R. O., 1983, *J. Nucl. Tech. Fusion*, 4, 3456.

[Sle98]   Sleep, N., & Zahnle, K., 1998, "Refugia from Asteroid Impacts on Early Mars and the Early Earth", *Journal of Geophysical Research*, 103, 28529.

[Slpa]   "SETI League Project Argus",
           See http://www.setileague.org.

[Sly85]   Slysh, V. I., 1985, "A Search in the Infrared to Microwave for Astroengineering Activity", *The Search for Extraterrestrial Life: Recent Developments*, ed. M. D. Papagiannis, D. Reidel Publishing Co., Dordrecht, 315-319.

[Smi96]   Smith, P. M., 1996, "Status of InP HEMT Technology for Microwave Receiver Applications", *IEEE-S Digest*, 5.

[Sny97]   Snyder, L. E., 1997, "The Search for Interstellar Gylcine", *Planetary and Interstellar Processes Relevant to the Origins of Life*, ed. D. C. B. Whittet, Kluwer Academic Publishers, Dordrecht, The Netherlands & New York, NY, 115.

[Spa90]   Spangler, S. R. S., & Gwinn, C. R., 1990, "Evidence for an Inner Scale to the Density Turbulence in the Interstellar Medium", *ApJ*, 353, L29.

[Sta97]   Staveley-Smith, L., 1997, "HI Multibeam Survey Techniques", *Astronomical Society of Australia*, 14, 111.

[Ste94]   Steffes, P. G., & DeBoer, D. R., 1994, "A SETI Search of Nearby Solar-type Stars at the 203 GHz Positronium Hyperfine Resonance", *Icarus*, 107, 215-218.

[Sto00]   Stootman, F., et al., 2000,"The Southern SERENDIP Project", *Bioastronomy '99 - A New Era in Bioastronomy*, Proceedings of a Conference held on the Kohala Coast, Hawaii, 2-6 August 1999, ASP Conference Series 213, 491.
                See http://seti.uws.edu.au.

[Str85]   Strickland, D., & Mourou, G., 1985, *Opt. Commun.*, 56, 219.

[Suc81]   Suchkin, G. L., & Tokarev, Y. V., 1981, presented at SETI-81 International Symposium held in Tallinn, Estonia.

[Sul78]   Sullivan, W. T. III, Brown, S., & Wetherhill, C., 1978, "Eavesdropping - The Radio Signature of the Earth", *Science*, 199, 377-388.

[Sul97]   Sullivan, W. T. III, Wellington, K. J., Shostak, G. S., Backus, P. R., & Cordes, J. M., 1997, "A Galactic Center Search for Extraterrestrial Intelligent Signals", *Astronomical and Biochemical Origins and the Search for Life in the Universe*, Proceedings at IAU Colloquium 161, , eds. C. B. Cosmovici, S. Bowyer, & D. Werthimer, Editrice Compositari, Bologna, 653.

[Svi92]   Sviatoslavsky, I. N., et al., 1992, *Fusion Technology*, 21, 1470.

[Swa00]   Swarup, G., 2000, "Pesek Lecture: SETI in India", *Paper #IAA-98-IAA.9.1.01*, presented at IAA meeting in 1998, Melbourne, Australia, published in a special issue of *Acta Astronautica*, 46, 10-12, 621-626.

[Tab94]   Tabak, M, et al., 1994, "Ignition and High Gain with Ultrapowerful Lasers", *Physics of Plasmas*, 1, 1626-1634.

[Tar79]   Tarter, J. C., Cuzzi, J., Black, D., Clark, T., Stull, M., & Drake, F. D., 1979,"SETI: High Sensitivity Search at NASA with High Speed Tape Recorders", *paper No. 79-A-43*, presented at 30th IAF Congress, Munich, Germany.

[Tar80]   Tarter, J. C., Black, D., Cuzzi, J., & Clark, T., 1980, "A High-Sensitivity Search for Extraterrestrial Intelligence at Lambda 18 cm", *Icarus*, 42, 136-144.

[Tar82]   Tarter, J. C., & Israel. F. P., 1982, "A Symbiotic Approach to SETI Observations: Use of Maps From the Westerbork Synthesis Radio Telescope", *Acta Astronautica*, 9, 415-419.

[Tar83]   Tarter, J. C., Duquet, R., Clark, T. A., & Lesyna, L., 1983, "Recent SETI Observations at Arecibo", *Acta Astronautica*, 10, 277.

[Tar85]   Tarter, J. C., 1985, "Statistics of 'Excess' Observatory Noise at the Nançay Telescope and Elsewhere", *paper No. IAA-85-473*, presented at 36th IAF Congress in Stockholm, Sweden.

[Tar85b]   Tarter, J. C., 1985, "SETI Observationsn Worldwide.", *The Search for Extraterrestrial Life: Recent Developments*, Proceedings of IAU Symposium 112, ed. M. D., Papagiannis, D., Reidel Publishing Company, 271-289.

[Tar95]   Tarter, J. C., 1995, "HRMS: Where We've Been, and Where We're Going", *Progress in the Search for Extraterrestrial Life*, ASP Conference Series 74, ed. S. Shostak, Astronomical Society of the Pacific, San Francisco, 456-469.

[Tar96]   Tarter, J. C., 1996, "Project Phoenix: the Australian Deployment", *The Search for Extraterrestrial Intelligence (SETI) in the Optical Spectrum II*, Proceedings of SPIE, eds., S. A., Kingsley & G. A., Lemarchand, 2704, 24.

[Tar97]   Tarter, J. C., 1997, "Results from Project Phoenix: Looking up from Down Under", *Astronomical and Biochemical Origins and the Search for Life in the Universe*, Proceedings at IAU Colloquium 161, eds. C. B. Cosmovici, S. Bowyer, & D. Werthimer, Editrice Compositari, Bologna, 633.

[Tar94]   Tarter, J. C., "SETI is Still Alive: Results from One Year of High Resolution Microwave Survey Observations and a Progress Report on Project Phoenix", *BAAS*, 54.01, 184.

[Tay93]   Taylor, J. H., Manchester, R. N., & Lyne, A. G., 1993, "Catalog of 706 posted 446 Pulsars", *ApJS*, 88, 529.

[Tay95]   Taylor, J. H., Manchester, R. N., Lyne, A. G., & Camilo, F., 1995, "Catalog of 706 posted 446 Pulsars."
             See http://pulsar.princeton.edu/catalog.shtml.

[Til95]   Tilgner, C. N., & Heinrichsen, I., 1995, "A Program to Search for Dyson Spheres With the Infrared Space Observatory", *paper No. IAA-95-IAA.9.1.11, IAF Congress*, Oslo, Norway.

[Tou00]   A. Tough, ed., 2000, *When SETI Succeeds: The Impact of High-Information Contact*, Foundation for the Future, Bellvue, Washington.

[Tow83]   Townes, C. H., 1983, "At What Wavelengths Should we Search for Signals from Extraterrestrial Intelligence?", Proceedings at NAS USA, 80, 1147.

[Tow97]   Townes, C. H., 1997, "Optical and Infrared SETI", *Astronomical and Biochemical Origins and the Search for Life in the Universe*, eds. C. B. Cosmovici, S. Bowyer, & D. Werthimer, Editrice Compositari, 585.

[Tro71]   Troitskii, V. S., Strodubtsev, A. M., Gershtein, L. I., & Rakhlin, V. L., 1971, "Search for Monochromatic 927-MHz Radio Emission from Nearby Stars", *Soviet AJ*, 15, 508.

[Tro75]   Troitskii, V. S., Bondar, L. N., & Strodubtsev, A. M., 1975, *Soviet Phys.-Usp.*, 17, 607.

[Tro82]   Troitskii, V. S., 1982, "Interview", *Leningradskaya Pravda*, 2 November 1982.

[Tur96]   Turon, C., 1996, "Hipparcos, A New Start for Astronomical and Astrophysical Topics", *Reviews of Modern Astronomy*, 9, 69.

[Vak98]   Vakoch, D. A., 1998, "The Dialogic Model: Representing Human Diversity in Messages to Extraterrestrials", *Acta Astronautica* 42.10 – 42.12, 705-710.

[Val83]   Valdes, F., & Freitas, R. A. Jr., 1983, "A Search for Objects Near the Earth-Moon Lagrarian Points", *Icarus*, 53, 453-457.

[Val86]   Valdes, F., & Freitas, R. A. Jr., 1986, "A Search for the Tritium Hyperfine Line from Nearby Stars", *Icarus*, 65, 152.

[Vae85]   Vallee, J. P., 1985, "Search for Strongly Polarized Radio Emission from E.T.I. and an Optimistic Approach to the Great Silence (Fermi's Paradox)", *The Search for Extraterrestrial Life: Recent Developments*, ed. M. D. Papagiannis, D. Reidel Publishing Co., Dordrecht, 321-325.

[Van97]   Van Wonterghem, B. M., et. al, 1997, "Performance of a Prototype for a Large-Aperture Multipass ND:glass Laser for Inertial Confienement Fusion" *Applied Optics*, 36, 4932-4953.

[Verit]   VERITAS.
            See http://Veritas.sao.arizona.edu/veritas/.

[Ves73]   Verschuur, G. L., 1973, "A Search for Narrow Band 21-cm Wavelenght Signals from Ten Nearby Stars", *Icarus*, 19, 329.

[Vog94]   Vogt, S. S., et al., 1994, "HIRES: the high-resolution echelle spectrometer on the Keck 10-m", *Instrumentation in Astronomy VIII*, Proceedings of SPIE, eds., D. L. Crawford, & E. R. Crain, 2198, 362.

[Wol92]   Wolszczan, A. F., & Rail, D., 1992, "A Planetary System Around the Millisecond Pulsar PSR1257+12", *Nature*, 255, 145.

[War00]   Ward, P. D., & Brownlee, D., 2000, *Rare Earth: Why Complex Life Is Noncommon in the Universe*, Copernicus Books.

[Wer96]   Werthimer, D., Donnelly, C., & Cobb, J., 1996, "SERENDIP SETI Project: New Instrumentation for SETI and Results of the Recent Arecibo Sky Survey", *The Search for Extraterrestrial Intelligence (SETI) in the Optical Spectrum II*, Proceedings of SPIE, eds., S. A., Kingsley & G. A., Lemarchand, 2704, 9.

[Wer96b]   Werthimer, D., Bowyer, S., Ng, D., Donnelly, C., Cobb, J., Lampton, M., & Airieau, S., 1996, "The Berkeley SETI Program: SERENDIP IV Instrumentation", *Poster Paper No. P4-20*, presented at 5th International Conference on Bioastronomy, IAU Colloquium 161, Capri, Italy.
            See http://seti.ssl.berkeley.edu/serendip/serendip.html.

[Wer97]   Werthimer, D., Nov. 1997, Private Communication.
            See http://seti.ssl.berkeley.edu/opticalseti/.

[Wer00]   Werthimer, D., Bowyer, S., Cobb, J., Ledofsky, M., & Lampton, M., 2000, "The SERENDIP IV Arecibo Sky Survey", *Bioastronomy 99: A new Era in Bioastronomy*, ASP Conference Series 213, eds. G. L. Lemarchand & K. Meech, 479.

[Wet94]   Wetherill, G. W., 1994, "Possible Consequences of Advance of Jupiters in Planetary Systems", *Ap&SS*, 212, 23.

[Whi96]   Whitmire, D. P., & Reynolds, R. T., 1996, "Circumstellar Habitable Zones: Astronomical Considerations", *Circumstellar Habitable Zones*, Proceedings of the First International Conference, ed. L. R. Doyle, Travis House Publications, Menlo Park, California.

[Wlk94]   Wilkinson, P. N., Narayan, R., & Spencer, R. E., 1994, "The Scatter-broadened Image of CYGNUS-X-3", *MNRAS*, 269, 67.

[Wil97]   Williams, D. M., Kasting, J.F F., & Wade, R. A., 1997, "Habitable Moons Around Extrasolar Giant Planets", *Nature*, 385, 234-236.

[Wil98]   Williams, D. M., Kasting, J. F., & Frakes, L. A., 1998, "Low-latitude Glaciation and Rapid Changes in Earth's Obliquity Explained by Obliquity - Oblateness Feedback", *Nature*, 396, 453.

[Woo94]   Woo, J. W., Clark, G.W., et al., 1994, "ASCA Measurements of the Grain-scattered X-ray Halos of Eclipsing Massive X-ray Binaries: VELA X-1 and Centaurus X-3", *ApJ Letters*, 436, L5.

[Wot00]   Wooten, H. A., 2000, An up-to-date list of detected molecules.
            See http://www.cv.nrao.edu/~awootten/allmols.html.

[Zad96]   Zadnik, M. G., Winterflood, J., et al., 1996, "Interstellar Communication Channel Search of Solar-type Targets Closer than 11 pc", *Poster paper No. P4-21*, presented at 5th International Conference on Bioastronomy, IAU Colloquium 161, Capri, Italy.

[Zom82]   Zombeck, M., 1982, *Handbook of Space Astronomy*, Cambridge University Press.

[Zuc85]   Zuckerman, B., 1985, "Preferred Frequencies for SETI Observations", *Acta Astronautica*, 12.2, 127.

Index